Spatial Econometric Methods in Agricultural Economics Using R

Editors

Paolo Postiglione

Professor in Economic Statistics, Department of Economic Studies
University "G. d'Annunzio" of Chieti-Pescara, Pescara, Italy

Roberto Benedetti

Professor in Economic Statistics, Department of Economic Studies
University "G. d'Annunzio" of Chieti-Pescara, Pescara, Italy

Federica Piersimoni

Senior Researcher
Italian National Institute of Statistics, Processes Design and
Frames Service, Rome, Italy

CRC Press
Taylor & Francis Group
Boca Raton   London   New York

CRC Press is an imprint of the
Taylor & Francis Group, an **informa** business

A SCIENCE PUBLISHERS BOOK

First edition published 2022
by CRC Press
6000 Broken Sound Parkway NW, Suite 300, Boca Raton, FL 33487-2742

and by CRC Press
2 Park Square, Milton Park, Abingdon, Oxon, OX14 4RN

© 2022 Taylor & Francis Group, LLC

CRC Press is an imprint of Taylor & Francis Group, LLC

Reasonable efforts have been made to publish reliable data and information, but the author and publisher cannot assume responsibility for the validity of all materials or the consequences of their use. The authors and publishers have attempted to trace the copyright holders of all material reproduced in this publication and apologize to copyright holders if permission to publish in this form has not been obtained. If any copyright material has not been acknowledged please write and let us know so we may rectify in any future reprint.

Except as permitted under U.S. Copyright Law, no part of this book may be reprinted, reproduced, transmitted, or utilized in any form by any electronic, mechanical, or other means, now known or hereafter invented, including photocopying, microfilming, and recording, or in any information storage or retrieval system, without written permission from the publishers.

For permission to photocopy or use material electronically from this work, access www.copyright.com or contact the Copyright Clearance Center, Inc. (CCC), 222 Rosewood Drive, Danvers, MA 01923, 978-750-8400. For works that are not available on CCC please contact mpkbookspermissions@tandf.co.uk

Trademark notice: Product or corporate names may be trademarks or registered trademarks and are used only for identification and explanation without intent to infringe.

ISBN: 978-1-498-76681-4 (hbk)
ISBN: 978-1-032-05370-7 (pbk)
ISBN: 978-0-429-15562-8 (ebk)

DOI: 10.1201/9780429155628

Typeset in Times New Roman
by Radiant Productions

Preface

The collection of spatial agricultural data and the spatial analysis of agriculture represent two issues of primary relevance for a large number of people. This book aims at supporting stakeholders to design spatial surveys for agricultural data and/or to analyse the geographically collected data. Hence, the book represents a comprehensive guide in methodological and empirical advanced techniques for practitioners.

This volume can also be considered as a primary tool for users from less developed countries, where agriculture is still the prevalent economic sector. Therefore, different contributions may guide one through the application of spatial survey methods, technologies developed in the past decades, such as remote sensing and GIS, and appropriate methods to analyse spatial agricultural data. Applied spatial analysts might also benefit from this work. In particular, a part of the book is devoted to the integration techniques used to merge agricultural data from different sources. Finally, both people from Academic institutions and National Statistical Offices may appreciate the occasion of deepening their knowledge of spatial techniques for agriculture.

Although the book could also represent a valued support on spatial methodologies in agriculture for graduate classes, the primary audience is mainly composed by researchers with some prior background in econometrics and spatial statistics.

The main objective of this book is to introduce agricultural economists to statistical approaches for the analysis of spatial data. The aim is to illustrate, for the main typologies of agricultural data, the most appropriate methods for the analysis, together with a description of available data sources and collection methods.

Spatial econometrics methods for different types of data are described and adopted with reference to typical analyses of agricultural economics. Topics such as spatial interpolation, point patterns, spatial autocorrelation, survey data analysis, small area estimation, regional data modelling, and spatial econometrics techniques are covered jointly with issues arising from the integration of several data types. Besides, the different phases of agricultural data collection, analysis, and integration are described in a simple way. The joint use of statistical methods, new technologies, and economic theory is treated considering the peculiarities of spatial data for a proper and efficient analysis of agricultural data.

Theoretical aspects of each model are described and complemented by examples on real data that are developed by using the open-source R software. The codes are available in the text, explained with details and in an intuitive way so that the readers can replicate these analyses on their own data. Moreover, any prior knowledge of the R programming environment is not assumed throughout the book.

The volume is organized in a number of review chapters on several specific themes. In particular, this book contains 13 Chapters, of which the first one can be considered as an introductory chapter, reviewing the main underlying concepts and presenting each contribution.

We would like to thank Alfredo Cartone for reading some parts of this book and for his support in the implementation of some R codes. Thanks also to Vijay Primlani of Science Publishers, CRC Press, for his continuous encouragement to complete this book. Finally, we are grateful to the individual chapter authors for their diligence in writing the documents. We are confident that their work will lead to new insights in the application of spatial econometric methods to agricultural data.

Rome, December 24th, 2020

Paolo Postiglione
Roberto Benedetti
Federica Piersimoni

Contents

CHAPTER 1
Basic Concepts

Paolo Postiglione,[1,*] *Roberto Benedetti*[1] and *Federica Piersimoni*[2]

1.1 Introducing space in agricultural economics

There is a great need for reliable data in agriculture. However, this information should be analysed through appropriate statistical methods to obtain evidence that can assist farmers, for example, to optimize farm returns, decrease unnecessary applications of fertilizers and pesticides, and preserve natural resources.

Standard statistical techniques often perform poorly when employed to agricultural data, due to its spatial nature. In fact, one of the common assumptions in traditional statistics is the independence and homogeneity among the observations, a hypothesis patently violated when applied to agricultural data. An agricultural variable could often display similar values in adjacent areas, leading to spatial clusters. In many cases, nearby fields have similar soil-type, climate, precipitation or an area cultivated with wheat may be close to other wheat-cultivated zones. Ignoring this *dependence* when analysing agricultural data may produce bias or inefficient estimates. For these reasons, the topic of statistical analysis of spatial data is worth a specific treatment.

For a long time, the analysis of geographically distributed phenomena in agricultural economics was carried out without the consideration of space as crucial information. The present book tries to fill this gap by highlighting potential applications of spatial analysis for agricultural facts. Indeed, space is very important in agricultural economics studies. The land is a crucial resource in agriculture and most of the data collected is spatially distributed. Besides, all agricultural activities are spatially located. However, the application of spatial models has grown to become important in applied agricultural economics only during the last few decades (Anselin 2002; Anselin and Bera 1998; Goodchild et al. 2000).

---
[1] "G. d'Annunzio" University of Chieti-Pescara, Department of Economic Studies.
[2] Italian National Institute of Statistics ISTAT, Processes Design and Frames Service.
  Emails: benedett@unich.it; piersimo@istat.it
* Corresponding author: postigli@unich.it

Generally speaking, the term *spatial* means that each unit has a geographical reference, i.e., we know where each case happens on a map. If the locations of these sites are observed and enclosed to the observations as labels, the resulting data is called spatial data. In spatial data analysis, the set of spatial locations are considered as essential information in the study. Our main idea is that location matters in agriculture and that the occurrences essentially follow the First Law of Geography, which according to Tobler (1970) states that: *"everything is related to everything else, but near things are more related than distant things"*.

A proper definition of new spatial data science methods is required in order to analyse agricultural data and to uncover interesting, useful, and non-trivial patterns. The first economist that explicitly claimed the importance of space in agricultural economics was von Thünen (1783–1850). Von Thünen developed the Isolated State model (von Thünen 1826), whose framework is considered to be the first serious application of spatial economics and economic geography in agriculture. His agricultural location theory conjectured that the optimal organisation of agricultural activities is based on location factors. Hence, these activities are arranged in concentric rings around a central consumers' town.

Recently, in many countries, the National Statistical Office geo-references the sampling frames of physical or administrative bodies used in agricultural surveys, not only with reference to the codes of a geographical taxonomy, but also adding data regarding the spatial position of each record.

Modern tools, such as GIS and remote sensing, are increasingly used in the monitoring of agricultural resources. For example, the developments in GIS technology offer growing opportunities to agricultural economics analysts dealing with large and detailed spatial databases, allowing them to combine spatial information from different sources and to produce different models, as well as tabular and graphic outputs.

The availability of these valuable sources of information makes the advanced models suggested in the spatial statistic and econometric literature applicable to agricultural economics.

More formally, spatial statistics is a field of spatial data analysis in which the observations are modelled using random variables. Ripley (1981) defines spatial statistics as *"the reduction of spatial patterns to a few clear and useful summaries"*, comparing such statistics *"with what might be expected from theories of how the pattern might have originated and developed"*.

Conversely, spatial econometrics is a branch of econometrics that deals with the modelling of spatial interaction and spatial heterogeneity in data analysis. The birth of this discipline can be traced back to the works of Paelinck and Klaassen (1979) and Anselin (1988). Following Anselin (1988), spatial econometrics can be defined as: *"the collection of techniques that deal with the peculiarities caused by space in the statistical analysis of regional science methods"*. In this book, we use a broad definition, referring to the term regional as those spatial units defined as areal regions, locations (i.e., points), and continuous units. Essentially, spatial econometrics represents a toolkit that allows for the rigorous treatment of data that is geographically distributed.

Interestingly, the analysis of agricultural yield data was also the motivation for the seminal paper by Whittle (1954), in which he analysed such data through two-dimensional stochastic models.

This book contains several contributions focused on spatial data and its use in monitoring agricultural resources, farms management, and regional markets. The theory of spatial methods is complemented by real and/or simulated examples implemented through the open-source software R.

The layout of this introductory chapter is as follows: in Section 1.2 the main typologies of spatial data are described. Section 1.3 contains some brief considerations about the R software. Section 1.4 outlines the contributions of this book, stressing the main evidences.

1.2 Spatial concepts: the essential

1.2.1 Spatial effects

Spatial econometrics aims to address in a formal way two effects that are typical of geo-referenced data: spatial dependence and spatial heterogeneity.

In recent years, a very extensive literature has stressed the role of spatial effects in many fields of statistical and econometric analysis (Anselin 1988; LeSage and Pace 2009; Benedetti et al. 2015; Kelejian and Piras 2017).

Spatial dependence may be defined as *"the propensity for nearby locations to influence each other and to possess similar attributes"* (Goodchild 1992, p.33). Empirical models that do not take spatial dependence and structural heterogeneities into account may show serious misspecification problems (see Chapters 9, 12, and 13 in this book).

Spatial dependence may also be referred to as the relationship among outcomes of a variable that is a result of the geographical position of their locations. It measures the similarity of variables within an area and the level of interdependence between the variables (Cliff and Ord 1981; Cressie 1993; Haining 2003). The procedures used to analyse patterns of spatial dependence vary according to the type of data.

For regional scientists, the economic counterpart of spatial dependence is the analysis of spillovers. Measuring the degree of spatial spillovers (LeSage and Pace 2009) and evaluating the extent of contagion (Debarsy et al. 2017) might help policy makers reach a more accurate comprehension of the agricultural phenomena.

The spatial analysis is often based on the definition of contiguity links that enable practitioners to entail geographical structures. These are defined in terms of a proximity matrix (the so-called **W**). Typically, the weights from the **W** matrix are non-stochastic and exogenous. The weights matrix is often defined with two areas defined as neighbours if they share a common border. In some cases, there is a need to capture and to model other forms of spatial proximity as hierarchical dependence and patterns of spatial competition (Haining 1990). However, another possible approach to define the elements of the weight matrix is in terms of similarity of one or more covariates (Conley and Topa 2002).

Surprisingly, in the spatial econometric and statistic literature, spatial heterogeneity, which is another relevant characteristic highlighted by spatial data, has been less investigated.

Spatial heterogeneity is connected to the absence of stability and it implies parameters to vary over space (Anselin 1988). The presence of a not constant relationship between a response variable and the covariates on a spatial unit has led to the introduction of spatially varying coefficients (Wheeler and Calder 2007), the geographically weighted regression (Fotheringham et al. 2002), the Bayesian regression models with spatially varying coefficient (Gelfand et al. 2003), and the local linear regression models (Loader 1999). In the field of linear estimation, spatial heterogeneity could lead to a serious problem of misspecification of the model (Postiglione et al. 2013).

The spatial heterogeneity can be classified into discrete heterogeneity and continuous heterogeneity. Continuous heterogeneity specifies how the regression coefficients change over space, as estimated, for example, through a local estimation process, as in the geographically weighted regression (GWR, Fotheringham et al. 2002). Discrete heterogeneity consists of a pre-specified set of spatial regimes or a predetermined group of spatial units (Anselin 1990; Postiglione et al. 2013), between which model coefficients are permitted to vary. For further details on spatial heterogeneity applications on agricultural data, see Chapters 8 and 9 in this book.

As highlighted by Postiglione et al. (2017), a new direction in the field of spatial analysis will be represented by the joint treatment of the two spatial effects: spatial dependence and spatial heterogeneity.

1.2.2 Types of spatial data

Spatial data refers to an observation on which we know the value of the variable and the location.

The variables, for example, may be univariate or multivariate, categorical or continuous. They may be based on an observational study, a well-designed experiment, or a sample survey.

The spatial domain, defined as the set of geographical coordinates, offers a potentially huge source of information for the process analysis. There are many different types of spatial data, and, as a consequence, different forms of spatial statistics are required.

According to the classification suggested by Cressie (1993), three types of spatial data can be identified:

- continuous or geostatistical data;
- lattice or areal data;
- point data.

In this classification, spatial data are distinguished by the nature of the spatial domain and not the size of the geographical unit under investigation (Cressie 1993; Schabenberger and Gotway 2005). The fundamental difference is so in the process that generated the data.

First, consider a spatial process in d dimensions as

$$\{x(\mathbf{z}): \mathbf{z} \in D \subset \mathbb{R}^d\} \tag{1.1}$$

where x represents the agricultural variable under investigation, detected at a location $\mathbf{z}$, defined using a $d \times 1$ vector of coordinates. If $d = 2$ the usual spatial process in a two-dimensional space is defined.

The geostatistical data is defined using a continuous domain D. In essence, with continuous data, there is an outcome for the variable of interest at any point across the territory under investigation D. The continuity is a property of the domain D, not of the variable being measured.

However, these data are usually measured in a discrete number of points defined in a bidimensional space in terms of the Cartesian coordinate $\mathbf{z}_i = (r_i, s_i)$. The point sampling may be chosen according to some design (e.g., random, stratified random, systematic, and spatial designs). See Chapters 2 and 3 for further details.

The points in D are non-stochastic. A spatial domain is said to be non-stochastic if it does not change from one realization of the spatial process to the next.

The term geostatistics refers to the analysis of continuous spatial variations. Geostatistics was defined by Matheron (1963) as "*the application of probabilistic methods to regionalized variables*". Diggle and Ribeiro (2007) identified three different scientific aims of geostatistics: model estimation (inference about the model parameters), prediction (inference about the unobserved values of the target variable), and hypothesis testing. Geostatistics can be also a valid support for optimising the data sampling plan. For further details about a possible application of geostatistics in yield prediction in agriculture, see Chapter 4.

In lattice data, the spatial domain D is fixed and discrete. The number of locations may be infinite, but they must be countable. Lattice data is often defined in terms of areal regions. The areas may be irregular in shape or form a regular grid. Remote sensed images are a typical example of spatial regular lattice data. In these images, the area is divided into a series of small rectangles, denoted as pixels. See Chapter 5 for some analysis on remote sensed data in agriculture.

Examples of irregular lattice data are most often based upon a partition of the territory into contiguous zones. In this case, the variables observed on spatial units identified by ZIP code, census tracts, provinces, or administrative regions.

The main investigation for areas concerns the relationship between values of neighbouring observations. The spatial analysis of areal data begins with the definition of the neighbourhood structure of the observations, then continues to measure the influence that observations have on their neighbours, and finally evaluates the significance of this influence. Examples of agricultural analyses on areal data are contained on Chapters 11 and 12 of this book.

The domain of geostatistical or lattice data is non-stochastic. Conversely, in spatial point patterns, the set of points changes according to each realization of the random process.

We are faced with point pattern data when each value of the variable refers to the location of a discrete object, the size of which is sufficiently small relative to the study area that it can be treated as a point (e.g., the location of agricultural farms or trees in a region).

The points may have additional information, called marks. In this case, we define the pattern as marked spatial point pattern; otherwise, it is denoted as unmarked.

In essence, spatial agricultural analysis may focus on how the points are distributed within the region (e.g., are the farms spatially clustered or random located?). In this case, the aim is to quantify the gap between the spatial distribution of observations and a completely random distribution in space. If the data presents a pattern more aggregated than a casual configuration, groups of points can be identified, and their importance measured.

Conversely, if we have a marked point pattern, the study may concern spatial analysis of the geographical location and the attribute available (e.g., how is the revenue of clustered/regularly located farms?). Point pattern analyses on agricultural data are presented in Chapter 7.

1.3 Using R software

The R software (http://www.r-project.org) is a statistical environment for the manipulation, analysis, and graphical representation of data. It is an interactive environment, wherein the commands produce an immediate response and provide for object-oriented programming.

The R software provides:

- an efficient data handling and storage capability;
- a collection of operators for calculations on arrays and matrices;
- a large and coherent collection of tools for statistical data analysis;
- graphical facilities for data analysis and display;
- a well-developed, simple and efficient programming language.

R was developed by Ross Ihaka and Robert Gentleman from the University of Auckland in the mid '90s' on the basis of the programming language S and another better-known commercial environment, S-PLUS (Ihaka and Gentleman 1996). The first official stable version was 1.0, released in 2000.

Unlike S-PLUS, this is a GNU-Software, that is, freely available under the constraints of the GPL (General Public License). R is available for Unix/Linux, Windows, and Macintosh platforms, and constitutes a real programming environment.

The codes are distributed through mirror sites, after compliance with the project standards has been checked in relation to the documentation as well. The analyses in R are organized in libraries (packages) which are implemented directly by the software developers. Packages are a collection of functions that are related to each other. Once R is installed, basic packages which only allow standard statistical analysis are immediately available, the other additional packages can be downloaded from mirror sites.

When you use the R program, it issues a prompt when it expects input commands. The default prompt is '>'.

The use of R has been simplified by the presence of R-studio (https://rstudio. com/products/rstudio/). RStudio is an integrated development environment (IDE) for R. It incorporates a console, syntax-highlighting editor, tools for plotting, history, debugging, and workspace management. RStudio is available in open source and commercial editions for Windows, Mac, and Linux, or in a browser connected to RStudio Server or RStudio Server Pro for Debian/Ubuntu, Red Hat/CentOS, and SUSE Linux.

The main problem for the R new users, however, is the knowledge and understanding of statistics. Differently from other commercial statistical software, where the lists of statistical methods are showed through windows or drop-down menus, R requires a priori understanding of the method that should be used. At a first glance, this may seem a huge drawback for using this software; but, in our opinion, this awareness moderates the possibility of a not appropriate use of statistical modelling.

In particular, R is a powerful computing tool that also supports geographical analysis and mapping for researchers interested in spatial analysis and mapping.

This book provides an introduction to the use of R for spatial statistical and econometric analysis for agricultural data. As aforementioned, the choice of using R as the statistical environment for these analyses derives from its particular properties.

1.4 The description of the book

One of the aims of this book is to stimulate and suggest research on the application of spatial statistical methods on agricultural data. The plan of the volume has been conceived to this end.

The book contains, in addition to this introductory Chapter, 12 chapters that attempt to cover multiple aspects of spatial analysis in agriculture. We are aware that this book cannot be considered an exhaustive and definitive reference for the topic, but we believe that the intentionally applicative approach and the extensive use of the R software may be very useful for a large audience of practitioners.

The first phase of a statistical analysis consists of data collection. Chapters 2 and 3 are devoted to this important issue.

In particular, Chapter 2 contains a review of spatial sampling designs. As noted, a spatial population, e.g., those represented by agricultural data, is characterized by dependence between units. As a consequence, neighbour units are likely to provide similar information and this aspect should be carefully considered in the definition of a sampling design. Since standard sampling designs do not consider the influence of spatial proximity between units, several spatial sampling designs have been introduced in the literature and described in this chapter. The R packages used are `sampling`, `Spbsampling`, `spsurvey`, and `BalancedSampling`.

Chapter 3 presents some issues related to the estimation phase of a survey. The main assumption is that data is coming from complex sampling design. The objective is to make design-based inference for a set of descriptive parameters of a finite population. Also, superpopulation models are considered, since the interest is in accounting for the spatial structure that very likely characterizes agricultural and environmental populations. Thus, a *model-assisted* approach is used to make inference (Särndal et al. 1992). The description of the estimation phase of sampling designs is supplemented by an illustration on real data coming from a survey on lakes in the Northeastern states of the US, using the R packages `sampling` and `survey`.

Chapter 4 describes the topic of yield prediction in agriculture, providing a comparison between the methods of regression kriging and random forest. Crop yields and prediction are becoming crucial activities for farmers as well as for consultants and agricultural organizations. For example, yield forecasting represents an important tool for developing agriculture operations and management and for regulating agriculture cultivation systems. The empirical analysis describes the application of regression kriging and random forest in the spatial estimation of winter wheat yield in the Southern Great Plains of the US, spanning across three states: Kansas, Oklahoma, and Texas. The R packages `caret` and `gstat` are used in the illustration.

Chapter 5 contains an interesting description of the problem of land pattern recognition. The production of land cover databases is noteworthy in the last few decades. These tools have been produced using aerial photos or satellite data. They are very important for managing agriculture food security, natural resources, and environment protection. The chapter also analyses different types of methods for the classification of remote sensed data. The application presents a comparison between different classification methods on Sentinel satellite data from Copernicus project of the European Space Agency and ground data provided by the Italian Ministry of Agriculture, using the R packages `randomForest`, `adabag`, and `glmnet`.

Chapter 6 describes different statistical systems in agriculture. The crucial information on agriculture is provided by the Census of Agriculture that is a key pillar for all national agricultural statistical systems. In some cases, this is the only information on the agricultural sector produced by many developing countries. The role of FAO and Eurostat in producing agricultural data is also emphasised in the chapter. This part of the book represents an interesting guide to different available agricultural data.

Chapter 7 presents the introductory aspects of point pattern analysis applied to agricultural data. In agricultural studies, the points, for example, could represent trees, farms, and animal nests. Using spatial point pattern analysis in agriculture, we may study, for instance, the distribution of farm sizes, and the reason why land concentration occurs in a given area. To this end, this chapter describes the introductive aspects of explorative analysis of spatial point patterns and offers a comprehensive description of possible applications on agricultural farms. The empirical exercise is performed using the R-package `spatstat`.

Chapter 8 focuses on detecting and modelling spatial effects in the analysis of the agricultural production at fixed locations, corresponding to farms. The agricultural output is modelled through a standard Cobb-Douglas production function. Exploratory spatial data analysis tools are presented in order to identify specific patterns of the points. Also, the technique of geographically weighted regression is described as a way to model continuous spatial heterogeneity and to develop local models. The main R-packages used in the application are `spdep` and `GWmodel`.

Chapter 9 analyses some aspects related to the spatial econometric modelling of farm data. The definition of the production function and, particularly, the spatial stochastic frontier model are introduced and estimated, using the R-packages `Benchmarking` and `ssfa`. Besides, a possible method to identify spatial regimes and contiguous spatial clusters are described for controlling unobserved spatial discrete heterogeneity. The library `spdep` and the function `skater` are used to this end.

Chapter 10 contains an overview of the most common areal interpolation methods. Areal interpolation aims at estimating variables for a set of target zones, based upon the known values in the set of source zones. The main focus is on the Bayesian interpolation method (BIM) introduced by Benedetti and Palma (1994). This technique will be illustrated as a special case of areal interpolation methods that accounts for spatial autocorrelation. The BIM is used to disaggregate data on the production per hectare of firms specialized in olive growing in Italy. The information is available at the regional level and estimated at the provincial level. The R-code to perform the procedure is available in the text.

Chapter 11 is devoted to a description of the topic of small area estimation. This includes the methods for obtaining more precise estimators in local/small areas, making use of the common characteristics of the areas. In particular, the authors analyse the most important area level models, which consider the spatial information, to provide estimates of agricultural and rural statistics at a local/small area level. The theory is complemented by R-illustration using a data set available in the `sae` R-package: the grape data set. This data set is based on the Italian Agricultural Census of year 2000 for the Italian region of Tuscany.

Chapter 12 presents research focusing on agricultural convergence. The β-convergence refers to a dynamic process in which poorer regions catch up with the richer ones. The importance of this analysis is justified by the relevance of agricultural policies. A cross-section application to β-convergence models for the agricultural sector in Europe is provided. The analysis focuses particularly on the consequences of spatial effects and their treatment in the R environment. To this end, several spatially augmented models are estimated using `spdep` and `spatialreg` R-packages.

Finally, Chapter 13 aims at contributing to the literature on spatial panel models, evidencing the main potential of analyses when applied to agricultural data. In particular, the chapter discusses the modern theoretical contributions on spatial panel model, with an illustration of the panel approach to stochastic frontier analysis for agricultural data. The R codes for the estimation are presented and commented on using the library `splm`. The empirical application concerns the analysis of production efficiency in Indonesian rice farming through the `RiceFarms` dataset, available in `splm`.

References

Anselin, L. 1988. *Spatial Econometrics: Methods and Models*. Dordrecht: Kluwer Academic Publisher.

Anselin, L. 1990. Spatial dependence and spatial structural instability in applied regression analysis. *Journal of Regional Science* 30: 185–207.

Anselin, L. and A. Bera. 1998. Spatial dependence in linear regression models with an introduction to spatial econometrics. pp. 237–289. *In*: Ullah, A. and D.E. Giles (eds.). *Handbook of Applied Economic Statistics*. Marcel Dekker, New York.

Anselin, L. 2002. Under the hood Issues in the specification and interpretation of spatial regression models. *Agricultural Economics* 27: 247–267.

Benedetti, R. and D. Palma. 1994. Markov random field-based image subsampling method. *Journal of Applied Statistics* 21: 495–509.

Benedetti, R., F. Piersimoni and P. Postiglione. 2015. *Sampling Spatial Units for Agricultural Surveys.* Springer: Berlin.

Cliff, A.J. and J.K. Ord. 1973. *Spatial Autocorrelation.* Pion: London.

Conley, T.G. and G. Topa. 2002. Socio-economic distance and spatial patterns in unemployment. *Journal of Applied Econometrics* 17: 303–327.

Cressie, N. 1993. *Statistics for Spatial Data.* Revised edition. New York: Wiley.

Debarsy, N., J.Y Gnabo and M. Kerkour. 2017. Sovereign wealth funds' cross-border investments: assessing the role of country-level drivers and spatial competition. *Journal of International Money and Finance* 76: 68–87.

Diggle, P.J. and P.J. Ribeiro. 2007. *Model-based Geostatistics.* Springer: New York.

Fotheringham, A.S., C. Brunsdon and M. Charlton. 2002. *Geographically Weighted Regression: The Analysis of Spatially Varying Relationships.* Chichester: Wiley.

Gelfand, A.E., H. Kim, C.F. Sirmans and S. Banerjee. 2003. Spatial modeling with spatially varying coefficient processes. *Journal of the American Statistical Association* 98: 387–396.

Goodchild, M.F. 1992. Geographical information science. *International Journal of Geographical Information Systems* 6: 31–45.

Goodchild, M.F., L. Anselin, R. Appelbaum and B. Harthorn. 2000. Toward spatially integrated social science. *International Regional Science Review* 23: 139–159.

Haining, R.P. 1990. *Spatial Data Analysis in the Social and Environmental Sciences.* Cambridge University Press, Cambridge.

Haining, R.P. 2003. *Spatial Data Analysis: Theory and Practice.* Cambridge University Press, Cambridge.

Ihaka, R. and R. Gentleman. 1996. R: A language for data analysis and graphics. *Journal of Computational and Graphical Statistics* 5: 299–314.

Kelejian, H. and G. Piras. 2017. *Spatial Econometrics.* Academic Press.

LeSage, J.P. and K. Pace. 2009. *Introduction to Spatial Econometrics.* Boca Raton: CRC Press.

Loader, C. 1999. *Local Regression and Likelihood.* New York: Springer.

Matheron, G. 1963. Principles of geostatistics. *Economic Geology* 58: 1246–66.

Paelinck, J. and L. Klaassen. 1979. *Spatial Econometrics.* Farnborough: Saxon House.

Postiglione, P., M.S. Andreano and R. Benedetti. 2013. Using constrained optimization for the identification of convergence clubs. *Computational Economics* 42: 151–174.

Postiglione, P., M.S. Andreano and R. Benedetti. 2017. Spatial clusters in EU productivity growth. *Growth and Change* 48: 40–60.

Ripley, B.D. 1981. *Spatial Statistics.* Wiley: New York.

Särndal, C.-E., B. Swensson and J. Wretman. 1992. *Model Assisted Survey Sampling.* Springer: Berlin.

Schabenberger, O. and C.A. Gotway. 2005. *Statistical Methods for Spatial Data Analysis.* CRC: Boca Raton, FL.

Tobler, W.R. 1970. A computer movie simulating urban growth in the Detroit region. *Economic Geography* 46: 234–240.

von Thünen, J.H. 1826. *Der isolirte Staat in Beziehung auf Landwirtschaft und Nationalökonomie.* Hamburg.

Wheeler, D.C. and C.A. Calder. 2007. An assessment of coefficient accuracy in linear regression models with spatially varying coefficients. *Journal of Geographical Systems* 9: 145–166.

Whittle, P. 1954. On stationary processes in the plane. *Biometrika* 41: 434–449.

CHAPTER 2
Spatial Sampling Designs

Francesco Pantalone[1,*] and *Roberto Benedetti*[2]

2.1 Introduction

Usually, a spatial population (i.e., units distributed over a region of interest) is characterized by dependence between units. This is due to the fact that those units are influenced by the same set of factors, which in turn could affect our hypothetical variable of interest. This is especially true in agricultural surveys, where units close together are influenced by the same soil fertility, weather, pollution, and other spatial factors. Correspondingly, contiguous units are likely to be similar and this aspect should be carefully considered in the definition of a sampling design.

A sampling design is a method to select a portion of the population of interest. Technically, given a finite population of interest $U = \{1, \ldots, N\}$, a sampling design is a probability function $p(s)$ that assigns a selection probability to each possible subset of the population U, which we call sample and we indicate by s. The set of samples that have a positive probability to be selected by a sampling design is called support of the sampling design. The probability function $p(s)$, in turn, defines the probabilities of the units in the population to be included in the sample. In particular, the probability that unit i is included in the sample is called *first-order inclusion probability*, denoted by $\pi_i = \sum_{s \ni i} p(s)$, and the probability that units i and j are included jointly in the sample is called *second-order inclusion probability*, denoted by $\pi_{ij} = \sum_{s \supset \{i,j\}} p(s)$. For more details on the principles of sampling designs see Tillé and Whilelm (2017). It is important to underline the difference between the sampling design $p(s)$, and the sampling algorithm. With the latter, we implement the sampling design and select samples that respect the probabilities given by the $p(s)$. In doing so, the algorithm usually uses selection probabilities during some steps, which are not to be confused with the already treated inclusion probabilities. For a deep treatment of sampling algorithms see Tillé (2006).

[1] University of Perugia, Department of Economics.
[2] "G. d'Annunzio" University of Chieti-Pescara, Department of Economic Studies.
  Email: benedett@unich.it
* Corresponding author: francesco.pantalone@studenti.unipg.it

Since standard sampling designs do not consider the spatial dependence between the units, several spatial sampling designs have been introduced in the literature over the last few years. For a review, see Benedetti et al. (2015, 2017c).

In this chapter, we focus on spatially balanced sampling designs, which are sampling designs that achieve samples well spread over the population of interest. The theoretical framework we consider is the design-based approach, which assumes that the variable of interest y is fixed and the only source of randomness comes from the selection of the sample. Moreover, we suppose we are interested in the total of a quantity of interest, say $t_y = \sum_{i=1}^{N} y_i$. In this framework, a well-known estimator for the total is the Horvitz-Thompson estimator (Horvitz and Thompson 1952), defined as $\hat{t}_{HT} = \sum_{i \in s} \frac{y_i}{\pi_i}$ with $\pi_i > 0 \; \forall i \in U$. Its variance is equal to:

$$V_{HT}(\hat{t}_{HT}) = Var_{HT}(\hat{t}_{HT}) = \sum \sum_{U} \left(\frac{\pi_{ij}}{\pi_i \pi_j} - 1 \right) y_i y_j$$

and is estimated by

$$\hat{V}_{HT}(\hat{t}_{HT}) = \sum \sum_{s} \frac{1}{\pi_{ij}} \left(\frac{\pi_{ij}}{\pi_i \pi_j} - 1 \right) y_i y_j.$$

The use of spatially balanced designs could improve the efficiency of the final estimates, in the sense that the obtained variance of the HT estimator could be lower with respect to the variance obtained by the use of some non-spatial sampling designs.

The outline of the chapter is as follows: the first sections are dedicated to some practical aspects, specifically the frame in Section 2.2 and the process of data collection in Section 2.3. Both of these aspects are important in the survey process, because they could lead to problems if not treated correctly. In Section 2.4, we provide a brief introduction to non-spatial sampling designs, specifically the simple random sampling, widely used as benchmark and as part of more sophisticated sampling designs; the systematic sampling, especially useful for its simplicity; the stratified sampling, which is a simple and efficient way to consider some structure of the population at the design level, and the CUBE method, for balanced samples. In Section 2.5, we focus on spatially balanced sampling designs. In particular, we analyse the generalized random tessellation stratified sampling (Steven and Olsen 2004), based on a function that maps two-dimensional space onto a one-dimensional space; the spatially correlated Poisson sampling (Grafström 2012), adaptation of Correlated Poisson Sampling (Bondesson and Thorburn 2008); the local pivotal method (Grafström et al. 2012), modification of pivotal method (Deville and Tillé 1998); product within distance and sum within distance (Benedetti and Piersimoni 2017a), based on a MCMC algorithm and a summary index of a distance matrix. Finally, we provide some concluding remarks.

2.2 Setting up the frame

An important task of any survey is the identification of the target population (e.g., people aged over 40, firms with more than 50 employees, etc.). Once the target population is identified, the next step is to gather information about units. In particular, a list of units is needed in order to perform a probability sampling design. We collect such information through the sampling frame, which is defined as any material or device used to obtain observational access to the finite population of interest (Särndal 1992). In the following, we list some of the properties a sampling frame should have (Colledge 2004):

- identification of the units through the use of a unique identifier code;
- completeness, i.e., the frame contains all the units in the population;
- accuracy, i.e., there are no repetitions of units;
- additional information about the units (when possible).

The quality of the sampling frame has a large impact on the quality of any survey.

In agricultural survey, defective sampling frames are a common source of non-sampling error. Therefore, definition and data collection of the frame is an important task and should be carried out with particular attention. To this end, we provide some guidelines for the definition and the construction of the frame. Roughly speaking, we can define two main steps: (i) definition of the statistical unit; and (ii) definition and construction of the frame. In the following, we analyse these steps with particular focus on agricultural surveys.

(i) Definition of the statistical unit

We can broadly divide statistical units into two categories:

- *legal bodies*, such as farms, households and businesses. Agricultural surveys are usually based on farms, but households and businesses are frequently used as well;
- *spatial units*, which are portions of land areas and they could be points, polygons, and lines. In agricultural surveys, points and polygons are of the main interest. These portions of areas can be set by identifiable physical boundaries, such as rivers or roads, using squared grid of map coordinates or superimposing their limits to the boundaries of the land of the agricultural holding.

(ii) Definition and construction of the frame

According to the type of statistical unit, a different frame is employed. In particular, when legal bodies are used as statistical units, list frames are defined, while in the case of spatial units, spatial frames are considered. The former is a list of agricultural holding addresses, along with other information, such as holding size, crops, livestock, and possibly more. The addresses can be obtained from previous agricultural censuses, administrative data sources, and farmer associations (Wallgren and Wallgren 2007; 2010). Among administrative data sources, it is worth mentioning the so-called cadastre, which is the up-to-date land records, usually owned by country institutions. Moreover, it is important to consider holdings under different legal status, such as cooperatives, government farms, and enterprises (FAO 1995).

A spatial frame is a list of non-overlapping areas. The preparation of this type of frame is particularly demanding and requires up-to-date graphical materials, such as maps, satellite images, and aerial photos.

2.3 Survey data processing

Once the units have been selected, the process of data collection begins. More specifically, we can identify three major processes: data collection, data editing, and validation process. In the data collection process, the selected units are observed. The main objectives of this process are three: (i) identification of statistical unit (e.g., farm, polygon, point); (ii) collection of data without introduction of endogenous bias (e.g., avoid to influence the response); (iii) consideration of all stages of the sampling and eventually a longitudinal structure in order to facilitate any future contact. Once the data is collected, it is necessary to detect any eventual missing or inconsistent data and account for it through data imputation. This is the task of the data editing process, which is usually fully automatic or computer-assisted (Atkinson and House 2010). Finally, a validation process evaluates the consistency and the quality of the data, and it is composed by all operations required to check if the achieved results meet the planned quality targets. Not only, this validation assesses if the quality of the data is sufficient for distributing the data to users, but it could also identify possible sources of error. Therefore, this process could determine relevant modifications to the future surveys in order to reduce eventual future errors.

2.4 Non-spatial sampling designs

In this section, we briefly review some of the non-spatial sampling designs. With the term non-spatial, we mean that these sampling designs have not been designed specifically for the use on spatial settings. Nevertheless, they can be extended to that use in some situations, and we provide insights on how to apply them, if possible. We first consider the Simple Random Sampling, the Systematic Sampling, and the Stratified Sampling. Such designs are widely used in practice, as standalone or as part of more complex sampling designs. Then we introduce the CUBE method, which is a way to select balanced samples. For each case, we provide R code in order to perform sampling selection. In doing that, we use a simulated population U available in the dataset `simul2` from package `Spbsampling` (Pantalone et al. 2020).

2.4.1 Simple random sampling

From a randomization point of view, simple random sampling (SRS) is one of the most basic sampling designs. In fact, given n as the sample size and N the population size, the SRS assigns to each sample of dimension n the same probability to be selected, given by $p(s) = \dfrac{1}{\binom{N}{n}}$.

When speaking of SRS, we need to differentiate between two types: with replacement (SRSWR), and without replacement (SRSWOR). In the former, the unit selected in a draw is then re-inserted in the population, allowing the possibility to have the same unit more than once in the sample; in the latter, once the unit is selected, it cannot be re-selected. The most widespread method in use is SRSWOR, both for reasons of efficiency, since the expected number of distinct units in the sample is constant and equal to n, and therefore variance of estimators of population total (and mean) is smaller compared to when using SRSWR; and for practical reasons, because in many surveys we may not want to observe the same unit more than once. Therefore, we focus on SRSWOR and we refer to it simply by SRS.

In this scenario, every unit in the population has the same first-order inclusion probability, constant and equal to the sampling fraction f,

$$\pi_i = \frac{n}{N} = f \ \forall i \in U,$$

while the second-order inclusion probabilities are equal to

$$\pi_{ij} = \frac{n}{N}\frac{n-1}{N-1} \forall i \neq j \in U.$$

Thus, the HT estimator of the total is given by

$$\hat{t}_{HT,SRS} = \sum_{i \in s} \frac{y_i}{\pi_i} = \frac{N}{n} \sum_{i \in s} y_i,$$

and the variance estimate by

$$\hat{V}_{HT}(\hat{t}_{HT,SRS}) = N^2 \frac{1-f}{n} S_y^2 = N^2 \left(\frac{1}{n} - \frac{1}{N}\right) S_y^2,$$

where $S_y^2 = \sum_{i \in s} \frac{(y_i - \bar{y})^2}{n-1}$.

By definition, SRS does not allow the use of auxiliary information in the design phase, therefore, no kind of population structure can be exploited. Indeed, the use of SRS is recommended when, for instance, the population lacks hierarchical or geographical structure. Also, many more complex sampling designs use in some steps a simple random sampling.

From a methodological point of view, SRS provides important insights on the analysis of complex sampling designs. In fact, it usually represents a benchmark in design effects comparison. Given $\hat{V}(\hat{t}_{HT,p})$ as the estimated variance of the HT estimator of the total when a sample of size n obtained by a generic sampling design p is used, and $\hat{V}(\hat{t}_{HT,SRS})$ is the estimated variance of the HT estimator of the total when a SRS of size n is used, the design effect is defined as the following variance ratio

$$deff(\hat{t}_{HT,p}, \hat{t}_{HT,SRS}) = \frac{\hat{V}(\hat{t}_{HT,p})}{\hat{V}(\hat{t}_{HT,SRS})}.$$

When this ratio is lower than 1, the use of the sampling design p leads to a gain in efficiency with respect to the use of SRS, otherwise there is a loss of efficiency. In addition, the design effect can be used to analyse different couples of estimators and sampling designs (usually referred as sampling strategy). Furthermore, following this approach, we could compare the efficiency of different estimators, given the use of the same sampling design.

We can perform SRS in the R environment by means of the function `srswor()` of the package `sampling` (Tillé and Matei 2016). The function requires as inputs the parameters n and N, which are the sample size and the population size, respectively.

The output is a vector of size N, where the i-th element is equal to 1 if the i-th unit is selected in the sample, 0 otherwise. As an example, we select a sample of dimension 100 through the following code.

```
# -- selection of the sample -- #
>N <- nrow(simul2) # population size
>n <- 100 # sample size
>pi <- rep(n/N, N) # inclusion probabilities
>set.seed(42)
>srs_sample <- which(srswor(n, N) == 1)
```

2.4.2 Systematic sampling

Systematic sampling consists of selecting sample units from a list of population units using a constant skip, i.e., every r-th unit is selected in the sample, where the start is random in order to achieve random samples. The list plays a crucial role. If it is shuffled randomly before the selection, the systematic sampling can be compared with SRS from a theoretical point of view. If it is ordered according to some criteria, the efficiency of the selection is greatly influenced by that. For example, if a population of agricultural firms is ordered according to yield crop, then the sample would reflect the yield crop distribution of the population better than by chance. With that being said, the use of systematic sampling can often encounter the practical disadvantage of a ratio N/n not integer, which implies that it is often impossible to find a step r such that we sample exactly n units (in univariate populations, a solution is proposed in Särndal et al. (1992, p. 6)). Moreover, due to the sampling mechanism, the majority of second order inclusion probabilities are equal to zero. Therefore, the standard estimator for the variance of HT cannot be used. For some possible solutions, see Särndal et al. (1992, p.83). The most used workaround is to assume as upper bound of the variance an SRS variance estimate in order to provide a cautionary estimate.

Systematic sampling can be used to consider spatial aspects of the population. In fact, using the coordinates of the units in the ordering phase could lead to a good spatial coverage. Moreover, to improve and increase the use of the spatial information, we can divide the population into equally spaced clusters. Cochran (1997) showed that this design has a lower variance than SRS if $\rho_{IC} < -n/[N(n-1)]$, where ρ_{IC} is the correlation between a pair of units in the same systematic sample, or intraclass correlation coefficient. The main difficulty when using the systematic design in a spatial context is given by the fact that only a regular grid of points or a set of regularly shaped polygons have a natural ordering that can be used to spread the sample over the study region.

The following code selects one sample of dimension 100.

```
# -- selection of the sample -- #
>N <- nrow(simul2) # population size
>n <- 100 # sample size
>pi <- rep(n/N, N) # inclusion probabilities
>set.seed(42)
>sys_sample <- which(UPsystematic(pi) == 1)
```

2.4.3 Stratified sampling

Taking into account the population structure during the selection process is an important task that can greatly benefit the efficiency of the final estimates. This process requires good auxiliary information, and in stratified sampling such information is used in order to create subgroups of the population, which are called *strata*. Afterwards, a sample is independently selected from each stratum. The auxiliary information should be used in such a way that the created strata have differences between them but similar on their inside. Indeed, it has been shown that stratified sampling is more efficient than SRS when the units within each stratum are as similar as possible, and the units in different strata are as different as possible (Cochran 1977).

The population U is partitioned into H strata, say $\{U_1, U_2,..., U_h,..., U_H\}$, which are exhaustive, i.e., $\bigcup_{h=1}^{H} U_h = U$, and non-overlapping, i.e., $U_h \cap U_r = \emptyset \ \forall h \neq r$. Therefore, given $\{N_1, N_2,..., N_H\}$ be the numbers of units of the population belonging to each stratum, then $\sum_{h=1}^{H} N_h = N$. From each stratum, a sample of dimension n_h is selected. Theoretically, any sampling design can be employed when sampling from the strata, and even the use of different sampling designs across the strata is allowed. Regarding point and variance estimation, since the random selection inside a stratum is independent from every other random selection in the other strata, the HT estimator and corresponding variance estimator are given simply by the sum of the estimates in each stratum. This allows a great deal of flexibility, because different sampling designs can be employed across the strata without the need of face complications on the estimation phase.

Here, we consider that an SRS is selected from each stratum. Therefore, first-order inclusion probability for unit i in stratum h is equal to

$$\pi_i = \frac{n_h}{N_h} = f_h \ \forall i \in U_h$$

and second-order inclusion probabilities are given by

$$\pi_{ij} = \frac{n_h}{N_h} \frac{n_h - 1}{N_h - 1} \forall i \neq j \in U_h \text{ and } \pi_{ij} = \frac{n_h n_m}{N_h N_m} \forall i \in U_h \ \forall j \in U_m, h \neq m.$$

Hence, we have the HT of a total as in the follow

$$\hat{t}_{HT,STR} = \sum_{i \in s} \frac{y_i}{\pi_i} = \sum_{h=1}^{H} \frac{N_h}{n_h} \sum_{i \in s_h} y_i$$

and the variance estimator equal to

$$\hat{V}_{HT}(\hat{t}_{HT,STR}) = \sum_{h=1}^{H} N_h^2 \frac{1 - f_h}{n_h} S_{y,h}^2,$$

where $S_{y,h}^2 = \sum_{i \in s_h} \frac{(y_i - \bar{y}_h)^2}{n_h - 1}$.

Another choice to face with stratified sampling is size allocation, i.e., how to determine n_h for each stratum. Two approaches are Neyman Allocation and Optimal Allocation. Here, we just give the idea behind these two methods and we refer to

Särndal (1992) for a deep treatment of the subject. Neyman allocation minimizes the variance of a sample estimate subject to a given sample size, and under this allocation n_h should be proportional to $N_h S_h$, where S_h is the population standard deviation of stratum h. Optimal allocation minimizes the variance of a sample estimate subject to a given total cost, and n_h should be proportional to $N_h S_h / \sqrt{C_h}$, where C_h is the cost of observing a unit i in the stratum h.

Spatial auxiliary information can be used in order to stratify the population, even though is not always clear how to efficiently perform such task. A common solution is to define the strata as subareas of the region of interest.

The package `sampling` (Tillé and Matei 2016) provides the function `strata()`, which performs a stratified sampling according to the main following parameters:

- `data`: population frame;
- `stratanames`: vector of stratification variables;
- `size`: vector of stratum sample sizes;
- `method`: method used inside the strata in order to select the units. The methods implemented are sampling without replacement (`srswor`), simple random sampling with replacement (`srswr`), Poisson sampling (`poisson`), systematic sampling (`systematic`).

In the following, we create two strata from the dataset, then select a stratified sample, and from each stratum we select an SRS.

```
# -- selection of the sample -- #
>N <- nrow(simul2) # population size
>n <- 100 # sample size
>pi <- rep(n/N, N) # inclusion probabilities
>stra <- as.numeric(simul2$z21 > 5)
>f <- cbind(simul2, stra)
>set.seed(42)
>str_sample <- strata(data = f, stratanames = "stra", size =
c(50, 50), method = "srswor")
```

2.4.4 The CUBE method

Balanced sampling can be viewed as a calibration method employed into the selection process, and it takes advantage of auxiliary information. Given $\mathbf{x}_i = (x_{i1}, x_{i2}, \ldots, x_{ip})$ as the vector of p auxiliary variables measured on unit i, a sample is called *balanced* respect to those variables if the HT estimates of the totals are equal to the known totals $\mathbf{t}_x = \sum_{i \in U} \mathbf{x}_i$, that is

$$\sum_{i \in s} \frac{\mathbf{x}_i}{\pi_i} = \mathbf{t}_x. \tag{2.1}$$

If auxiliary variables used are correlated with the variable of interest, a balanced sample could lead to gain in efficiency estimates. Note that balanced sampling generalises other sampling designs. For example, a sampling design with fixed sample size n is balanced on the variable $x_i = \pi_i$, while a stratified sampling is balanced on the indicator variables of strata.

The CUBE method (Deville and Tillé 2004) selects balanced samples with any number of auxiliary variables, while a set of prescribed first-order inclusion probabilities is satisfied (even unequal probabilities are allowed). The algorithm starts from the vector $\boldsymbol{\pi} = (\pi_1, \pi_2, \ldots, \pi_N)$, and at each step at least one component is changed to 0 or 1. This change is made in such a way that the prescribed inclusion probabilities are satisfied, and the balancing equations (2.1) are satisfied as well. In particular, the method is composed of two phases. In the first one, the *flight phase*, the balancing equations are satisfied exactly. The second one, the *flight phase*, takes place in the eventuality that in the flight phase no sample satisfies the equations precisely, and it slightly relaxes the constraints as consequence. In this case we achieve a sample that approximately satisfies the balancing equations (2.1).

The implementation in the R environment is provided by the package BalancedSampling (Grafström and Lisic 2016), through the function cube(). Parameters needed are prob, vector of first order inclusion probabilities, and Xbal, matrix of auxiliary balancing variables. Note that, in order to have a fixed sample size n, we need to insert the first order inclusion probabilities that sum to n as first column of the matrix Xbal. The output of the function is a vector containing the labels of the units selected. In the following, we select a sample by the CUBE method using the spatial coordinates of the units as balancing variables.

```
# -- selection of the sample -- #
>N <- nrow(simul2) # population size
>n <- 100 # sample size
>pi <- rep(n/N, N) # inclusion probabilities
>X <- cbind(pi, simul2$x, simul2$y)
>set.seed(42)
>cube_sample <- cube(prob = pi, Xbal = X)
```

2.5 Spatially balanced sampling designs

Over the last decade, spatial data has increased substantially. New technologies allow geo-coded data to be obtained easily and, consequently, official statistical offices have started to geo-reference their registers. When dealing with spatial populations, we should consider the spatial structure. In fact, units that are spatially distributed usually show spatial dependence. In agricultural research, surveys are routinely used to gather data. The observed units are typically geo-referenced and it would, therefore, be desirable to take into account the spatial distribution when selecting the sample. *Spatially balanced sampling designs* address the task of taking into consideration the spatial dependence of the units in the population. In particular, these designs select samples well spread over the population of interest. Technically, a well spread sample has a number of selected units on every part of the study region close to what is expected on average (Grafström and Lundström 2013). The idea is that a spread sample could capture the spatial heterogeneity of the population, which in turn could improve the efficiency of estimates compared to the efficiency of estimates achieved by data obtained from non-spatial sampling design (e.g., SRS). We review two theoretical reasons that clarify this idea, that are the *lemma decomposition* and the use of the *anticipated variance*.

The decomposition lemma (Knottnerus 2003, p.87) states that, given a sample of size n drawn by a sampling design $p(s)$, the constant and unknown population variance of the variable $\breve{y}$ is given by

$$\sigma_{\breve{y}}^2 = V_S(\bar{\breve{y}}_S) + \frac{n-1}{n} E_S(S_{\breve{y},S}^2),$$

where $\breve{y}$ is the expanded vector of y/π_i, $V_S(\bar{\breve{y}}_S)$ is the variance between samples of HT estimator of the mean according to the design $p(s)$ and $E_S(S_{\breve{y},S}^2)$ is the expectation of the sample variances of $\breve{y}$. From this result, we can observe that HT estimator could be more efficient when the first-order inclusion probabilities are such that the y/π_is are constant or proportional to y, and/or the design $p(s)$ increases the expected within sample variance. The latter suggests that the $p(s)$ should be proportional or more than proportional to the sample variance S^2. Since this variance is unknown because it is relative to the unobserved variable of interest y, we should have auxiliary information related to it in order to exploit this theoretical insight. In the spatial setting, such information can be provided by the distance between units, as pointed out in the spatial interpolation literature (Ripley 1981; Cressie 1993).

Another way to understand the possible consequences of a well spread sample is by modelling the population. To this end, we introduce a spatial model that *(i)* links the variable of interest with some auxiliary information; *(ii)* considers the spatial correlation. From the model, we compute the *anticipated variance* (Isaki and Fuller 1982), from now on AV, of the HT estimator. In particular, the AV of a parameter of interest θ and its estimator $\hat{\theta}$ is given by

$$AV(\hat{\theta} - \theta) = E_m \left[E_S \left[(\hat{\theta} - \theta)^2 \right] \right] - \left[E_m \left[E_S \left[\hat{\theta} - \theta \right] \right] \right]^2,$$

where E_m denotes expectation with respect to the model and E_S denotes expectation with respect to the sample design.

The following linear model is assumed to hold for y_i, given the known auxiliary variables $\mathbf{x}_i$, for each unit in the population:

$$\begin{cases} y_i = \mathbf{x}_i'\boldsymbol{\beta} + \epsilon_i \\ E_m(\epsilon_i) = 0 \\ Var_m(\epsilon_i) = \sigma_i^2 \\ Cov_m(\epsilon_i\ \epsilon_j) = \sigma_i\sigma_j\rho_{ij} \end{cases} \tag{2.2}$$

where Var_m and Cov_m denote variance and covariance with respect to the model, respectively, $\boldsymbol{\beta}$ is a vector of regression coefficients, ϵ_i is a zero-mean random variable with variance σ_i^2 and ρ_{ij} is an autocorrelation parameter, for $i \neq j$ and such that $\rho_{ij} = 1$ if $i = j$. Under the model (2.2), the AV of the HT estimator of the total t_y is given by (Grafström and Tillé 2013)

$$AV(t_{y,HT}) = E_S E_m(t_{y,HT} - t_y)^2 = E_S \left[\left(\sum_{i \in s} \frac{x_i}{\pi_i} - \sum_{i \in U} \mathbf{x}_i \right)' \boldsymbol{\beta} \right]^2 + \sum_{i \in U} \sum_{j \in U} \sigma_i\sigma_j\rho_{ij} \frac{\pi_{ij} - \pi_i\pi_j}{\pi_i\pi_j}, \tag{2.3}$$

From equation (2.3), we observe that the uncertainty about the estimate can be split into two terms. The first term can be reduced selecting a sample that is balanced on **x**, while the second term of the equation can be reduced exploiting the spatial information of the population. Indeed, if ρ_{ij} increases as the distance between the units decreases, then selecting units far apart reduces the second term. Therefore, a sample well spread over the region of interest could potentially improve the HT estimation.

For the HT variance estimation, we cannot rely to the general variance estimator presented in the introduction since these sampling designs usually set the second order inclusion probabilities of nearby units to zero. Moreover, the computation of such probabilities can be prohibitive. To overcome this problem, we can employ the *local mean variance estimator* (Steven and Olsen 2003), or a generalization of that estimator by Grafström and Schelin (2014). Indeed, both proposals do not involve the use of the π_{ij}s. With a different approach, Benedetti et al. (2017b) proposed a model-based estimation of the variance in non-measurable designs.

The spatial balance of a sample can be measured by the *spatial balance index* (*SBI*, Steven and Olsen 2004), which is an index based on the use of Voronoi polygons. For a given sample *s*, a Voronoi polygon is constructed for each sample unit *i*, and it includes all population units closer to *i* than to any other sample unit *j*. For the *i*-th Voronoi polygon, v_i is defined as the sum of the inclusion probabilities of the units it contains. The index is then

$$SBI(s) = Var(v_i) = \frac{1}{n}\sum_{i \in s}(v_i - 1)^2.$$

Since for a spatially balanced sample $v_i \approx 1 \; \forall i \in s$ (Steven and Olsen 2004), then the closer *SBI* is to zero, the better the spread.

In the next sections, we will review the main spatially balanced sampling designs. The theory is presented along with illustrations in R code. For the sake of the illustration, we select samples from the same simulated population *U* we used in the previous section and, in order to investigate the spatial balance, we compute the corresponding SBI by means of the sbi() function, which can be found in the package Spbsampling (Pantalone et al. 2020). This function requires three parameters: dis, distance matrix that contains all the distances of all the pairs of units in the population, pi, vector of first order inclusion probabilities, and s, vector of labels of the units selected in the sample. Here, we compute the *SBI* for the samples we have already selected in the previous section. In this way, we can compare the results in term of spatial balance between the non-spatial and the spatial sampling designs.

```
# -- computation of sbi -- #
>d <- as.matrix(dist(cbind(simul2$x, simul2$y)))
>sbi(dis = d, pi = pi, s = srs_sample)
[1] 0.389899

# -- computation of sbi -- #
>d <- as.matrix(dist(cbind(simul2$x, simul2$y)))
```

```
>sbi(dis = d, pi = pi, s = sys_sample)
[1] 0.3810101

# -- computation of sbi -- #
>d <- as.matrix(dist(cbind(simul2$x, simul2$y)))
>sbi(dis = d, pi = pi, s = str_sample$ID_unit)
[1] 0.4539394

# -- computation of sbi -- #
>d <- as.matrix(dist(cbind(simul2$x, simul2$y)))
>sbi(dis = d, pi = pi, s = cube_sample)
[1] 0.4791919
```

2.5.1 *Generalized random tessellation stratified sampling*

Generalized random tessellation stratified (GRTS) sampling is the first spatially sampling design introduced in the literature (Stevens and Olsen 2004). Some preliminary works can be found in Stevens (1997), who derived inclusion and joint inclusion functions for several grid-based precursor designs to GRTS, and introduced the multiple-density, nested, random-tessellation stratified (MD-NRTS) design.

Core of GRTS is a function that maps two-dimensional space into one-dimensional space while preserving some spatial relationships. Specifically, the units located within a geographic region are placed on a line by means of a *hierarchical randomization process*, from where a systematic sample is then selected. In particular, a *quadrant-recursive* function is used through the ordering process of the units on the line. This process starts with a 2×2 square grid put randomly over the region of interest, with the resulting cells placed randomly in a line. Then, for each cell the same process is repeated, and the resulting sub-cells are randomly ordered inside the original cell (at this second step, 16 cells are put in a line). The process continues until at most one population unit occurs in a cell, and the generated random order is used to place the units on the line, which therefore has length N. A systematic sample of dimension n is then selected from the line. Indeed, the line is divided into N/n length segments, and a starting point r is selected randomly from $(0, N/n)$. Then, every $(r + iN/n)$th point, for $i = 1,\ldots,n-1$, is selected, where if the point occurs within one of the units, then that unit is selected (Brewer and Hanif 1983). If unequal probabilities are prescribed, then each point is given a length proportional to its inclusion probability. This procedure allows to achieve spatially balanced samples that respect given inclusion probabilities. Even though we have focused on area sampling, the GRTS can be implemented for point, linear features and not contiguous phenomena as well. However, there is the possibility to lose some spatial relationships during the hierarchization process, for instance it can happen that units close in the space may be far apart in the one-dimensional space. Finally, note that this method cannot be applied when units have more than two coordinates, i.e., when the space domain is given by $D \subset \mathbb{R}^h$, with $h > 2$.

GRTS is implemented on the R package `spsurvey` (Kincaid and Olsen 2016) through the function `grts()`. In order to ease the illustration, we present a function which wraps the original. Inputs are given by p, vector of first-order inclusion probabilities, and x and y, vector of first and second coordinates of the units, respectively. The output of this function is a vector of length *N* that contains 0 when the unit is not selected and 1 when is selected. In the following we select one sample, which is then plotted in Figure 2.1.

```
>GRTS <- function(p, x, y)
+ {
+ N <- length(p)
+ n <- round(sum(p))
+ index <- 1:N
+ s <- rep(0,times = N)
+ att <- data.frame(x = x, y = y, mdcaty = p, ids = index)
+ design <- list(None = list(panel = c(Panel1 = n),
+ seltype = "Continuous", caty.n = c("Caty 1" = n), over = 0))
>res <- grts(design, DesignID = "Site", SiteBegin = 1,
+ type.frame = "finite", src.frame = "att.frame",
+ in.shape = NULL, sp.object = NULL, att.frame = att, id = NULL,
+ xcoord = "x", ycoord = "y", stratum = NULL, mdcaty = "mdcaty",
+ startlev = NULL, maxlev = 11, maxtry = 1000,
+ shift.grid = TRUE, do.sample = rep(TRUE, length(design)),
+ shapefile = FALSE, prjfilename = NULL, out.shape = "sample")
+ s[res$ids] <- 1
+ s
}
```

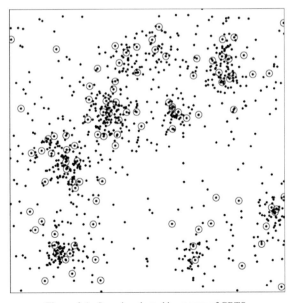

Figure 2.1. Sample selected by means of GRTS.

```
>N <- nrow(simul2) # population size
>n <- 100 # sample size
>pi <- rep(n/N, N)

# -- selection of the sample -- #
>set.seed(42)
>grts_sample <- which(GRTS(p = pi, x = simul2$x,
+ y = simul2$y) == 1)
Stratum: None
Current number of levels: 4
Current number of levels: 5
Current number of levels: 6
Final number of levels: 6

# -- computation of sbi -- #
>d <- as.matrix(dist(cbind(simul2$x, simul2$y)))
>sbi(d, p, grts_sample)
[1] 0.2290909

# -- plot -- #
>par(mar = c(1,1,1,1), xaxs = "i", yaxs = "i")
>plot(simul2$x, simul2$y, axes = F, cex = 0.5, pch = 19,
+ xlim = c(0, 1), ylim = c(0, 1))
>points(simul2$x[grts_sample], simul2$y[grts_sample],
+ pch = 1, cex = 2)
>box()
```

2.5.2 *Spatially correlated Poisson sampling*

The spatially correlated Poisson sampling (SCPS, Grafström 2012) is an adaptation of the correlated Poisson sampling (Bondesson and Thorburn 2008). The latter is an unequal probability sampling design that has been introduced for real time sampling, where a sampler visits the units one by one and it must decide at that moment if the unit is sampled or not. It is based on a list sequential criterion, where at step $t = 1,...,$ N the outcome of the t-th unit in the list is decided, i.e., in the first step, a decision about the first unit is taken, in the second step, a decision about the second unit is taken, and so on. The indicator variable I_t is used in order to record the outcome and it is set to 1 if the unit t has been selected, 0 otherwise. The decision at step t is taken according to a vector of selection probabilities $\pi^{(t)} = \pi_1^{(t)},...,\pi_N^{(t)}$, which are updated at each step after the decision has been made. The algorithm starts at step 1, where the vector of selection probabilities is set equal to the vector of inclusion probabilities, i.e., $\pi^{(0)} = \pi_1,..., \pi_N$, and the decision about the first unit is taken, which is therefore selected with probability $\pi_1^{(0)} = \pi_1$. At the general step t, the values for $I_1,..., I_{t-1}$ are known and the unit t is included with probability $\pi_t^{(t-1)}$. Afterwards, the vector of selection probabilities $\pi^{(t)}$ is updated. For unit t, $\pi_t^{(k)} = I_t$ for $k \geq t$, while for the units $i = t + 1,..., N$ the selection probabilities are updated according to the following rule

$$\pi_i^{(t)} = \pi_i^{(t-1)} - \left(I_t - \pi_t^{(t-1)}\right) w_t^{(i)}$$

where $w_t^{(i)}$ are weights given by unit t to the other units $i = t + 1,\dots N$. The algorithm continues until the N-th step, where the vector of selection probabilities has been modified N times and it consists of N indicator variables, that is the sample. It is important to underline that the selection probabilities are updated step by step in such a way that the prescribed inclusion probabilities are satisfied. Moreover, in order to have selection probabilities between zero and one, the weights need to satisfy the following constraint

$$-\min\left(\frac{1-\pi_i^{(t-1)}}{1-\pi_t^{(t-1)}}, \frac{\pi_i^{(t-1)}}{\pi_t^{(t-1)}} \right) \le w_t^{(i)} \le \min\left(\frac{\pi_i^{(t-1)}}{1-\pi_t^{(t-1)}}, \frac{1-\pi_i^{(t-1)}}{\pi_t^{(t-1)}} \right) \tag{2.4}$$

The weights are used to create negative or positive correlation between the indicator variables, since positive weights usually give negative correlation and negative weights usually give positive correlation.

The adaption introduced by the SCPS regards the weights, which are computed taking into account the distance between the units through a distance function $d(t,j)$. The author proposed two sets of weights: *maximal weights* and *Gaussian preliminary weights*. The first approach at each step t gives as much weight as possible to the closest unit $i = t + 1, t + 2,\dots N$, then as much weight as possible to the second closest unit, and so on. In case of equal distances, the weight is distributed equally on those units that have the same distance. The weights have to sum up to 1. This procedure implies that the sample size is locally balanced. In fact, for any subset A of all units within some subregion, the number of selected units is close to the sum of the corresponding inclusion probabilities, $\sum_{i \in A} I_i \approx \sum_{i \in A} \pi_i$.

In the second approach, the weights are regulated by a Gaussian distribution centered at the position of unit t and, hence, chosen by

$$w_t^{(i)*} \propto exp\left(-\left(\frac{d(i,t)}{\sigma} \right)^2 \right), i = t+1, t+2, \dots, N \tag{2.5}$$

where σ is a parameter that controls the spread of the weights and it can be chosen as the average of the distances between each unit and its closest neighbour. Weights defined by (2.5) do not always ensure that (2.4) is satisfied, and when this happens, they are cut off in order to (2.4) to hold (hence the name *preliminary* weights).

In both approaches, the order of the units in the list does not affect the general properties of SCPS. Important features of this method are given by the possibility to handle equal and unequal first-order inclusion probabilities, and to deal with n coordinates, i.e., space in $\mathbb{R}^n$.

The design is implemented on the R package `BalancedSampling` (Grafström and Lisic 2016) by the function `scps()`, which has only two parameters: `prob`, the first-order inclusion probabilities, and `x`, the matrix of auxiliary information, which in our case we use for the coordinates of the units. The output is a vector containing the labels of the selected units. In the following an illustration of the usage, with the selected sample plotted in Figure 2.2.

```
> N <- nrow(simul2) # population size
> n <- 100 # sample size
```

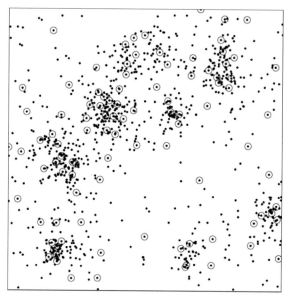

Figure 2.2. Sample selected by means of SCPS.

```
> pi <- rep(n/N, N) # first order inclusion probabilities
> X <- cbind(simul2$x, simul2$y) # matrix of coordinates
# -- selection of the sample -- #
> set.seed(42)
> scps_sample <- scps(prob = pi, x = X)
# --------------------------- #
# -- computation of sbi -- #
> d <- as.matrix(dist(cbind(simul2$x, simul2$y)))
> sbi(dis = d, pi = pi, s = scps_sample)
[1] 0.1622222
# ---------------------- #
# -- plot -- #
> par(mar = c(1, 1, 1, 1), xaxs = "i", yaxs = "i")
> plot(simul2$x, simul2$y, axes = F, cex = 0.5, pch = 19,
+ xlim = c(0, 1), ylim = c(0, 1))
+ points(simul2$x[scps_sample], simul2$y[scps_sample],
+ pch = 1, cex = 2)
> box()
# --------- #
```

2.5.3 *Local pivotal method*

The *local pivotal method* (LPM, Grafström et al. 2012) is based on the pivotal method (Deville and Tillé 1998), which is an unequal probability sampling design that works as follows. At each step, the inclusion probabilities are updated for a couple of units in such a way that the sampling outcome is decided for at least one of the units. Once a unit is selected, it cannot be chosen again. The procedure continues until a decision has been taken for all the units in the population, which can take at most N steps. The probabilities at each step are updated according to the following rules. Let i and j the couple of units chosen, with corresponding probabilities inclusion π_i and π_j. If $\pi_i + \pi_j < 1$, then

$$(\pi_i', \pi_j') = \begin{cases} (0, \pi_i + \pi_j) \text{ with probability} \dfrac{\pi_j}{\pi_i + \pi_j} \\[4mm] (\pi_i + \pi_j, 0) \text{ with probability} \dfrac{\pi_i}{\pi_i + \pi_j} \end{cases}$$

otherwise, if $\pi_i + \pi_j \geq 1$, then

$$(\pi_i', \pi_j') = \begin{cases} (1, \pi_i + \pi_j - 1) \text{ with probability} \dfrac{1 - \pi_j}{2 - \pi_i - \pi_j} \\[4mm] (\pi_i + \pi_j - 1, 1) \text{ with probability} \dfrac{1 - \pi_i}{2 - \pi_i - \pi_j} \end{cases}$$

The way the units are chosen at each step determines the properties of the sampling design. In the original paper, one proposed solution is to randomly choose them (Deville and Tillé 1998). In this case, the pivotal method has high entropy (Grafstrom 2010).

The adaptation introduced in the LPM consists of choosing the units in a different way in order to achieve spatial balance. In fact, the units are chosen randomly between nearby units. There are two versions of this design, referred by LPM1, which obtains more spatially balanced samples, and LPM2, which does not obtain the same level of spread of the LPM1 but is faster and simpler. In fact, the number of expected number of computations for LPM1 is in the worst case proportional to N^3, and proportional to N^2 in the best case, while for LPM2 is always proportional to N^2.

As in the case of the SCPS design, LPM provides the possibility to handle equal and unequal first-order inclusion probabilities and to deal with space in $\mathbb{R}^n$.

Both versions of this sampling design are implemented in the R package BalancedSampling (Grafström and Lisic 2016) by the functions lpm1() and lpm2(), respectively. Both functions use as inputs the parameter p for the first-order inclusion probabilities, and the parameter x for the matrix of auxiliary variables. The output is a vector of the labels of the units selected. In the following, we select a sample for each version of this sampling design (Figure 2.3).

```
>N <- nrow(simul2) # population size
>n <- 100 # sample size
>pi <- rep(n/N, N) # first order inclusion probabilities
>X <- cbind(simul2$x, simul2$y) # matrix of coordinates
# -- selection of the sample -- #
>set.seed(42)
>lpm1_sample <- lpm1(prob = pi, x = X)
```

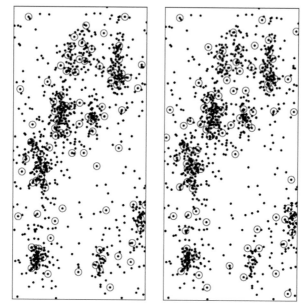

Figure 2.3. Sample selected by means of LPM1 on the left, sample selected by means of LPM2 on the right.

```
# ---------------------------- #
# -- computation of sbi -- #
>d <- as.matrix(dist(cbind(simul2$x, simul2$y)))
>sbi(dis = d, pi = pi, s = lpm1_sample)
[1] 0.1159596
# ---------------------- #
# -- selection of the sample -- #
>set.seed(42)
>lpm2_sample <- lpm2(prob = pi, x = X)
# ---------------------------- #
# -- computation of sbi -- #
>sbi(dis = d, pi = pi, s = lpm2_sample)
[1] 0.1632323
# --------------------- #
# -- plot -- #
>par(mfrow = c(1, 2), mar = c(1, 1, 1, 1), xaxs = "i",
+ yaxs = "i")
>plot(simul2$x, simul2$y, axes = F, cex = 0.5, pch = 19,
+ xlim = c(0, 1), ylim = c(0, 1))
>points(simul2$x[lpm1_sample], simul2$y[lpm1_sample], pch
= 1, + cex = 2)
>box()
>plot(simul2$x, simul2$y, axes = F, cex = 0.5, pch = 19,
+ xlim = c(0, 1), ylim = c(0, 1))
```

```
>points(simul2$x[lpm2_sample],     simul2$y[lpm2_sample],
pch = 1, + cex = 2)
>box()
```

2.5.4 Product within distance—sum within distance

Product within distance and sum within distance (Benedetti and Piersimoni 2017a) are two sampling designs based on an iterative procedure that relies on a distance matrix $\mathbf{D}_U = \{d_{ij}; i = 1,\dots, N; j = 1,\dots, N\}$, which contains all the distances of all the pairs of units in the population, and an index $M(\mathbf{D}_S)$ of the distance matrix of a given sample that summarizes the distances of the sample units, where the higher the overall distance, the higher the index. The distances between the units can be computed with any metric and in any dimension, for example, Euclidean, Manhattan distance or similarity measure, and this provides the flexibility to handle different spatial settings. Given $\mathbf{s}^{(t)} = \{l_1, l_2,\dots, l_n\}$ as a vector of unit labels for the t-th iteration, the algorithm starts at iteration $t = 0$ where a simple random sample of size n is selected. Then, for each iteration t, the following steps are performed:

1. two units, i and j, are selected randomly. One unit is selected from inside the current configuration, i.e., $i \in \mathbf{s}^{(t)}$, the other one is selected from outside the current configuration, i.e., $i \notin \mathbf{s}^{(t)}$;

2. a new configuration $\mathbf{s}_e^{(t)}$ is defined, and it is given by exchanging the two units selected in the previous step. In this way, $\mathbf{s}^{(t)}$ and $\mathbf{s}_e^{(t)}$ differ only for one unit;

3. the current configuration $\mathbf{s}^{(t)}$ is updated to $\mathbf{s}_e^{(t)}$ or it is retained. Specifically, the update occurs with a probability given by the following acceptance rule

$$p = min\left[1, \left(\frac{M\left(D_{S_e^{(t)}}\right)}{M\left(D_{S^{(t)}}\right)}\right)^{\beta}\right] \qquad (2.6)$$

with β known constant. The acceptance rule gives more probability to sample that are relatively more spread, since $M(D_{S_e^{(t)}})$ is higher when the overall distances on the configurations is higher. Moreover, the parameter β regulates the amount of spread of the sample: the higher it is, the more spread the sample will be.

4. steps 1, 2 and 3 are repeated N times.

The algorithm runs for $t = 1,\dots T$ iterations, for a total of $T \times N$ steps, unless at iteration $t < T$ no update occurred, in that case, the algorithm is stopped. The number T is given in advance: the higher, the better the convergence, the slower the algorithm. The authors proposed a value of 10 as a good compromise. This procedure can be seen as an MCMC-based algorithm, where at each step a Markov-chain is run and given a configuration, a new one is selected or rejected according to the Metropolis-Hastings criterion acceptance rule (Robert and Casella 2010). A random outcome from a multivariate probability $p(s)$ proportional to the summary index M of the distance matrix is then generated. Two indexes M are proposed

$$M_0(D_s) = \prod_{i:s_i=1} \prod_{j\neq i:s_j=1} d_{ij}, \qquad (2.7)$$

$$M_1(D_s) = \sum_{i;s_i=1} \sum_{j;s_j=1} d_{ij}. \tag{2.8}$$

When (2.7) is used, the sampling design is referred to as *product within distance* (*PWD*), while when (2.8) is used, the sampling design is referred to as *sum within distance* (*SWD*).

With regard to inclusion probabilities, the distance matrix plays a key role. Indeed, in order to achieve constant equal inclusion probabilities equal to n/N, the distance matrix has to be standardized, since different standardizations can lead to different set of inclusion probabilities. When we use the *SWD*, we iteratively constrain the row (or column) sums of the matrix to known totals (e.g., 1), while when we use the *PWD* we iteratively constrain the row (or column) sums of the logarithmic transformed matrix to known totals (e.g., 0). Moreover, at each iteration of the standardization, an average with its transpose is performed, so that the symmetry of the matrix is conserved. For more details, see Benedetti and Piersimoni (2017a).

These methods are implemented on the R package Spbsampling (Pantalone et al. 2020). The functions performing the sampling designs are pwd() and swd(), which share the same following parameters:

- dis: distance matrix $N \times N$ that specifies how far all the pairs of units in the population are;
- n: sample size;
- beta: level of spread, with default value equal to 10.
- nrepl: number of samples to draw, default equal to 1;
- niter: maximum number of iterations for the algorithm, default equal to 10.

The output is a list with components s, a matrix nrepl x n that contains the nrepl selected samples, each of them stored in a row, and iterations, the number of iterations run by the algorithm.

Regarding matrix standardization, the package provides a function to be used along pwd(), which is called stprod(), and a function to be used with swd(), that is stsum(). These functions have the same parameters:

- mat: matrix to be standardized;
- con: vector of constraints;
- differ: maximum accepted difference with the constraint, with default equal to 1e-15;
- niter: maximum number of iterations, set to a default value of 1000.

Both functions have as output a list with components mat, which is the standardized matrix, iterations, that is the number of iterations run by the algorithm, and conv, the convergence reached.

The code for the *PWD* design is the following. Note that we standardize the distance matrix in order to get equal inclusion probabilities n/N. Selected sample is showed in Figure 2.4.

```
>N <- nrow(simul2) # population size
>n <- 100 # sample size
```

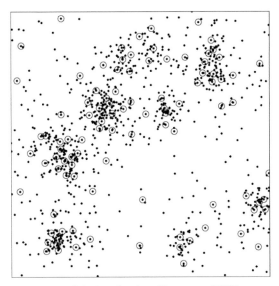

Figure 2.4. Sample selected by means of *PWD*.

```
>pi <- rep(n/N, N) # first order inclusion probabilities
# -- standardization of the distance matrix -- #
>d <- as.matrix(dist(cbind(simul2$x, simul2$y)))
>s_d <-stprod(mat = d, con = rep(0, N))$mat
# ----------------------------------------- #
# -- selection of the sample -- #
>set.seed(42)
>pwd_sample <- pwd(dis = s_d, n = n)$s
# --------------------------- #
# -- computation of sbi -- #
>sbi(dis = d, pi = pi, s = pwd_sample)
[1] 0.0589899
# ----------------------- #
# -- plot -- #
>par(mar = c(1, 1, 1, 1), xaxs = "i", yaxs = "i")
>plot(simul2$x, simul2$y, axes = F, cex = 0.5, pch = 19, xlim =
+ c(0, 1), ylim = c(0, 1))
>points(simul2$x[pwd_sample], simul2$y[pwd_sample], pch = 1,
+ cex = 2)
>box()
```

In the following, the code for the *SWD* design is given. As before, we standardize the distance matrix. Selected sample is depicted in Figure 2.5.

```
>N <- nrow(simul2) # population size
>n <- 100 # sample size
>pi <- rep(n/N, N) # first order inclusion probabilities
# -- standardization of the distance matrix -- #
```

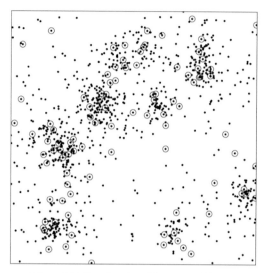

Figure 2.5. Sample selected by means of *SWD*.

```
>d <- as.matrix(dist(cbind(simul2$x, simul2$y)))
>s_d <-stsum(mat = d, con = rep(1, N))$mat
# ------------------------------------------- #
# -- selection of the sample -- #
>set.seed(42)
>swd_sample <- swd(dis = s_d, n = n)$s
# --------------------------- #
# -- computation of sbi -- #
>sbi(dis = d, pi = pi, s = swd_sample)
[1] 0.2870707
# ---------------------- #
# -- plot -- #
>par(mar = c(1, 1, 1, 1), xaxs = "i", yaxs = "i")
>plot(simul2$x, simul2$y, axes = F, cex = 0.5, pch = 19, xlim =
c(0, 1), ylim = c(0, 1))
>points(simul2$x[swd_sample], simul2$y[swd_sample], pch  =
1,
+ cex = 2)
>box()
```

2.6 Conclusion

Populations distributed over the space are characterized by spatial dependence that has to be accounted for. The major challenge in the design phase is how to incorporate and use the spatial information in the mechanism selection. In fact, exploiting such information could lead to great improvement on the efficiency of final estimates. Since standard sampling designs do not offer the possibility to incorporate and account for any spatial structure, several spatial sampling designs have been introduced in the literature. In this chapter, we have focused on spatially balanced sampling designs. Firstly, we have seen some theoretical reasons in support of the idea that a well spread sample could capture spatial heterogeneity and, therefore, improve the final estimates. Secondly, we have reviewed the main spatially balanced sampling designs, along with illustrations on how to implement these methods through the R statistical programming language.

References

Benedetti, R., F. Piersimoni and P. Postiglione. 2015. *Sampling Spatial Units for Agricultural Surveys.* Springer: Berlin, Heidelberg.

Benedetti, R. and F. Piersimoni. 2017a. A spatially balanced design with probability function proportional to the within sample distance. *Biometrical Journal* 59: 1067–1084.

Benedetti, R., G. Espa and E. Taufer. 2017b. Model-based variance estimation in non-measurable spatial designs. *Journal of Statistical Planning and Inference* 181: 52–61.

Benedetti, R., F. Piersimoni and P. Postiglione. 2017c. Spatially balanced sampling: A review and a reappraisal. *International Statistical Review* 85: 439–454.

Bondesson, L. and D. Thorburn. 2008. A list sequential sampling method suitable for real-time sampling. *Scandinavian Journal of Statistics* 35: 466–483.

Cochran, W.G. 1977. *Sampling Techniques.* Wiley: New York.

Colledge, M. 2004. OECD/UNESCAP/ADB. Workshop on assessing and improving statistical quality: measuring the non-observed economy. Bangkok, 11–14 May 2004.

Deville, J.C. and Y. Tillé. 2004. Efficient balanced sampling: the cube method. *Biometrika* 91: 893–912.

FAO. 1995. *Conducting Agricultural Censuses and Surveys.* FAO Statistical Development Series 6. FAO, Rome.

Grafström, A. 2010. Entropy of unequal probability sampling designs. *Statistical Methodology* 7: 84–97.

Grafström, A. 2012. Spatially correlated Poisson sampling. *Journal of Statistical Planning and Inference* 142: 139–147.

Grafström, A., N.L.P. Lundström and L. Schelin. 2012. Spatially balanced sampling through the pivotal method. *Biometrics* 68: 514–520.

Grafström, A. and Y. Tillé. 2013. Doubly balanced spatial sampling with spreading and restitution of auxiliary totals. *Environmetrics* 24: 120–131.

Grafström A. and L. Schelin. 2014. How to select representative samples. *Scandinavian Journal of Statistics* 41(2): 277–290.

Grafström, A. and J. Lisic. 2016. *BalancedSampling: Balanced and Spatially Balanced Sampling.* R package version 1.5.2, https://CRAN.Rproject.org/package=BalancedSampling.

Grafström, A. and N.L.P. Lundström. 2013. Why well spread probability samples are balanced. *Open Journal of Statistics* 3: 36–41.

Horvitz, D.G. and D.J. Thompson. 1952. A generalization of sampling without replacement from a finite Universe. *Journal of the American Statistical Association* 47: 663–685.

Isaki, C.T. and W.A. Fuller. 1982. Survey design under the regression superpopulation model. *Journal of the American Statistical Association* 77: 89–96.

Johnson, C., R. Masson and M. Trosset. 2005. On the diagonal scaling of Euclidean distance matrices to doubly stochastic matrices. *Linear Algebra and Its Applications* 397: 253–264.

Kincaid, T.M. and A.R. Olsen. 2016. *Spsurvey: Spatial Survey Design and Analysis.* R package version 3.3.

Knottnerus, P. 2003. *Sample Survey Theory: Some Pythagorean Perspectives.* Springer, New York.

Pantalone, F., R. Benedetti and F. Piersimoni. 2020. *Spbsampling: Spatially Balanced Sampling.* R package version 1.3.4, URL https://CRAN.R-project.org/package=Spbsampling.

Robert, C.P. and G. Casella. 2010. *Introducing Monte Carlo Methods with R.* Springer.

R Core Team. 2018. *R: A Language and Environment for Statistical Computing.* R Foundation for Statistical Computing, Vienna, Austria. URL https://www.R-project.org.

Särndal, C.E., B. Swensson and J. Wretman. 1992. *Model Assisted Survey Sampling.* Springer: New York.

Stevens, D.L., Jr. 1997. Variable density grid-based sampling designs for continuous spatial populations. *Environmetrics* 8: 167–195.

Stevens, D.L. Jr. and A.R. Olsen. 2003. Variance estimation for spatially balanced samples of environmental resources. *Environmetrics* 14(6): 593–610.

Stevens, D.L. Jr. and A.R. Olsen. 2004. Spatially balanced sampling of natural resources. *Journal of the American Statistical Association* 99: 262–278.

Tillé, Y. 2006. *Sampling Algorithms.* Springer, New York.

Tillé, Y. and A. Matei. 2016. *Sampling: Survey Sampling.* R package version 2.8. https://CRAN.R-project.org/package=sampling.

Tillé, Y. and M. Wilhelm. 2017. Probability sampling designs: Principles for choice of design and balancing. *Statistical Science* 32: 176–189.

Wallgren, A. and B. Wallgren. 2007. *Register-based Statistics: Administrative Data for Statistical Purposes.* Wiley, Chichester.

Wallgren, A. and B. Wallgren. 2010. Using administrative registers for agricultural statistics. pp. 27–44. *In*: Benedetti, R., M. Bee, G. Espa and F. Piersimoni (eds.). *Agricultural Survey Methods.* Wiley, Chichester.

CHAPTER 3
Including Spatial Information in Estimation from Complex Survey Data

Francesco Pantalone[1],* and *Maria Giovanna Ranalli*[2]

3.1 Introduction

This chapter focuses on the estimation phase of a survey. We assume that we have data coming from a possibly complex sampling design and that we are interested in doing inference for a set of descriptive parameters of a finite population. We focus on estimation of finite population totals and (differentiable) functions thereof. We consider a design-based framework for inference, that is, the population values of the variables of interest are considered as fixed quantities, so that randomness comes into play only from the sampling design. Since we are interested in accounting for the spatial structure that very likely characterizes agricultural and environmental populations, we also consider superpopulation models in our inferential framework. These models *assist* our estimation procedures in the hope of providing more accurate estimates. However, we only consider design-consistent estimators that are robust to model misspecifications: these estimators show an improved efficiency if the model well describes the population values of the variable of interest, but they are not biased in the case of model failure. Therefore, we consider a *model-assisted* approach to inference in the sense of Särndal et al. (1992) and not a model-based approach.

In order to improve in efficiency over the basic Horvitz-Thompson design-based estimator, we do not only need a good model, but also relevant auxiliary information. Population level auxiliary information can have different sources and can take different levels of detail. Section 3.5 discusses the different types and roles of auxiliary information for model-assisted inference. According to the available

[1] University of Perugia, Department of Economics.
[2] University of Perugia, Department of Political Science.
  Email: giovanna.ranalli@unipg.it
* Corresponding author: francesco.pantalone@studenti.unipg.it

auxiliary information, there are several alternative choices for the assisting model in the definition of the final estimator. We move from simple linear regression models and well-known regression estimators to more complex assisting models in order to better capture the spatial structure of agriculture and environmental population data. Variance estimation is also discussed in order to provide tools to evaluate the precision of estimates. When available, analytic variance estimator formulas should be used to avoid computationally intensive procedures. Resampling methods, such as bootstrap and jackknife, are useful alternatives when the quality of analytic variance formulas is deemed unreliable and when the parameters of interest are (possibly complicated) non-linear or non-differentiable functions of population totals. We will not discuss resampling methods here. See Wolter (1985) and Mashreghi et al. (2016) for an introduction to resampling methods and an updated treatment of bootstrap, respectively.

Estimators and variance estimators are introduced; then, their computation is illustrated on real data coming from a survey on lakes in the Northeastern states of the U.S. conducted by the Environmental Protection Agency. Numerical illustrations are carried out using two leading packages for survey sampling and estimation in the R software (R Core Team 2019): the package `sampling` (Tillé and Matei 2016) and the package `survey` (Lumley 2019), along with *ad hoc* functions reported in the Appendix.

The structure of the chapter is as follows. In Section 3.2 we introduce notation and unbiased estimation from a design-based perspective, then in Section 3.3 we describe the data used to illustrate the methods in this chapter. We dedicate Section 3.4 to the computation of basic estimators, Section 3.5 to discuss the role of auxiliary information, and Section 3.6 to the framework of model-assisted estimation. In Section 3.7 we show the first class of estimators analyzed in the chapter, based on linear models, while Sections 3.8 and 3.9 are dedicated to more sophisticated estimators which involve generalized linear models for binary data and geoadditive models to include geographical information in a more flexible way, respectively. Finally, we provide concluding remarks in Section 3.10.

3.2 Notation and basic estimators

Usually, real finite populations are quite complex to model, especially if we are concerned with estimating parameters for more than one variable of interest. The advantage of the design-based and, to some extent, the model-assisted approach to inference, is the possibility to avoid choosing and relying on a statistical model for each variable of interest. Instead, we assume the randomization principle, which states that randomness is coming only from the selection of the sample, and we construct estimators that rely on this principle. In this setting, for the population of interest $U = \{1,...,i,..., N\}$, we assume that the values of the variable of interest, y_i, $i = 1,..., N$, are fixed and unknown quantities. A sample s is a subset of U selected using a sampling design $p(s)$, which is a probability function that assigns a probability to each possible sample from U. The sample is represented through an indicator random variable I_i, that takes value 1 when unit i is selected in the sample, and 0 otherwise. The probability of selecting unit i in the sample is the expectation of I_i, it is called *first-order inclusion probability* and is denoted by π_i. The probability of selecting unit i and j jointly in the sample is the expectation of I_iI_j, it is called *second-order inclusion probability* and is denoted by π_{ij}. Let $\Delta_{ij} = \pi_{ij} - \pi_i\pi_j$.

For simplicity now, consider the case in which we are interested in estimating the population total of y, denoted by $t_y = \sum_{i=1}^{N} y_i$. In this framework, an unbiased estimator of t_y is the Horvitz and Thompson (1952) estimator, HT from now on, given by

$$HT(y) = \sum_{i \in s} \frac{y_i}{\pi_i}.$$

This estimator is unbiased as long as all the units in the population have a positive first order inclusion probability. This is a simple estimator, yet an unbiased one. The price to pay for this simplicity is that, in situations where there is low correlation between the variable of interest and the inclusion probabilities, the estimator can be inefficient. In fact, the variance of $HT(y)$ is given by

$$V(HT(y)) = \sum_{i=1}^{N}\sum_{j=1}^{N} \Delta_{ij} \frac{y_i}{\pi_i} \frac{y_j}{\pi_j},$$

where $\pi_{ij} = \pi_{ii}$ if $i = j$, and it can be estimated by

$$\hat{V}_{HT}(HT(y)) = \sum_{i \in s}\sum_{j \in s} \frac{y_i}{\pi_i} \frac{y_j}{\pi_j} \frac{\Delta_{ij}}{\pi_{ij}}.$$

The population mean of y is given by t_y/N and can be estimated using $HT(y)/N$. The variance is given by $V_{HT}(HT(y))/N^2$ and can be estimated by $\hat{V}_{HT}(HT(y))/N^2$. When the population size N is not known, a common alternative is the *Hajek estimator, HJ(y)*, that is given by $HT(y)/HT(1)$, where $HT(1) = \sum_s \pi_i^{-1} = \hat{N}$ denotes the HT estimate of the population size. Note that the Hajek estimator is recommended even when N is known. In fact, the Hajek estimator is more efficient than the Horvitz-Thompson estimator in the case of weak or negative correlation between π_i and y_i and when the sampling design is such that $\sum_s \pi_i^{-1} \neq N$ by definition.

3.3 The U.S. North-eastern Lakes survey

The data used throughout the chapter come from a survey conducted by the U.S. Environmental Protection Agency through the Environmental Monitoring and Assessment Program (EMAP), between 1991 and 1996, on lakes in the North-eastern states of the US. We refer to Messer et al. (1991) and to Larsen et al. (2001) for a description of the EMAP and of the North-eastern lakes survey. In particular, the survey is based on a population of 21,026 lakes, from which 334 lakes were surveyed, some of which visited several times during the study period. Figure 3.1 displays the area of interest and the sample locations. These lakes were selected by means of a complex sampling design based on the use of a hexagonal grid frame. See Larsen et al. (1993) for a description of the sampling design. The total number of measurements is 551. If multiple measurements are available for the same lake, we average these in order to obtain one measurement per lake sampled.

The major variable of interest from the survey is the acidity of the lake, measured through the acid neutralizing capacity (ANC), which is defined as the ability of the

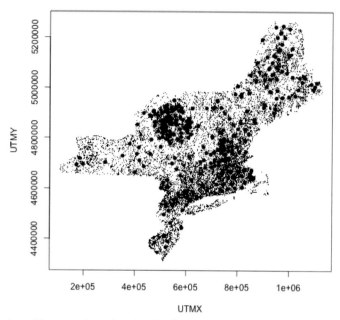

Figure 3.1. Area of interest and sample points. Smaller dots represent the lakes in the frame, larger dots represent the sampled lakes.

water to buffer acid. An ANC value less than 0 μeq/L indicates that the water has lost all its ability to buffer acid. Surface waters with ANC values below 200 μeq/L are considered at risk of acidification, and values less than 50 μeq/L are considered at high risk. Moreover, some variables are available for each lake in the population, such as elevation, x-geographical coordinate and y-geographical coordinate of the centroid of each lake in the UTM coordinate system, and a categorical variable for eco-region.

Regarding variance estimation, since second order inclusion probabilities are not available from this dataset, we cannot use the standard variance estimator and we need a workaround in order to achieve a measure of accuracy for the estimates. To this end, we treat the complex sampling design as a stratified sample taken with replacement. In particular, we use 14 strata corresponding to groups of spatial clusters of lakes that were used in the original sampling design. These clusters were used to ensure spatial distribution of the sampled lakes over the region of interest. For more details on the construction of the spatial clusters, see Larsen et al. (1993). Let H be the number of strata, n_h the number of sample observations within stratum h, $p_i = n_h^{-1}\pi_i$ and s_h the set of sampled units that fall in stratum h, then we write the variance estimator for the total as $\sum_{h \in H} S_h^2$, where S_h^2 is the estimated within-stratum weighted residual variance for stratum h. In the case of $HT(y)$, the estimated variance is given by

$$\hat{V}(HT(y)) = \sum_{h \in H} \frac{1}{n_h(n_h - 1)} \sum_{i \in s_h} \left(\frac{y_i}{p_i} - \sum_{j \in s_h} \frac{y_j}{\pi_j} \right)^2,$$

while for the Hajek estimator of the population mean we have

$$\hat{V}(HJ(y)) = \frac{1}{\hat{N}^2} \sum_{h \in H} \frac{1}{n_h(n_h - 1)} \sum_{i \in s_h} \left(\frac{y_i - HJ(y)}{p_i} - \sum_{j \in s_h} \frac{y_j - HJ(y)}{\pi_j} \right)^2.$$

Breidt et al. (2007) follow the same approach and provide more details on this. In order to implement variance estimation, we provide in the Appendix an *ad hoc* function, called `var_est()`, specifically designed for this dataset, which computes the S_h^2's and provides the variance estimate for the estimate of a population total.

To carry out estimation, we have two datasets available: the first one, `lakes.fr`, contains the information on the frame of the population, while the second one, `lakes`, contains information on the sample. As mentioned before, in the former we have the following variables available for each lake in the population:

- `UTMX`, *x*-geographical coordinate of the centroid of the lake, in UTM coordinate system;
- `UTMY`, *y*-geographical coordinate of the centroid of the lake, in UTM coordinate system;
- `GRID.ELEV.M`, elevation of the lake;
- `ECO.NUM`, classification code for seven eco-regions.

In the sample dataset, we have the variables from the frame and all the data coming from the survey. In particular, we will focus on the following variables:

- `ANC`, Acid Neutralizing Capacity, measured in µeq/L;
- `MG`, Magnesium, measured in µeq/L;
- `CL`, Chloride, measured in µeq/L;
- `SO4`, Sulfate, measured in µeq/L;
- `ACRISK`, binary response that indicates if the lake is considered at risk of acidification or not, 1 being at risk, 0 being not at risk;
- `WGT.3X`, basic design weight, that is π_i^{-1};
- `CLUSTER.1`, spatial cluster, stratification variable.

3.4 Basic estimators from the North-eastern lakes survey data

In this section, we compute the basic estimators and corresponding standard errors introduced previously. As target of inference, we consider `ANC` and the binary variable `ACRISK`. In particular, we focus on the estimation of the population total and mean of these variables. We use the function `HTestimator()` from package `sampling` (Tillé and Matei 2016) to compute $HT(y)$. This function requires two arguments: y, the variable of interest (or variables if we supply a vector), and `pik`, the inclusion probabilities. For variance estimation, we use the above mentioned `var_est()` function reported in the Appendix. It requires two arguments as well, the variable of interest and the basic design weight.

To begin, we load the required library and we attach the dataset.

```
>library(sampling)
>N <- nrow(lakes.fr)
>n <- nrow(lakes)
>attach(lakes)
>N_hat <- HTestimator(rep(1, n), 1/WGT.3X)
```

The following code computes the estimator of the total, of the mean and the corresponding standard errors for the variable ANC.

```
# -- HT estimator of the total of ANC -- #
>ht <- HTestimator(ANC, 1/WGT.3X)
>ht
          [,1]
[1,] 11924749
# -- SE of HT total of ANC -- #
>se_ht <- sqrt(var_est(ANC, WGT.3X))
>se_ht
[1] 1812230
# -- HJ estimator of the mean of ANC -- #
>hj <- HTestimator(ANC, 1/WGT.3X) /
+ HTestimator(rep(1, n), 1/WGT.3X)
>hj
         [,1]
[1,] 555.941
# -- SE of HJ mean of ANC -- #
>se_hj <- sqrt(var_est(ANC - c(hj), WGT.3X) / (N_hat ^ 2)
>se_hj
          [,1]
[1,] 65.67598
```

Analogously, the following code computes the estimates for the variable ACRISK.

```
# -- HT estimator of the total of ACRISK -- #
>ht <- HTestimator(ACRISK, 1/WGT.3X)
>ht
           [,1]
[1,] 8025.708
# -- SE of HT total of ACRISK -- #
>se_ht <- sqrt(var_est(ACRISK, WGT.3X))
>se_ht
[1] 1234.464
# -- HJ estimator of the mean of ACRISK -- #
>hj <- HTestimator(ACRISK, 1/WGT.3X) /
+HTestimator(rep(1, n), 1/WGT.3X)
```

```
>hj
          [,1]
[1,] 0.3741647
# -- SE of HJ mean of ACRISK -- #
>se_hj <- sqrt(var_est(ACRISK - c(hj), WGT.3X)/(N_hat ^ 2))
>se_hj
          [,1]
[1,] 0.04523369
```

These values can be interpreted as follows: the estimated average value of ANC across all lakes is 555.9 µeq/L, with a confidence interval for this estimate given by (427.2, 684.6), using a normal approximation for the distribution of the HT estimator. The estimates for ACRISK provide us with the information that approximately 8026 lakes are at risk of acidification, i.e., 37% of the lakes. The confidence intervals for these estimates are (5606.1, 10445.2) and (28%, 46%), respectively.

3.5 Auxiliary information

In order to consider alternative estimators to the basic HT and Hajek, we need to have auxiliary information available to help increase efficiency. Auxiliary information can have different sources, can take different forms and can be available at several different levels of granularity. We can identify two extreme settings: (*i*) information in the form of population totals for auxiliary variables, and (*ii*) individual level covariate information for each unit in the population, usually coming from the sampling frame. The former setting is most common in many countries for household surveys, for which auxiliary information is imported from different sources, such as administrative aggregate data. In this case, the specification of the model is essentially limited to linear models with few exceptions. In addition, issues with temporal misalignment of auxiliary data may further reduce the predictive power of this information. Setting (*ii*), on the other hand, is very detailed and allows one to work also with generalized, non-linear, nonparametric, mixed and additive models. It is typical of household surveys and of business surveys in countries for which individual level administrative registers are usually well-established, such as the Nordic European countries. Several intermediate settings can be envisioned for the availability of auxiliary information, particularly if the population is naturally clustered.

In order to properly account for the spatial structure of agricultural and environmental populations, we usually need a type (*ii*) level of detail for auxiliary information, that is the case of the North-eastern lakes survey considered in this chapter. Availability of population level individual covariates has recently increased, thanks to the improved practice of integration of administrative sources. In addition, useful information to model spatial dependence comes from satellite geo-referenced data. The level of detail of this information can range from the ideal one, in which geo-referenced information is available for units that coincide with population units, to settings in which information is available for larger areas that represent clusters of population units. This latter situation can still be handled with the methods illustrated in this chapter, although the predictive power of this auxiliary information is expected to be reduced by the coarser level of definition of the units for which information is available. There is a further setting in which population units of interest are defined by geographical areas, auxiliary information is available for a different geography, and the two geographies are not nested within one another. In this case, we have spatial misalignment for which more complex, usually Bayesian Hierarchical, models should be envisioned that are not dealt with in this chapter.

3.6 Model-assisted estimation

In order to improve the efficiency of estimates in the presence of useful auxiliary information, we can use the *model-assisted* approach to build new estimators of the population total and mean. The idea behind this approach is to build a model that well describes the relationship between the values of the variable of interest and those of auxiliary variables for which we have population level information. This model is used to obtain predicted values of the variable of interest for both sampled and non-sampled units. The population total of these predicted values would be unbiased (with respect both to the design *and* the model) for the population total of *y* if the model were correctly specified. A bias adjustment is envisioned to robustify such population total of predictions to model misspecifications, so that the final estimator is consistent and approximately unbiased with respect to the sampling design, irrespective of the goodness of the model used.

To illustrate the model-assisted approach, we follow the path proposed in the review paper by Breidt and Opsomer (2017), where it is introduced by means of the *difference estimator*. Let $\mathbf{x}$ be a vector of auxiliary variables and assume, for the moment, that we know the value taken by $\mathbf{x}$ on all units in the population (*complete auxiliary information*). Now, suppose we have a model $m(\cdot)$, independent from the sample, for predicting y_i from $\mathbf{x}_i$. The Authors refer to $m(\cdot)$ as a *method*: this is particularly useful in order to give to the assisting model the appropriate role of being just a tool to obtain population level predictions for the variable of interest. This is also useful to remove the emphasis on the role of the model in the properties of the final estimator. Given the predictions $m(\mathbf{x}_i)$, for $i = 1,\dots, N$, that we obtain from the use of this method with auxiliary information, the difference estimator is defined as

$$DIFF(y;m) = \sum_{i=1}^{N} m(\mathbf{x}_i) + \sum_{i \in s} \frac{y_i - m(\mathbf{x}_i)}{\pi_i} = \sum_{i=1}^{N} m(\mathbf{x}_i) + HT(y - m).$$

This estimator is exactly unbiased, regardless of the quality of the method used, since

$$E[DIFF(y;m)] = \sum_{i=1}^{N} m(\mathbf{x}_i) + E[HT(y - m)] = t_m + t_y - t_m = t_y,$$

where expectation is with respect to the sampling design and t_m denotes the population total of the predictions. Moreover, since this method does not depend on the sample, this total is not random and the design variance of the difference estimator follows immediately from the design variance of the HT estimator:

$$V(DIFF(y;m)) = V(HT(y - m)) = \sum_{i=1}^{N} \sum_{j=1}^{N} \Delta_{ij} \frac{y_i - m(\mathbf{x}_i)}{\pi_i} \frac{y_j - m(\mathbf{x}_j)}{\pi_j}.$$

From this expression, it is clear that the better the method used to obtain predictions, i.e., the better the model describes the relationship between the variable of interest and the auxiliary variables, the smaller the residuals and, therefore, the lower the variance. At the same time, if the model is misspecified, in the sense that the method used provides predictions with corresponding residuals that do not improve the efficiency of the estimator, the difference estimator is still design unbiased. In this sense, the approach based on the difference estimator is only model-assisted and not model-based. The estimated variance is given by

$$\hat{V}(DIFF(y;m)) = \hat{V}(HT(y-m)) = \sum_{i \in s} \sum_{j \in s} \frac{\Delta_{ij}}{\pi_{ij}} \frac{y_i - m(\mathbf{x}_i)}{\pi_i} \frac{y_j - m(\mathbf{x}_j)}{\pi_j}.$$

As mentioned in Section 3.3, we do not know second order inclusion probabilities for the North-eastern lakes survey; therefore, we write the variance estimator in our application as

$$\hat{V}(DIFF(y;m)) = \sum_{h \in H} S_h^2,$$

in which S_h^2 can be calculated as

$$S_h^2 = \frac{1}{n_h(n_h-1)} \sum_{i \in s_h} \left(\frac{e_i}{p_i} - \sum_{j \in s_h} \frac{e_j}{p_j} \right)^2$$

where $e_i = y_i - m(\mathbf{x}_i)$ (see, e.g., Särndal et al. 1992, p.421–422).

In practice, we seldom have a method $m(\cdot)$ that does not depend on sample data. We may have a functional form that is independent from the sample. However, such functional form unavoidably depends on a series of parameters, so that most often we use sample data in order to estimate such parameters. Then, in the model-assisted approach we generally use an estimator of $m(\cdot)$ in $DIFF(y; m)$, which we denote by $\hat{m}(\cdot)$. The rest of the chapter is devoted to illustrating different alternative specifications of the functional form of $m(\cdot)$ and describes how to obtain the corresponding estimate $\hat{m}(\cdot)$ from sample data. We will look at different working models and different ways to introduce the spatial information in the definition of the method $m(\cdot)$, trying to achieve better predictions in order to improve the efficiency of the final estimator.

Since we use sample data to obtain $\hat{m}(\cdot)$, the final estimator will no longer be exactly unbiased. However, we will consider estimators that can be shown to be, under mild conditions, asymptotically design unbiased and/or design consistent for the total of y. Usually, to achieve this, the census level estimate $m_N(\cdot)$ of $m(\cdot)$ is defined and derived, as if the values of y on all units in the finite population were observed. If conditions can be established that ensure that $\hat{m}(\cdot)$ is design consistent (asymptotically design unbiased) for $m_N(\cdot)$, then the final estimator will be design consistent (asymptotically design unbiased) for the population total of y. We will not discuss in detail the theoretical properties of the different estimators considered, but we will refer to the appropriate literature. A general set of conditions is provided in Breidt and Opsomer (2017). Finally, note that we can estimate the population mean of y by $DIFF(y; m)/N$ and estimate the variance by $\hat{V}(DIFF(y; m))/N^2$.

3.7 Linear models

The *Generalized Regression Estimator* (Särndal et al. 1992, Chapter 6), in short GREG, is the most known model-assisted survey estimator. It is inspired by linear multiple regression so that

$$m(\mathbf{x}_i) = \mathbf{x}_i' \boldsymbol{\beta}.$$

Following the discussion in the previous section, if the entire population were observed, the parameters $\boldsymbol{\beta}$ would be estimated via ordinary least squares, obtaining the following predictor

$$m_N(\mathbf{x}_i) = \mathbf{x}_i' \boldsymbol{\beta}_N = \mathbf{x}_i' \left(\sum_{j=1}^{N} \mathbf{x}_j \mathbf{x}_j' \right)^{-1} \sum_{j=1}^{N} \mathbf{x}_j y_j.$$

For the sake of simplicity, we will only consider the simpler case of homoschedastic regression models, although heteroschedasticity can be easily accounted for by using weighted least squares in the estimation of $\boldsymbol{\beta}$. In practice, we have a sample s, and we replace the finite population totals we find in $m_N(\mathbf{x}_i)$ by the corresponding HT estimators

$$\hat{m}(\mathbf{x}_i) = \mathbf{x}_i' \hat{\boldsymbol{\beta}} = \mathbf{x}_i' \left(\sum_{j \in s} \frac{\mathbf{x}_j \mathbf{x}_j'}{\pi_j} \right)^{-1} \sum_{j \in s} \frac{\mathbf{x}_j y_j}{\pi_j}.$$

Last, plugging the estimated method $\hat{m}$ into the difference estimator, we obtain the GREG as

$$DIFF(y, \mathbf{x}_i' \hat{\boldsymbol{\beta}}) = \sum_{i=1}^{N} \mathbf{x}_i' \hat{\boldsymbol{\beta}} + \sum_{i \in s} \frac{y_i - \mathbf{x}_i' \hat{\boldsymbol{\beta}}}{\pi_i}.$$

A nice property of the GREG is that it can be written as a linear combination of the observed *y*-values:

$$DIFF(y, \mathbf{x}_i' \hat{\boldsymbol{\beta}}) = \sum_{i \in s} \frac{y_i - \mathbf{x}_i' \hat{\boldsymbol{\beta}}}{\pi_i} + \sum_{i=1}^{N} \mathbf{x}_i' \hat{\boldsymbol{\beta}} = \sum_{i \in s} \frac{1}{\pi_i} \left\{ 1 + (t_x - HT(\mathbf{x}))' \left(\sum_{j \in s} \frac{\mathbf{x}_j \mathbf{x}_j'}{\pi_j} \right)^{-1} \mathbf{x}_i \right\} y_i$$

$$= \sum_{i \in s} \omega_{is} y_i,$$

where $t_x = \sum_{i=1}^{N} \mathbf{x}_i$. The weights ω_{is} have three important features: (*i*) they are a modification of the basic ones $\{\pi_i^{-1}\}$ and they take into account the auxiliary information on $\mathbf{x}$; (*ii*) they only depend on $\{\mathbf{x}_i\}_{i \in s}$ and the population $\mathbf{x}$-totals, without having the necessity to know the values of $\mathbf{x}$ for each unit in the population (aggregate auxiliary information); (*iii*) they do not depend on the variable of interest y, which in turn implies that we can use these weights for any variable of interest in a multi-purpose survey. A possible drawback with these weights is given by the possibility that they can take very large or even negative values.

We deal with the fit of the model using the R base function `lm()`, and the function `predict()` for predicting values for all units in the frame. Note that we specify the argument `weights` in order to use the basic weights in the fit procedure, i.e., we use a weighted least square procedure to obtain the estimates $\hat{\beta}$. We start with a simple model that takes into account just the elevation of the lakes.

```
# -- method m(.) -- #
>mod_greg1 <- lm(formula = ANC ~ GRID.ELEV.M, weights = WGT.3X,
+data = lakes)
# -- predictions over U -- #
>m_u <- predict(mod_greg1, newdata = lakes.fr)
# -- sample level residuals -- #
>e <- ANC - predict(mod_greg1, newdata = lakes)
# -- GREG of the total of ANC -- #
>greg1_tot <- sum(m_u) + HTestimator(e, 1/WGT.3X)
>greg1_tot
          [,1]
[1,] 12166922
# -- SE of GREG total of ANC -- #
>se_greg1_tot <- sqrt(var_est(e, WGT.3X))
>se_greg1_tot
[1] 1404950
# -- GREG of the mean of ANC -- #
>greg1_mean <- greg1_tot/N
>greg1_mean
          [,1]
[1,] 578.6608
# -- SE of GREG mean of ANC -- #
>se_greg1_mean <- sqrt(var_est(e, WGT.3X)/(N^2))
>se_greg1_mean
[1] 66.81964
```

Keeping in mind that we want to achieve good predictions in order to improve the GREG estimates, i.e., reduce their variance, we now introduce models which exploit auxiliary spatial information in order to improve the prediction power. Therefore, in the second model, along with elevation, we consider the lakes coordinates as regressors as well.

```
# -- method m(.) -- #
>mod_greg2 <- lm(formula = ANC ~ GRID.ELEV.M + UTMX + UTMY,
+weights = WGT.3X, data = lakes)
# -- predictions over U -- #
>m_u <- predict(mod_greg2, newdata = lakes.fr)
# -- residuals -- #
>e <- ANC - predict(mod_greg2, newdata = lakes)
# -- GREG of the total of ANC -- #
>greg2_tot <- sum(m_u) + HTestimator(e, 1/WGT.3X)
>greg2_tot
```

```
         [,1]
[1,] 12438601
# -- SE of GREG total of ANC -- #
>se_greg2_tot <- sqrt(var_est(e, WGT.3X))
>se_greg2_tot
[1] 1236668
# -- GREG of the mean of ANC -- #
>greg2_mean <- greg2_tot/N
>greg2_mean
          [,1]
[1,] 591.5819
# -- SE of GREG mean of ANC -- #
>se_greg2_mean <- sqrt(var_est(e, WGT.3X)/(N ^ 2))
>se_greg2_mean
[1] 58.81613
```

Other than the spatial coordinates, we have another variable in the frame with information on the spatial structure of the population. In fact, ECO.NUM is a variable that describes the eco-region that characterizes the lake, with seven levels. In the third model we use this categorical variable using the factor function.

```
# -- method m(.) -- #
>mod_greg3 <- lm(formula = ANC ~ GRID.ELEV.M + UTMX + UTMY +
+factor(ECO.NUM), weights = WGT.3X, data = lakes)
# -- predictions over U -- #
>m_u <- predict(mod_greg3, newdata = lakes.fr)
# -- residuals -- #
>e <- ANC - predict(mod_greg3, newdata = lakes)
# -- GREG of the total of ANC -- #
>greg3_tot <- sum(m_u) + HTestimator(e, 1/WGT.3X)
>greg3_tot
          [,1]
[1,] 12138858
# -- SE of GREG total of ANC -- #
>se_greg3_tot <- sqrt(var_est(e, WGT.3X))
>se_greg3_tot
[1] 1196749
# -- GREG of the mean of ANC -- #
>greg3_mean <- greg3_tot/N
>greg3_mean
          [,1]
[1,] 577.3261
# -- SE of GREG mean of ANC -- #
>se_greg3_mean <- sqrt(var_est(e, WGT.3X)/(N ^ 2))
>se_greg3_mean
```

```
[1] 56.91759
# ---------------------------- #
```

As our final model specification, in order to catch a more complex structure in space, we consider also an interaction term between the spatial coordinates.

```
# -- method m(.) -- #
>mod_greg4 <- lm(formula = ANC ~ GRID.ELEV.M + UTMX + UTMY +
+factor(ECO.NUM) + UTMX : UTMY, weights = WGT.3X,
+data = lakes)
# -- predictions over U -- #
>m_u <- predict(mod_greg4, newdata = lakes.fr)
# -- residuals -- #
>e <- ANC - predict(mod_greg4, newdata = lakes)
# -- GREG of the total of ANC -- #
>greg4_tot <- sum(m_u) + HTestimator(e, 1/WGT.3X)
>greg4_tot
          [,1]
[1,] 11669962
# -- SE of GREG total of ANC -- #
>se_greg4_tot <- sqrt(var_est(e, WGT.3X))
>se_greg4_tot
[1] 1142566
# -- GREG of the mean of ANC -- #
>greg4_mean <- greg4_tot/N
>greg4_mean
          [,1]
[1,] 555.0253
# -- SE of GREG mean of ANC -- #
>se_greg4_mean <- sqrt(var_est(e, WGT.3X)/(N^2))
>se_greg4_mean
[1] 54.34064
```

Analyzing the models and corresponding estimates, we can see that, as expected, when the model considers more spatial information, the standard error decreases: this is due to the more predictive power achieved by the considered models.

Another way to obtain the GREG estimates is through the R package survey (Lumley 2019). This package is well-known in the survey field and it is quite useful in all situations where the sampling design is not too complex, because the package can handle the major standard sampling designs, such as stratified sampling, cluster sampling, and so on. For these reasons, we now illustrate briefly how to use it. For a complete overview of the package, see Lumley (2010). We focus on the GREG estimator and specifically we replicate the setting of the first model we have seen in the previous section. The first step is the definition of the survey design. Since our complex sampling design is not implemented in the package, we need to approximate it by the closest one provided by the package. We consider a stratified unequal probabilities sampling design, where the strata are given by the variable CLUSTER.1 and the probabilities by the reciprocal of WGT.3X. The function used to define the survey design object is svydesign. For this function the major parameters are:

- id: formula specifying (eventual) clusters. ~ 1 indicates no clusters;
- strata: stratifying variable;
- probs: inclusion probabilities of the units in the sample;
- data: sample data.

We can define our sampling design as follows:

```
>library(survey)
>dstrat <- svydesign(id = ~ 1, strata = ~ CLUSTER.1,
+probs = 1/lakes$WGT.3X, data = lakes)
```

Note that with the symbol ~ we indicate that the variables are inside the dataset supplied by the parameter data. Now, since we have a survey design object, we can proceed with the estimation through the function calibrate, which has the following major parameters:

- design: survey design object which contains information about the sampling design;
- formula: auxiliary variables used in the estimation;
- population: vector of population totals of the auxiliary variables.

The code needed for our purpose is:

```
>cal.e <- calibrate(dstrat, formula = ~ GRID.ELEV.M,
+population = c(`(Intercept)` = nrow(lakes.fr),
+GRID.ELEV.M = sum(lakes.fr$GRID.ELEV.M)))
```

Finally, we get the estimates for the population total with the function svytotal, with the first argument indicating the variable of interest and the second containing the results of calibrate, as follows:

```
>svytotal(~ ANC, cal.e)
          total      SE
ANC 12166922 1483696
```

We can note that the value of the estimate is the same as that of greg1 considered in the previous section. The estimate of the standard error is different, because the package uses a different approximation to the variance in this context of unknown second order inclusion probabilities.

3.8 Generalized linear models

In this section we explore the possibility of using a generalized linear model (McCullagh and Nelder 1989) for our method. In particular,

$$m(\mathbf{x}_i) = h(\mathbf{x}_i'\boldsymbol{\beta})$$

where h is the inverse of the link function in a generalized linear model. For example,

$$m(\mathbf{x}_i) = \frac{\exp(\mathbf{x}_i'\boldsymbol{\beta})}{1+\exp(\mathbf{x}_i'\boldsymbol{\beta})}$$

in a logistic regression model for binomial responses, and

$$m(\mathbf{x}_i) = \exp(\mathbf{x}_i'\boldsymbol{\beta})$$

for a log-linear model for Poisson responses. This class of models allows us to deal with binary response variables. Therefore, we focus on total and mean estimation of the binary response ACRISK and use a logistic model. The function used to fit such models is glm(). As in the case of the linear models analyzed in Section 3.7, we use the basic weights in the fit procedure, through the parameter weights provided by the function. In this case, we provide the basic weights normalized in such a way the mean of the weights is equal to 1, in order to ease the convergence of the algorithm used by the function. Moreover, in order to specify the binomial family, we need to use the parameter family. The caveat here is to set this parameter to quasi-binomial, and not to binomial, in order to avoid a warning about non-integer number of successes. This is due to the behavior of glm(), since for a binomial GLM prior weights are used to give the number of trials when the response is the proportion of successes. Note that the quasi-binomial family gives the same point estimates and standard errors of the binomial family. In the function predict(), the parameter type = "response" sets the predictions to the same scale of the response variable. For variance estimation we still use the function var_est() employed before. Finally, the specification of the linear predictor is the same as that considered for the linear models in Section 3.7.

```
# -- method m(.) -- #
>mod_glm1 <- glm(formula = ACRISK ~ GRID.ELEV.M, family =
"quasibinomial", +weights = WGT.3X/mean(WGT.3X), data = lakes)

# -- predictions over U -- #
>m_u <- predict(mod_glm1, type = "response", newdata = lakes.fr)

# -- residuals -- #
>e <- ACRISK - predict(mod_glm1, type = "response", newdata =
lakes)

# -- estimate of the total of ACRISK -- #
>glm1_tot <- sum(m_u) + HTestimator(e, 1/WGT.3X)
>glm1_tot
          [,1]
[1,] 8181.957

# -- SE of the estimate of the total of ACRISK -- #
>se_glm1_tot <- sqrt(var_est(e, WGT.3X))
>se_glm1_tot
[1] 975.9185

# -- estimate of the mean of ACRISK -- #
>glm1_mean <- glm1_tot/N
```

```
>glm1_mean
           [,1]
[1,] 0.3891352
# -- SE of the estimate of the mean of ACRISK -- #
>se_glm1_mean <- sqrt(var_est(e, WGT.3X)/(N^2))
>se_glm1_mean
[1] 0.04641484
# -- method m(.) -- #
>mod_glm2 <- glm(formula = ACRISK ~ GRID.ELEV.M + UTMX + UTMY,
+family = "quasibinomial", weights = WGT.3X/mean(WGT.3X),
data = lakes)
# -- predictions over U -- #
>m_u <- predict(mod_glm2, type = "response", newdata = lakes.fr)
# -- residuals -- #
>e <- ACRISK - predict(mod_glm2, type = "response", newdata =
lakes)
# -- estimate of the total of ACRISK -- #
>glm2_tot <- sum(m_u) + HTestimator(e, 1/WGT.3X)
>glm2_tot
          [,1]
[1,] 8285.054
# -- SE of the estimate of the total of ACRISK -- #
>se_glm2_tot <- sqrt(var_est(e, WGT.3X))
>se_glm2_tot
[1] 852
# -- estimate of the mean of ACRISK -- #
>glm2_mean <- glm2_tot/N
>glm2_mean
           [,1]
[1,] 0.3940385
# -- SE of the estimate of the mean of ACRISK -- #
>se_glm2_mean <- sqrt(var_est(e, WGT.3X)/(N^2))
>se_glm2_mean
[1] 0.04052126
# -- method m(.) -- #
>mod_glm3 <- glm(formula = ACRISK ~ GRID.ELEV.M + UTMX + UTMY +
+factor(ECO.NUM), family = "quasibinomial",
+weights = WGT.3X/mean(WGT.3X), data = lakes)
# -- predictions over U -- #
>m_u <- predict(mod_glm3, type = "response", newdata = lakes.fr)
# -- residuals -- #
>e <- ACRISK - predict(mod_glm3, type = "response", newdata =
lakes)
```

```
# -- estimate of the total of ACRISK -- #
>glm3_tot <- sum(m_u) + HTestimator(e, 1/WGT.3X)
>glm3_tot
          [,1]
[1,] 7967.892
# -- SE of the estimate of the total of ACRISK -- #
>se_glm3_tot <- sqrt(var_est(e, WGT.3X))
>se_glm3_tot
[1] 859.2785
# -- estimate of the mean of ACRISK -- #
>glm3_mean <- glm3_tot/N
>glm3_mean
          [,1]
[1,] 0.3789542
# -- SE of the estimate of the mean of ACRISK -- #
>se_glm3_mean <- sqrt(var_est(e, WGT.3X)/(N^2)
>se_glm3_mean
[1] 0.04086743
# -- method m(.) -- #
>mod_glm4 <- glm(formula = ACRISK ~ GRID.ELEV.M + UTMX + UTMY
+ +factor(ECO.NUM) + UTMX : UTMY, family = "quasibinomial",
+weights = WGT.3X/mean(WGT.3X), data = lakes)
# -- predictions over U -- #
>m_u <- predict(mod_glm4, type = "response", newdata = lakes.fr)
# -- residuals -- #
>e <- ACRISK - predict(mod_glm4, type = "response", newdata =
lakes)
# -- estimate of the total of ACRISK -- #
>glm4_tot4 <- sum(m_u) + HTestimator(e, 1/WGT.3X)
>glm4_tot
          [,1]
[1,] 7720.197
# -- SE of the estimate of the total of ACRISK -- #
>se_glm4_tot <- sqrt(var_est(e, WGT.3X))
>se_glm4_tot
[1] 725.8333
# -- estimate of the mean of ACRISK -- #
>glm4_mean <- glm4_tot/N
>glm4_mean
          [,1]
[1,] 0.3671738
# -- SE of the estimate of the mean of ACRISK -- #
>se_glm4_mean <- sqrt(var_est(e, WGT.3X)/(N^2))
```

```
>se_glm4_mean
[1] 0.03452075
# ----------------------------------------------- #
```

A summary of the estimates obtained is reported in the first four rows of Table 3.1. Here, it is enough to notice that, as for estimation of the population total for ANC, the standard error decreases when the spatial information is included in the model.

Table 3.1. Estimates for the population mean of ANC and ACRISK with different working methods identified by the set of variables included. ANC uses linear and additive regression models, while ACRISK uses logistic regression linear and additive models.

	ANC mean		ACRISK mean	
Estimator/variables in the model	**Estimate**	**Standard error**	**Estimate**	**Standard error**
HJ (y)	555.9	65.6	37.4%	4.5%
GRID.ELEV.M	578.6	66.8	38.9%	4.6%
GRID.ELEV.M + UTMX + UTMY	591.5	58.8	39.0%	4.0%
GRID.ELEV.M + UTMX + UTMY + factor (ECO.NUM)	577.3	56.9	37.0%	4.0%
GRID.ELEV.M + UTMX + UTMY + factor (ECO.NUM) + UTMX : UTMY	555.0	54.3	36.7%	3.4%
GRID.ELEV.M + s (UTMX, UTMY, bs = "tp")	600.0	35.8	38.9%	2.4%
GRID.ELEV.M + s (UTMX, UTMY, bs = "tp") + factor (ECO.NUM)	551.8	34.8	36.2%	2.2%

3.9 Geoadditive models

The flexibility of the difference estimator allows us to use more complex than (generalized) linear models as a tool to define the method $m(\cdot)$. Here, we focus on the use of penalized spline regression, which is a non-parametric regression method, in order to better capture nonlinearities in the spatial structure of the data. In fact, penalized splines can be used in a multivariate context, but here we are interested in bivariate smoothing, in order to capture the relationship between the location in space of the units space and the response variable. Breidt et al. (2005) introduced the use of penalized splines regression into the model-assisted approach and Cicchitelli and Montanari (2012) applied that approach in a spatial context. Ruppert et al. (2003) provide an excellent introduction to smoothing with penalized splines and, in particular, they discuss bivariate smoothing, together with its extension that can handle the inclusion of other continuous or categorical covariates and a non-normal response called *geoadditive model* (Kamman and Wand 2003). For the sake of simplicity, we will only briefly introduce the methodology for bivariate smoothing with penalized splines. The interested reader can find more details in the abovementioned references.

Let $\mathbf{x}_i = (x_{i1}, x_{i2})$ be a pair of spatial coordinates and $\kappa_1,..., \kappa_k$ a set of *knot* locations such that the $K \times K$ matrix $\Omega = \{(\|\kappa_k - \kappa_l\|)^2 \, log(\|\kappa_k - \kappa_l\|) \}_{k,l=1,...,K}$ is non singular, where $\|.\|$ is the Euclidean norm. Then, for each location $\mathbf{x}_i$ we define K pseudo-covariates given by $[z_1(\mathbf{x}_i),\ldots, z_K(\mathbf{x}_i)] = [\breve{z}_1(\mathbf{x}_i),\ldots, \breve{z}_K(\mathbf{x}_i)]\Omega^{-1/2}$, where $\breve{z}_k(\mathbf{x}_i) = (\|\mathbf{x}_i - \kappa_k\|)^2 \, log(\|\mathbf{x}_i - \kappa_k\|)$, for $i = 1,\ldots N$ and $k = 1,\ldots, K$. We define our method $m(\cdot)$ as a bivariate smoother function in space given by the following spline regression

$$m(\mathbf{x}_i) = m(x_{i1}, x_{i2}) = \beta_0 + \beta_1 x_{i1} + \beta_2 x_{i2} + u_1 z_1(\mathbf{x}_i) + \cdots + u_K z_K(\mathbf{x}_i),$$

with $\beta_0, \beta_1, \beta_2, u_1,\ldots, u_k$ and σ_ε^2 model parameters. The model accounts for the spatial component of the population by means of β_1, β_2 and through the coefficients $u_1,\ldots, u_k$ (note that $z_1(\mathbf{x}_i),\ldots, z_K(\mathbf{x}_i)$ contain spatial information). In the hypothetical situation where we know the response variable for units in the population, we could estimate the model by a penalized least-square criterion, minimizing the function

$$\sum_{i=1}^{N}[y(\mathbf{x}_i) - \beta_0 \beta_1 x_{i1} - \beta_2 x_{i2} - u_1 z_1(\mathbf{x}_i) - \cdots - u_K z_K(\mathbf{x}_i)]^2 + \lambda \sum_{k=1}^{K} u_k^2$$

with respect to $\beta_0, \beta_1, \beta_2, u_1,\ldots, u_k$, and with the term $\lambda \sum_{k=1}^{K} u_k^2$ used to regulate the variation of the coefficients $u_1,\ldots, u_k$. Note that with larger values of λ, the fit approaches the least-square regression line while, with smaller values of λ, the fit is more wiggly and accommodates departures of the spatial structure of the data from a plane. For a fixed value of λ, the census level solution is given by

$$\begin{bmatrix} \boldsymbol{\beta}_N \\ \mathbf{u}_N \end{bmatrix} = \begin{bmatrix} \begin{bmatrix} \mathbf{X'X} & \mathbf{X'Z} \\ \mathbf{Z'X} & \mathbf{Z'Z} \end{bmatrix} + \lambda \mathbf{D} \end{bmatrix}^{-1} \begin{bmatrix} \mathbf{X'} \\ \mathbf{Z'} \end{bmatrix} \mathbf{y}$$

where $\boldsymbol{\beta}_N = [\beta_0, \beta_1, \beta_2]'$; $u_N = [u_1,\ldots, u_K]'$; $\mathbf{X}$ is a $N \times 3$ matrix with i-th row $[1, x_{i1}, x_{i2}]$ for $i = 1,\ldots, N$; $\mathbf{Z}$ is a $N \times K$ matrix with ik-th entry equal to $z_k(\mathbf{x}_i)$ for $k = 1,\ldots, K$ and $i = 1,\ldots, N$; $\mathbf{D} = blockdiag[0_{3\times 3}, \mathbf{I}_K]$ with $\mathbf{I}_K$ $K \times K$ identity matrix; $\mathbf{y} = [y(\mathbf{x}_1),\ldots, y(\mathbf{x}_N)]'$. In the end, the population level vector of fitted values is given by $\mathbf{S}_\lambda \mathbf{y}$, where $\mathbf{S}_\lambda$ is the smoothing matrix and is given by

$$\mathbf{S}_\lambda = [\mathbf{X}, \mathbf{Z}] \begin{bmatrix} \begin{bmatrix} \mathbf{X'X} & \mathbf{X'Z} \\ \mathbf{Z'X} & \mathbf{Z'Z} \end{bmatrix} + \lambda \mathbf{D} \end{bmatrix}^{-1} \begin{bmatrix} \mathbf{X'} \\ \mathbf{Z'} \end{bmatrix}.$$

Then, $m_N(\mathbf{x}_i)$ is the i-th component of $\mathbf{S}_\lambda \mathbf{y}$. Now, since $\boldsymbol{\beta}_N$ and $\mathbf{u}_N$ are unknown population quantities, we estimate them using sample data. Given the matrices $\mathbf{X}_s$ and $\mathbf{Z}_s$, obtained as submatrices of $\mathbf{X}$ and $\mathbf{Z}$ consisting of the rows for which $i \in s$, and the matrix $\Pi_s = diag(1/\pi_i)$, consistent estimators are given by

$$\begin{bmatrix} \hat{\boldsymbol{\beta}} \\ \hat{\mathbf{u}} \end{bmatrix} = \begin{bmatrix} \begin{bmatrix} \mathbf{X}_s'\Pi_s\mathbf{X}_s & \mathbf{X}_s'\Pi_s\mathbf{Z}_s \\ \mathbf{Z}_s'\Pi_s\mathbf{X}_s & \mathbf{Z}_s'\Pi_s\mathbf{Z}_s \end{bmatrix} + \lambda \mathbf{D} \end{bmatrix}^{-1} \begin{bmatrix} \mathbf{X}_s' \\ \mathbf{Z}_s' \end{bmatrix} \Pi_s \mathbf{y}_s,$$

and, as a consequence, the estimated smoothing matrix can be written as

$$\hat{\mathbf{S}}_\lambda = [\mathbf{X}, \mathbf{Z}] \begin{bmatrix} \begin{bmatrix} \mathbf{X}_s'\Pi_s\mathbf{X}_s & \mathbf{X}_s'\Pi_s\mathbf{Z}_s \\ \mathbf{Z}_s'\Pi_s\mathbf{X}_s & \mathbf{Z}_s'\Pi_s\mathbf{Z}_s \end{bmatrix} + \lambda \mathbf{D} \end{bmatrix}^{-1} \begin{bmatrix} \mathbf{X}_s' \\ \mathbf{Z}_s' \end{bmatrix} \Pi_s.$$

Finally, we can use this matrix to obtain the population vector of estimated predictions as $\hat{\mathbf{S}}_\lambda \mathbf{y}_s$, and $\hat{m}(\mathbf{x}_i)$ is the i-th component of this vector.

Regarding the implementation in R, here the only modification needed is due to the use of a function for estimating the bivariate spline and the corresponding geoadditive model. We use the function gam(), from the package mgcv (Wood 2017). The usage of this function is quite similar to lm(), the only additional part is the use of the parameter s inside the formula in order to specify a smooth function of the covariates instead of a linear function. Specifically, s(UTMX, UTMY, bs = "tp") indicates that we have a bivariate function on UTMX and UTMY, and that we use a *thin-plate* set of basis functions (defined by bs = "tp"). Different sets of basis functions can be specified. See Wood (2017) for details on this. We consider two geoadditive models. The first one considers the elevation of the lake, GRID. ELEV.M, along with the bivariate smooth term over UTMX and UTMY. In the second one we add to the first model the categorical variable ECO.NUM. With that said, we proceed to the estimate of the total and the mean of the variable ANC.

```
>library(mgcv)
# -- method m(.) -- #
>ms_anc1 <- gam(formula = ANC ~ GRID.ELEV.M + s(UTMX, UTMY, bs
= "tp"), +weight = WGT.3X, data = lakes)
# -- predictions over U -- #
>m_u <- predict(ms_anc1, type = "response", newdata = lakes.fr)
# -- residuals -- #
>e <- ANC - predict(ms_anc1, type = "response", newdata =
lakes)
# -- estimate of the total of ANC -- #
>spl1_anc_tot <- sum(m_u) + HTestimator(e, 1/WGT.3X)
>spl1_anc_tot
          [,1]
[1,] 12615785
# -- SE of the estimate of the total of ANC -- #
>se_spl1_anc_tot <- sqrt(var_est(e, WGT.3X))
>se_spl1_anc_tot
[1] 754456
# -- estimate of the mean of ANC -- #
>spl1_anc_mean <- spl1_anc_tot/N
>spl1_anc_mean
          [,1]
[1,] 600.0088
# -- SE of the estimate of the mean of ANC -- #
>se_spl1_anc_mean <- sqrt(var_est(e, WGT.3X)/(N^2))
>se_spl1_anc_mean
[1] 35.88205
# -- method m(.) -- #
>ms_anc2 <- gam(formula = ANC ~ GRID.ELEV.M + s(UTMX, UTMY, bs
= "tp") + +factor(ECO.NUM), weight = WGT.3X, data = lakes)
```

```
# -- predictions over U -- #
>m_u <- predict(ms_anc2, type = "response", newdata = lakes.fr)
# -- residuals -- #
>e <- ANC - predict(ms_anc2, type = "response", newdata =
lakes)
# -- estimate of the total of ANC -- #
>spl2_anc_tot <- sum(m_u) + HTestimator(e, 1/WGT.3X)
>spl2_anc_tot
          [,1]
[1,] 11602391
# -- SE of the estimate of the total of ANC -- #
>se_spl2_anc_tot <- sqrt(var_est(e, WGT.3X))
>se_spl2_anc_tot
[1] 733490.5
# -- estimate of the mean of ANC -- #
>spl2_anc_mean <- spl2_anc_tot/N
>spl2_anc_mean
          [,1]
[1,] 551.8116
# -- SE of the estimate of the mean of ANC -- #
>se_spl2_anc_mean <- sqrt(var_est(e, WGT.3X)/(N^2))
>se_spl2_anc_mean
[1] 34.88493
```

Considering the estimation of the population mean and total of ANC, we note that we have achieved a further improvement in efficiency. In fact, the variance in this case is lower than the variance of the GREG coming from model number 4 in Section 3.7, which was the most efficient estimator.

Now, we consider the binary response ACRISK. To accommodate the nature of this variable we use again the quasibinomial family option and the standardized weights in the fit.

```
# -- method m(.) -- #
>ms_acrisk1 <- gam(formula = ACRISK ~ GRID.ELEV.M +
+s(UTMX, UTMY, bs = "tp"), family = "quasibinomial",
+weight = WGT.3X/mean(WGT.3X), data = lakes)
# -- predictions over U -- #
>m_u <- predict(ms_acrisk1, type = "response", newdata =
lakes.fr)
# -- residuals -- #
>e <- ACRISK - predict(ms_acrisk1, type = "response", newdata
= lakes)
# -- estimate of the total of ACRISK -- #
>spl1_acrisk_tot <- sum(m_u) + HTestimator(e, 1/WGT.3X)
>spl1_acrisk_tot
```

```
           [,1]
[1,] 8184.316
# -- SE of the estimate of the total of ACRISK -- #
>se_spl1_acrisk_tot <- sqrt(var_est(e, WGT.3X))
>se_spl1_acrisk_tot
[1] 509.5923
# -- estimate of the mean of ACRISK -- #
>spl1_acrisk_mean <- spl1_acrisk_tot/N
>spl1_acrisk_mean
           [,1]
[1,] 0.3892474
# -- SE of the estimate of the mean of ACRISK -- #
>se_spl1_acrisk_mean <- sqrt(var_est(e, WGT.3X)/(N ^ 2))
>se_spl1_acrisk_mean
[1] 0.02423629
# -- method m(.) -- #
>ms_acrisk2 <- gam(formula = ACRISK ~ GRID.ELEV.M +
+s(UTMX, UTMY, bs = "tp") + factor(ECO.NUM), family =
"quasibinomial", +weight = WGT.3X/mean(WGT.3X), data = lakes)
# -- predictions over U -- #
>m_u <- predict(ms_acrisk2, type = "response", newdata =
lakes.fr)
# -- residuals -- #
>e <- ACRISK - predict(ms_acrisk2, type = "response", newdata
= lakes)
# -- estimate of the total of ACRISK -- #
>spl2_acrisk_tot <- sum(m_u) + HTestimator(e, 1/WGT.3X)
>spl2_acrisk_tot
           [,1]
[1,] 7615.335
# -- SE of the estimate of the total of ACRISK -- #
>se_spl2_acrisk_tot <- sqrt(var_est(e, WGT.3X))
>se_spl2_acrisk_tot
[1] 479.212
# -- estimate of the mean of ACRISK -- #
>spl2_acrisk_mean <- spl2_acrisk_tot/N
>spl2_acrisk_mean
           [,1]
[1,] 0.3621866
# -- SE of the estimate of the mean of ACRISK -- #
>se_spl2_acrisk_mean <- sqrt(var_est(e, WGT.3X)/(N ^ 2))
>se_spl2_acrisk_mean
[1] 0.0227914
```

Table 3.1 reports all the estimates for the mean of ANC and of ACRISK. Here, it is important to notice that the standard error for the estimates that include a bivariate smoothing term in space is considerably smaller than that of the other estimators, by this providing evidence that the spatial structure of the data is more complex than a plane in space, even after accounting for the other covariates.

3.10 Concluding remarks

Estimation in agricultural and environmental surveys and in surveys in general can effectively take advantage of auxiliary information. In particular, the model-assisted approach incorporates auxiliary information by means of a working model or method $m(\cdot)$, in order to improve the efficiency over the basic HT estimator, while maintaining design consistency in case of model misspecification. A strength of the illustrated approach is flexibility. In fact, there is no restriction on the type of methods we can use. In the last decades, statistical learning techniques have also been used in order to account for non-linear and complex dependence structures. For example, Breidt and Opsomer (2000) used local polynomial regression, Breidt et al. (2005) introduced penalized splines, Breidt et al. (2007) considered additive models, Montanari and Ranalli (2005) proposed Neural Networks, and Toth and Eltinge (2011) presented regression trees. In this chapter, we have analyzed some of these working models that are associated to the so-called difference estimator for both continuous and binary variables of interest. In particular, we have reviewed some models that could be useful in the spatial context, starting from a basic linear regression model that leads to the well-known GREG estimator and, then, progressively introducing more complex models.

The proposed approach based on the difference estimator can be implemented using the following steps: choose a method $m(\cdot)$, obtain a census-level solution $m_N(\cdot)$, use sample data and basic design weights in order to achieve a sample based estimate $\hat{m}(\cdot)$ and corresponding residuals, compute the difference estimator and the corresponding variance estimate. In fact, even for the most complex model we have considered—a geoadditive model for a binary response with a bivariate smoother for latitude and longitude and a linear term for continuous and categorical covariates—the steps followed to compute the estimate of the population total and mean are essentially the same: (*i*) fit the model on the sample data accounting for the basic design weights, (*ii*) compute predicted values for all units in the population, (*iii*) compute the sample residuals, (*iv*) compute the estimator of the total as the sum of the population predictions plus the Horvitz-Thompson estimate of the population total of the residuals; (*v*) variance estimation is simply given by the variance estimator of the Horvitz-Thompson estimator of the population total (mean) of the residuals.

Although we have focused on total and mean estimation, more complex parameters, such as the distribution function and population quantiles, can be investigated (Breidt et al. 2007). Finally, a multipurpose approach can be envisioned. Table 3.2 reports the estimates for other variables of interest also considered in Breidt et al. (2007) that use the same models considered for ANC. The code is reported in the software Appendix, where the presence of outlying (leverage) observations is handled by setting to zero the corresponding weight in the first step (*i*) of the estimation procedure.

Table 3.2. Estimates for the population mean of other variables of interest from the survey—MG, CL and SO4—using the different working methods identified by the set of variables included. See the Appendix for a description of the code used. Linear and additive regression models are used for all the survey variables.

Estimator/variables in the model	MG mean		CL mean		SO4 mean	
	Estimate	**Standard error**	**Estimate**	**Standard error**	**Estimate**	**Standard error**
HJ (*y*)	418.5	196.6	1315.7	995.0	333.0	138.9
GRID.ELEV.M	443.4	199.7	1381.8	1010.0	359.5	141.1
GRID.ELEV.M + UTMX + UTMY	455.5	199.3	1977.9	1009.8	368.0	137.6
GRID.ELEV.M + UTMX + UTMY + factor(ECO.NUM)	454.5	199.2	1373.8	1011.7	369.0	136.1
GRID.ELEV.M + UTMX + UTMY + factor(ECO.NUM) + UTMX : UTMY	440.7	200.0	1350.5	1013.4	362.5	136.4
GRID.ELEV.M + s(UTMX, UTMY, bs = "tp")	456.2	196.6	1413.8	1010.4	363.5	123.0
GRID.ELEV.M + s(UTMX, UTMY, bs = "tp") + factor(ECO.NUM)	413.6	197.9	1334.0	1012.3	328.2	125.5

Software appendix

1. Variance estimation

```
var_est <- function(y, w)
{
#
# estimation of total variance
# y : variable for which we estimate the total variance
# w : weight
#
n_var <- table(lakes$CLUSTER.1)
n_h <- rep(0, n)
for(i in 1:length(n_var))
{
n_h[lakes$CLUSTER.1 == i] <- n_var[i]
}
y_w <- y * w * n_h
S2h <- tapply(y_w, lakes$CLUSTER.1, var)/n_var
sum(S2h)
}
```

2. Estimation for variables MG, SO4, CL

```
# ------ MG basic estimates ------ #
N_hat <- HTestimator(rep(1, n), 1/WGT.3X)
```

```
ht <- HTestimator(MG, 1/WGT.3X)
sd_ht <- sqrt(var_est(MG, WGT.3X))
hj<-HTestimator(MG, 1/WGT.3X)/HTestimator(rep(1, n), 1/WGT.3X)
sd_hj <- sqrt(var_est(MG - c(hj), WGT.3X)/(N_hat ^ 2))
# ----------------------------- #
# ------ CL basic estimates ------ #
N_hat <- HTestimator(rep(1, n), 1/WGT.3X)
ht <- HTestimator(CL, 1/WGT.3X)
sd_ht <- sqrt(var_est(CL, WGT.3X))
hj <- HTestimator(CL, 1/WGT.3X)/HTestimator(rep(1, n), 1/
WGT.3X)
sd_hj <- sqrt(var_est(CL - c(hj), WGT.3X)/(N_hat ^ 2))
# ----------------------------- #
# ------ SO4 basic estimates ------ #
N_hat <- HTestimator(rep(1, n), 1/WGT.3X)
ht <- HTestimator(SO4, 1/WGT.3X)
sd_ht <- sqrt(var_est(SO4, WGT.3X))
hj <- HTestimator(SO4, 1/WGT.3X)/HTestimator(rep(1, n), 1/
WGT.3X)
sd_hj <- sqrt(var_est(SO4 - c(hj), WGT.3X)/(N_hat ^ 2))
# ----------------------------- #
# ------ MG GREG estimates ------ #
out <- 160
WGT.3X0 <- WGT.3X
WGT.3X0[out] <- 0
mod_greg1_mg <- lm(formula = MG ~ GRID.ELEV.M, weights =
WGT.3X0, data = lakes)
m_u <- predict(mod_greg1_mg, newdata = lakes.fr)
e <- MG - predict(mod_greg1_mg, newdata = lakes)
greg1_tot_mg <- sum(m_u) + HTestimator(e, 1/WGT.3X)
sd_greg1_tot_mg <- sqrt(var_est(e, WGT.3X))
greg1_mean_mg <- greg1_tot_mg/N
sd_greg1_mean <- sqrt(var_est(e, WGT.3X)/(N ^ 2))
mod_greg2_mg <- lm(formula = MG ~ GRID.ELEV.M + UTMX + UTMY,
weights = WGT.3X0, data = lakes)
m_u <- predict(mod_greg2_mg, newdata = lakes.fr)
e <- MG - predict(mod_greg2_mg, newdata = lakes)
greg2_tot_mg <- sum(m_u) + HTestimator(e, 1/WGT.3X)
sd_greg2_tot_mg <- sqrt(var_est(e, WGT.3X))
greg2_mean_mg <- greg2_tot_mg/N
sd_greg2_mean_mg <- sqrt(var_est(e, WGT.3X)/(N ^ 2))
mod_greg3_mg <- lm(formula = MG ~ GRID.ELEV.M + UTMX + UTMY +
factor(ECO.NUM), weights = WGT.3X0, data = lakes)
m_u <- predict(mod_greg3_mg, newdata = lakes.fr)
e <- MG - predict(mod_greg3_mg, newdata = lakes)
greg3_tot_mg <- sum(m_u) + HTestimator(e, 1/WGT.3X)
```

```
sd_greg3_tot_mg <- sqrt(var_est(e, WGT.3X))
greg3_mean_mg <- greg3_tot_mg/N
sd_greg3_mean_mg <- sqrt(var_est(e, WGT.3X)/(N ^ 2))
mod_greg4_mg <- lm(formula = MG ~ GRID.ELEV.M + UTMX + UTMY +
factor(ECO.NUM) + UTMX * UTMY, weights = WGT.3X0, data = lakes)
m_u <- predict(mod_greg4_mg, newdata = lakes.fr)
e <- MG - predict(mod_greg4_mg, newdata = lakes)
greg4_tot_mg <- sum(m_u) + HTestimator(e, 1/WGT.3X)
sd_greg4_tot_mg <- sqrt(var_est(e, WGT.3X))
greg4_mean_mg <- greg4_tot_mg/N
sd_greg4_mean_mg <- sqrt(var_est(e, WGT.3X)/(N ^ 2))
# ----------------------------- #
# ------ CL GREG estimates ------ #
out <- 160
WGT.3X0 <- WGT.3X
WGT.3X0[out] <- 0
mod_greg1_cl <- lm(formula = CL ~ GRID.ELEV.M, weights =
WGT.3X0, data = lakes)
m_u <- predict(mod_greg1_cl, newdata = lakes.fr)
e <- CL - predict(mod_greg1_cl, newdata = lakes)
greg1_tot_cl <- sum(m_u) + HTestimator(e, 1/WGT.3X)
sd_greg1_tot_cl <- sqrt(var_est(e, WGT.3X))
greg1_mean_cl <- greg1_tot_cl/N
sd_greg1_mean <- sqrt(var_est(e, WGT.3X)/(N ^ 2))
mod_greg2_cl <- lm(formula = CL ~ GRID.ELEV.M + UTMX + UTMY,
weights = WGT.3X0, data = lakes)
m_u <- predict(mod_greg2_cl, newdata = lakes.fr)
e <- CL - predict(mod_greg2_cl, newdata = lakes)
greg2_tot_cl <- sum(m_u) + HTestimator(e, 1/WGT.3X)
sd_greg2_tot_cl <- sqrt(var_est(e, WGT.3X))
greg2_mean_cl <- greg2_tot_cl/N
sd_greg2_mean_cl <- sqrt(var_est(e, WGT.3X)/(N ^ 2))
mod_greg3_cl <- lm(formula = CL ~ GRID.ELEV.M + UTMX + UTMY +
factor(ECO.NUM), weights = WGT.3X0, data = lakes)
m_u <- predict(mod_greg3_cl, newdata = lakes.fr)
e <- CL - predict(mod_greg3_cl, newdata = lakes)
greg3_tot_cl <- sum(m_u) + HTestimator(e, 1/WGT.3X)
sd_greg3_tot_cl <- sqrt(var_est(e, WGT.3X))
greg3_mean_cl <- greg3_tot_cl/N
sd_greg3_mean_cl <- sqrt(var_est(e, WGT.3X)/(N ^ 2))
mod_greg4_cl <- lm(formula = CL ~ GRID.ELEV.M + UTMX + UTMY +
factor(ECO.NUM) + UTMX * UTMY, weights = WGT.3X0, data = lakes)
m_u <- predict(mod_greg4_cl, newdata = lakes.fr)
e <- CL - predict(mod_greg4_cl, newdata = lakes)
greg4_tot_cl <- sum(m_u) + HTestimator(e, 1/WGT.3X)
sd_greg4_tot_cl <- sqrt(var_est(e, WGT.3X))
```

```
greg4_mean_cl <- greg4_tot_cl/N
sd_greg4_mean_cl <- sqrt(var_est(e, WGT.3X)/(N^2))
# ---------------------------- #
# ------ SO4 GREG estimates ------ #
out <- 160
WGT.3X0 <- WGT.3X
WGT.3X0[out] <- 0
mod_greg1_so4 <- lm(formula = SO4 ~ GRID.ELEV.M, weights =
WGT.3X0, data = lakes)
m_u <- predict(mod_greg1_so4, newdata = lakes.fr)
e <- SO4 - predict(mod_greg1_so4, newdata = lakes)
greg1_tot_so4 <- sum(m_u) + HTestimator(e, 1/WGT.3X)
sd_greg1_tot_so4 <- sqrt(var_est(e, WGT.3X))
greg1_mean_so4 <- greg1_tot_so4/N
sd_greg1_mean <- sqrt(var_est(e, WGT.3X)/(N^2))
mod_greg2_so4 <- lm(formula = SO4 ~ GRID.ELEV.M + UTMX + UTMY,
weights = WGT.3X0, data = lakes)
m_u <- predict(mod_greg2_so4, newdata = lakes.fr)
e <- SO4 - predict(mod_greg2_so4, newdata = lakes)
greg2_tot_so4 <- sum(m_u) + HTestimator(e, 1/WGT.3X)
sd_greg2_tot_so4 <- sqrt(var_est(e, WGT.3X))
greg2_mean_so4 <- greg2_tot_so4/N
sd_greg2_mean_so4 <- sqrt(var_est(e, WGT.3X)/(N^2))
mod_greg3_so4 <- lm(formula = SO4 ~ GRID.ELEV.M + UTMX + UTMY +
factor(ECO.NUM), weights = WGT.3X0, data = lakes)
m_u <- predict(mod_greg3_so4, newdata = lakes.fr)
e <- SO4 - predict(mod_greg3_so4, newdata = lakes)
greg3_tot_so4 <- sum(m_u) + HTestimator(e, 1/WGT.3X)
sd_greg3_tot_so4 <- sqrt(var_est(e, WGT.3X))
greg3_mean_so4 <- greg3_tot_so4/N
sd_greg3_mean_so4 <- sqrt(var_est(e, WGT.3X)/(N^2))
mod_greg4_so4 <- lm(formula = SO4 ~ GRID.ELEV.M + UTMX + UTMY +
factor(ECO.NUM) + UTMX * UTMY, weights = WGT.3X0, data = lakes)
m_u <- predict(mod_greg4_so4, newdata = lakes.fr)
e <- SO4 - predict(mod_greg4_so4, newdata = lakes)
greg4_tot_so4 <- sum(m_u) + HTestimator(e, 1/WGT.3X)
sd_greg4_tot_so4 <- sqrt(var_est(e, WGT.3X))
greg4_mean_so4 <- greg4_tot_so4/N
sd_greg4_mean_so4 <- sqrt(var_est(e, WGT.3X)/(N^2))
# ---------------------------- #
# ------ MG splines estimates ------ #
out <- 160
WGT.3X0 <- WGT.3X
WGT.3X0[out] <- 0
ms_mg1 <- gam(formula = MG ~ GRID.ELEV.M + s(UTMX, UTMY, bs =
"tp"), weight = WGT.3X0, data = lakes)
```

```
m_u <- predict(ms_mg1, type = "response", newdata = lakes.fr)
e <- MG - predict(ms_mg1, type = "response", newdata = lakes)
spl1_mg_tot <- sum(m_u) + HTestimator(e, 1/WGT.3X)
se_spl1_mg_tot <- sqrt(var_est(e, WGT.3X))
spl1_mg_mean <- spl1_mg_tot/N
se_spl1_mg_mean <- sqrt(var_est(e, WGT.3X)/(N^2))
ms_mg2 <- gam(formula = MG ~ GRID.ELEV.M + s(UTMX, UTMY, bs =
"tp") + factor(ECO.NUM), weight = WGT.3X0, data = lakes)
m_u <- predict(ms_mg2, type = "response", newdata = lakes.fr)
e <- MG - predict(ms_mg2, type = "response", newdata = lakes)
spl2_mg_tot <- sum(m_u) + HTestimator(e, 1/WGT.3X)
se_spl2_mg_tot <- sqrt(var_est(e, WGT.3X))
spl2_mg_mean <- spl2_mg_tot/N
se_spl2_mg_mean <- sqrt(var_est(e, WGT.3X)/(N^2))
# ------------------------------- #
# ------ CL splines estimates ------ #
out <- 160
WGT.3X0 <- WGT.3X
WGT.3X0[out] <- 0
ms_cl1 <- gam(formula = CL ~ GRID.ELEV.M + s(UTMX, UTMY, bs =
"tp"), weight = WGT.3X0, data = lakes)
m_u <- predict(ms_cl1, type = "response", newdata = lakes.fr)
e <- CL - predict(ms_cl1, type = "response", newdata = lakes)
spl1_cl_tot <- sum(m_u) + HTestimator(e, 1/WGT.3X)
se_spl1_cl_tot <- sqrt(var_est(e, WGT.3X))
spl1_cl_mean <- spl1_cl_tot/N
se_spl1_cl_mean <- sqrt(var_est(e, WGT.3X)/(N^2))
ms_cl2 <- gam(formula = CL ~ GRID.ELEV.M + s(UTMX, UTMY, bs =
"tp") + factor(ECO.NUM), weight = WGT.3X0, data = lakes)
m_u <- predict(ms_cl2, type = "response", newdata = lakes.fr)
e <- CL - predict(ms_cl2, type = "response", newdata = lakes)
spl2_cl_tot <- sum(m_u) + HTestimator(e, 1/WGT.3X)
se_spl2_cl_tot <- sqrt(var_est(e, WGT.3X))
spl2_cl_mean <- spl2_cl_tot/N
se_spl2_cl_mean <- sqrt(var_est(e, WGT.3X)/(N^2))
# ------------------------------- #
# ------ SO4 splines estimates ------ #
out <- 160
WGT.3X0 <- WGT.3X
WGT.3X0[out] <- 0
ms1_so4 <- gam(formula = SO4 ~ GRID.ELEV.M + s(UTMX, UTMY, bs =
"tp"), weight = WGT.3X0, data = lakes)
m_u <- predict(ms1_so4, type = "response", newdata = lakes.fr)
e <- SO4 - predict(ms1_so4, type = "response", newdata = lakes)
spl1_so4_tot <- sum(m_u) + HTestimator(e, 1/WGT.3X)
se_spl1_so4_tot <- sqrt(var_est(e, WGT.3X))
```

```
spl1_so4_mean <- spl1_so4_tot/N
se_spl1_so4_mean <- sqrt(var_est(e, WGT.3X)/(N^2))
ms2_so4 <- gam(formula = SO4 ~ GRID.ELEV.M + s(UTMX, UTMY, bs =
"tp") + factor(ECO.NUM), weight = WGT.3X0, data = lakes)
m_u <- predict(ms2_so4, type = "response", newdata = lakes.fr)
e <- SO4 - predict(ms2_so4, type = "response", newdata = lakes)
spl2_so4_tot <- sum(m_u) + HTestimator(e, 1/WGT.3X)
se_spl2_so4_tot <- sqrt(var_est(e, WGT.3X))
spl2_so4_mean <- spl2_so4_tot/N
se_spl2_so4_mean <- sqrt(var_est(e, WGT.3X)/(N^2))
# -------------------------------- #
```

References

Breidt, F.J., G. Claeskens and J.D. Opsomer. 2005. Model-assisted estimation for complex surveys using penalised splines. *Biometrika* 92: 831–846.

Breidt, F.J. and J.D. Opsomer. 2000. Local polynomial regression estimators in survey sampling. *The Annals of Statistics* 28: 1026–1053.

Breidt, F.J. and J.D. Opsomer. 2017. Model-assisted survey estimation with modern prediction techniques. *Statistical Science* 32: 190–205.

Breidt, F.J., J.D. Opsomer, A.A. Johnson and M.G. Ranalli. 2007. Semiparametric model-assisted estimation for natural resource surveys. *Survey Methodology* 33: 35–44.

Larsen, D.P., T.M. Kincaid, S.E. Jacobs and N.S. Urquhart. 2001. Designs for evaluating local and regional scale trends. *Bioscience* 51: 1049–1058.

Larsen, D.P., K.W. Thornton, N.S. Urquhart and S.G. Paulsen. 1993. Overview of survey design and lake selection. EMAP Surface Waters 1991 Pilot Report. Larsen, D.P. and S.J. Christie (eds.). *Technical Report EPA/620/R 93/003, U.S. Environmental Protection Agency*.

Lumley, T. 2010. *Complex Surveys: A Guide to Analysis Using R*. Wiley.

Lumley, T. 2019. *Survey: Analysis of Complex Survey Samples*. R package version 3.35-1. https://cran.r-project.org/web/packages/survey/index.html.

Kammann, E.E. and M.P. Wand. 2003. Geoadditive models. *Journal of the Royal Statistical Society: Series C* (Applied Statistics) 52: 1–18.

Mashreghi, Z., D. Haziza and C. Léger. 2016. A survey of bootstrap methods in finite population sampling. *Statistics Surveys* 10: 1–52.

McCullagh, P. and J.A. Nelder. 1989. *Generalized Linear Models*. Chapman and Hall: London.

Messer, J.J., R.A. Linthurst and W.S. Overton. 1991. An EPA program for monitoring ecological status and trends. *Environmental Monitoring and Assessment* 17: 67–78.

Montanari, G.E. and M.G. Ranalli. 2005. Nonparametric model calibration estimation in survey sampling. *Journal of the American Statistical Association* 100: 1429–1442.

R Core Team. 2019. R: A language and environment for statistical computing. R Foundation for Statistical Computing, Vienna, Austria. URL https://www.R-project.org/.

Ruppert, D., M. Wand and R. Carroll. 2003. *Semiparametric Regression*. Cambridge University Press: Cambridge.

Sarndal, C., B. Swensson and J. Wretman. 1992. *Model Assisted Survey Sampling*. Springer, New York.

Tillé, Y. and A. Matei. 2016. *Sampling: Survey Sampling. R Package Version 2.8*. https://cran.r-project.org/web/packages/sampling/index.html.

Toth, D. and J.L. Eltinge. 2011. Building consistent regression trees from complex sample data. *Journal of the American Statistical Association* 106: 1626–1636.

Wand, M. 2018. *SemiPar: Semiparametic Regression. R Package Version 1.0-4.2*. https://cran.r-project.org/web/packages/SemiPar/index.html.

Wolter, K.M. 1985. *Introduction to Variance Estimation*. Springer-Verlag: New York.

Wood, S.N. 2017. *Generalized Additive Models: An Introduction with R* (2nd edition). Chapman and Hall/CRC.

CHAPTER 4
Yield Prediction in Agriculture
A Comparison Between Regression Kriging and Random Forest

Eugenia Nissi[1],* and *Annalina Sarra*[2]

4.1 Introduction

Agriculture plays a vital and significant role in the worldwide economy. Over the last few years, this sector has faced important challenges: population pressure, climate change, and food security. With respect to the first concern, owing to the increase in world population, forecasted to reach 9.9 billion in 2050 (Buttafuoco and Lucà 2016) and to the resulting reduction of land availability, farmers are obliged to produce "more from less land".

As a result, both increasing and predicting crop yields are becoming more and more essential activities for farmers as well as for consultants and agriculture related organizations.

The crop forecasting is defined by the Food and Agriculture Organization of the United Nations (FAO) as "*the art of predicting crop yield and production before the harvest actually takes place, typically a couple of months in advance*" (Gommes et al. 1996).

There are various noteworthy evidences which show the benefits associated with early crop yield prediction. Specifically, yield forecasting represents an important tool for developing agriculture operations and management, for regulating agriculture cultivation systems, for food security (Horie et al. 1992), for analysing global trends, for making informed management and financial decisions, for avoiding the losses during any kinds of unfavourable conditions, like those due to climate change, and for evaluating crop-area insurance contracts (Nadolnyak et al. 2008). The prediction

---

[1] "G. d'Annunzio" University of Chieti-Pescara, Department of Economic Studies.
[2] "G. d'Annunzio" University of Chieti-Pescara, Department of Philosophical, Pedagogical and Quantitative Economic Sciences.
Email: asarra@unich.it
* Corresponding author: eugenia.nissi@unich.it

of crop yield is extremely challenging because, due to the complexity of agro-systems, a number of factors and their interactions must be taken into account.

The outcome of crop yield prediction is primarily dependent on environmental parameters, such as sunlight, soil water, rainfall and humidity, and temperature-related factors; other factors that may affect a selected crop production are related to land capability, planting practices (fertilizer application and irrigation) and soil chemical properties. In the past, most farmers have relied on their long-term experience to figure out the prediction associated with crop yield.

The existing research trends have highlighted that there are several methods of yield forecasting. Many scholars propose Crop Simulation Models (CSM) to study the relationship between crop growth and environment (see, among others, Brisson et al. 2003; Franko et al. 2007; Nendel et al. 2011; Stöckle et al. 2003). These models can generate results under various conditions. After nearly 40 years of development, the CSM shifted from a simple qualitative simulation of crop growth to more complex simulations of the whole growth process in a certain area and contexts. Besides, owing to the heterogeneity of crop growth systems and the uncertainty surrounding many input parameters, stochastic models are required (Basso et al. 2001). However, many of the CSM models entail intensive data and calibration requirements for having accurate crop growth simulations that make their use tricky (van Ittersum et al. 2013). Accordingly, a large literature has used statistical models to provide reasonable crop yield predictions. Statistical models include mainly time-series, panel, and cross-section models (Lobell and Burke 2010).

Traditional regression approaches (simple or multiple linear regression), in which historical data on crop yields and agro-meteorological parameters are used to calibrate regression equations, are rather common in many yields forecast researches and programs (Lobell and Burke 2009). Several studies have pointed to a number of advantages and drawbacks of the statistical models applied in the field of agriculture crop yield predictions (Tao and Zhang 2010; Zhang et al. 2010; Schlenker and Lobell 2010). For instance, the limited data requirements and the assessment of model uncertainties are the most substantial advantages of the regression models. Conversely, statistical models can hardly extrapolate values beyond the model parameterization limits.

In the recent years, many researches support that data mining technique algorithms can be profitable when applied to a wide range of agriculture problems and, in particular, with regard to reasonable crop yield predictions. The common tasks from standard data mining are adapted for dealing with spatial relations in agriculture data. In this chapter, we carried out an experimental study to compare the potentiality of two hybrid spatial interpolation techniques for the spatial estimation of the crop yield: regression kriging (RK) and random forest (RF). As a case study, we interpolated the winter wheat (*Triticum aestivum* L.) yields in the Southern U.S. Great Plains.

The chapter is organized as follows. Section 4.2 provides a brief review of the most interesting researches which employ data mining techniques in agriculture. Section 4.3 covers the basic features of regression kriging and random forest framework, giving special emphasis to the uncertainty quantification in random forest and the performance measures in guiding the interpretation of the results. Section 4.4 introduces the study area and describes the data, then the results are given in Section 4.5. Within Sections 4.4 and 4.5, details of R code are also provided. The concluding remarks of this chapter are finally summarized in Section 4.6.

4.2 Data mining techniques in agriculture: A brief literature survey

Data Mining (DM), also known as "Knowledge Discovery in Databases", refers to a set of techniques and methodologies aimed at extracting knowledge from existing large datasets to discover meaningful patterns and rules and establish useful classifications (Fayyad et al. 1996; Klosgen and Zytkow 2002).

DM is an inter-disciplinary area which builds a bridge among database technology, machine learning, statistics, fuzzy logic, and pattern recognition (Han et al. 2006). By ease of use and the possibility of presenting complex results in a simple fashion, DM techniques have gained a lot attention in the agriculture field (Miller et al. 2009). The potential of applying data mining techniques in agriculture are discussed, for instance, in Raorane and Kulkarni (2013) and Kalpana et al. (2014). In agriculture research, it is possible to find a growing number of DM promising applications, set up for different purposes, like yield prediction, management zone delineation, soil classification, weather forecasting, disease detection, and pesticide and fertilizer optimization.

In what follows, we briefly highlight few interesting researches related to some of the above-mentioned areas, while the interested reader can refer to Wu et al. (2008) for a review of the most used DM techniques.

As for yield prediction, the study of Kaul et al. (2005) shows, for example, how an artificial neural network model can deliver more reliable and accurate crop yield forecasting compared to other methods. Gonzalez-Sanchez et al. (2014) carry out a broad research to examine the predictive ability of some machine learning techniques for crop yield prediction. Specifically, multiple linear regression, regression trees, artificial neural network, support vector machines, and k-nearest neighbours' algorithms are compared for seeking the most accurate technique for forecasting the issue at hand. In Supriya (2017), the problem of predicting the crop yield is formalized as a classification rule, where Naïve Bayes and k-nearest neighbours are used. Everingham et al. (2016) exploit the properties of a random forest algorithm to investigate the prediction of sugarcane.

Another classic task in agriculture, known as delineation of management zones, can be accomplished via DM techniques. In this respect, the work of Kitchen et al. (2005) presents an advanced approach for delineating spatially coherent regions, representing subfield regions with homogenous characteristics. This objective is achieved by running a fuzzy c-means clustering.

Along these lines of research, Arango et al. (2016) explore the potentialities of using a methodology based on machine learning algorithms and satellite data to automatically delimitate cultivable lands. Bhargavi and Jyothi (2009) propose an analysis of soil profile experimental datasets using a Naïve Bayes classifier. Another study by Verheyen et al. (2001) is aimed at clustering soils in combination with GPS-based technologies via k-means clustering.

In the context of pest control decision making program, the study of Bi and Chen (2010) has proved the effectiveness of the Bayesian network method as a valid tool for crop disease. Later, Bhagawati et al. (2015) recognize the important role of artificial neural networks in development of precise forecasting and forewarning models of plant diseases.

Other studies found that the association rule mining could be potentially useful for elucidating hidden patterns and associates between different climates and crop productions (Dhanya and Kumar 2009).

4.3 Methodology

Over the past few years, a number of methods for spatial interpolation have been proposed and applied. They range from deterministic models, such as inverse distance weighting, spline, and radial basis functions, to stochastic approaches to spatial interpolation, developed to account for the uncertainty in the resulting outcomes. Geostatistical modelling (Goovaerts 1997; Cressie 2015) provides a convenient approach to spatial interpolation combining deterministic and stochastic components.

Geostatistics translates one of the key principles of geography, *"Everything is related to everything else, but near things are more related than distant things"* (Tobler 1970), into a mathematical model, through the spatial autocorrelation functions or variograms.

In the traditional literature of geostatistical models, the target variable Y, which can be the occurrence, the quantity and/or state geographical phenomena, can be modelled as the sum of a deterministic mean (μ) and a stochastic residual (e):

$$Y(s) = \mu(s) + e(s)$$

Spatial interpolation is concerned with the prediction of the target variable Y, at new and unsampled location (s_0), relying on training data, $y(s_i)$, $i = 1,\ldots, n$ where $s_i \in D$, n is the number of observed locations and D is the geographical domain.

In what follows, we provide a brief methodological comparison between regression kriging and random forest regression kriging, employed in this research for spatial prediction and comparative evaluation of wheat yield prediction.

4.3.1 Regression kriging

Kriging is the mainstream geostatistical technique for generating spatial predictions of the values of a target variable at any location, based on a number of observations in the same area.

Kriging enjoys the benefit of including the knowledge about spatial autocorrelation, in the variable of interest, in modelling and prediction, and provides a spatially explicit measure of prediction uncertainty (Matheron 1969; Oliver and Webster 2014).

An extension of kriging that combines the multiple linear regression and kriging is the so-called regression kriging (RK), first devised by Odeh et al. (1994). Through this hybrid approach, the predictions of the target variable are made via a deterministic method, i.e., regression model with covariate information, and a stochastic method, where the spatial autocorrelation of the model residuals is determined with a variogram.

Formally, the following equation is used to predict the value of Y at unsampled location (s_0) $(\hat{y}_{s_0})$:

$$\hat{y}_{s_0} = \hat{\mu}_{s_0} + \hat{e}_{s_0} \tag{4.1}$$

In Eq. 4.1, $\hat{\mu}_{s_0}$ is the fitted deterministic part and $\hat{e}_{s_0}$ is the interpolated residual.

The basic framework of spatial interpolation through RK relies on a linear regression, used to fit the trend, and on kriging to interpolate the residual $\hat{e}$. By adopting the matrix notation, the RK can be represented as follows:

$$\mathbf{y} = \mathbf{q}' \, \boldsymbol{\beta} + \boldsymbol{\varepsilon} \tag{4.2}$$

$$\hat{y}_{(s0)} = \mathbf{q}_0' \, \hat{\boldsymbol{\beta}} + \boldsymbol{\lambda}_0' \mathbf{e} \tag{4.3}$$

where $\mathbf{e}$ is the vector of n regression residuals, q_0 is the vector of p auxiliary variables at s_0, $\hat{\boldsymbol{\beta}}$ is the vector of $p + 1$ estimated model coefficients, $\boldsymbol{\lambda}_0$ is the vector of n kriging weights. The regression coefficients $\hat{\boldsymbol{\beta}}$ are estimated from the sample by some fitting methods. Specifically, to take the spatial autocorrelation between individual observations into account, it is possible to solve the regression coefficients, using the following generalized least square estimation (Cressie 2015):

$$\hat{\boldsymbol{\beta}}_{GLS} = (\mathbf{q}' \, \mathbf{C}^{-1} \mathbf{q})^{-1} \, \mathbf{q}' \, \mathbf{C}^{-1} \mathbf{y} \tag{4.4}$$

where $\mathbf{q}$ is the matrix of the auxiliary variables, $\mathbf{C}$ is the $n \times n$ covariance matrix of the residuals, and $\mathbf{y}$ is the vector of the sampled response observations.

Using the matrix notation, the predicted value at location s_0 (i.e., $\hat{y}_{(s0)}$) can be written as:

$$\hat{y}_{s_0} = \mathbf{q}_0' \, \hat{\boldsymbol{\beta}}_{GLS} + \boldsymbol{\lambda}_0' \, (\mathbf{y} - \mathbf{q}\hat{\boldsymbol{\beta}}_{GLS}) \tag{4.5}$$

As pointed out in Christensen (2001), the prediction model has an error that reflects the position of a new location in both geographical and feature space. The RK procedure has been fruitfully adopted over the past years for producing spatial predictions in a variety of environmental fields, like physical geography, soil science, agriculture, climatology, epidemiology, and natural hazard monitoring (Li and Heap 2011).

4.3.2 Random forest for spatial prediction

RF is a machine learning approach, first described in Breiman (2001). As the name suggests, random forest takes advantage of an ensemble of decision trees, constructed for classification or regression purposes. In this study, we only consider RF for regression. Formally, an RF model is defined as a collection of regression trees $\{Tb : b = 1,..., B\}$ each built from a bootstrap sample of the data set $\{Y, X\}$.

Given the predictor values $\mathbf{x} = (x_1,..., x_p)$, the prediction at a new site is obtained by averaging the predictions made by each tree in the ensemble:

$$\hat{f}(x) = \frac{1}{B} \sum_{b=1}^{B} T_b(x) \tag{4.6}$$

Such averaging has the effect of stabilizing predictions because individual trees have low bias but high variance.

Being a completely non-parametric procedure, RF makes no assumptions about the relationships between the predictor variables and the response. In the standard

RF framework, the geographical space is not included in the model and the sampling locations are ignored during the estimation of model parameters. Recently, RF algorithm has acquired a good reputation as a potential method for predicting values across a region and creating interpolated surfaces.

One way to account for spatial autocorrelation in RF modelling is through the specification of a Random Forest Regression Kriging (see, among others, Hengl et al. 2015; Fayad et al. 2016).

In analogy with regression kriging, a prediction for a new site is given by summing the RF prediction and the kriging prediction of RF residuals.

Moving along these lines of research, one can consider the general framework proposed by Hengl et al. (2018) to make RF applicable to a spatial statistics problem expressed as:

$$Y(s) = f(X_G, X_R, X_P) \tag{4.7}$$

In Eq. (4.7), X_G denotes the covariates accounting for geographical proximity between observations, X_R are the surface reflectance covariates, i.e., usually spectral bands of remote sensing images, and X_P refers to the process-based covariates.

Note that the predictions would be similar to ordinary kriging, when RF for spatial data is fitted using only the geographical covariates (X_G), whereas, if all covariates in Eq. 4.7 are used, RF for spatial data would be like regression kriging.

4.3.3 *Parameters and uncertainty quantification in random forest*

The implementation of RF requires to tune some inner parameters. First and foremost, in order to make RF implementable, it is necessary to set the *minimum size of a node*, used to determine the terminal nodes. Another important parameter that controls the RF is the *forest size*, that is the total number of trees grown. A default choice for the forest size parameter is 500; actually, this parameter does not really need to be fine-tuned but it is advised to set a computationally feasible large number to achieve a higher accuracy.

The algorithm also depends on the number of predictors to randomly consider per split. In practice, as suggested in Wright and Ziegler (2015), this number is either the number of predictors or the square root of the total predictors/covariates.

It is common belief that the implemented default values for these parameters lead to satisfactory empirical performance in prediction. In contrast, Scornet (2018) stresses the absence of theoretical justification for these default values and provides insights about how to choose parameters in random forest procedure.

Another critical issue in RF modelling regards the uncertainty quantification of random forest predictions. In this respect, different approaches are noteworthy. Wager et al. (2014) rely on jackknife-after-bootstrap methods for estimating standard errors in RF predictions.

Their procedure evaluates the average variability between RF predictions built on the whole training set and the RF predictions (Mentch and Hooker 2016) with the jackknife training sets that exclude one observation pair iteratively. Later et al. (2016) demonstrate that formal statistical inference procedures are possible within the context of supervised ensemble learning.

By training a multitude of trees on strict subsample combinations of the training set and averaging their results, the authors demonstrate that the ensembles can be viewed as U-statistics which are proven to be asymptotically normal. Accordingly, it is possible to compute confidence intervals for predictions and formally test the significance of features.

Instead, reasonable estimates of prediction uncertainty for random forest regression models are obtained by Coulston et al. (2016), following a Monte Carlo approach.

In literature, great attention is also paid to the algorithm proposed by Meinshausen (2006) which is based on the quantile RF. The quantile RF allows one to estimate the quantiles of the distribution of the response at prediction points by inferring the full conditional distribution of a target variable.

Consequently, Meinshausen's technique can be used to make prediction intervals and not for confidence intervals because the empirical conditional cumulative distribution function does not provide information on the uncertainty of the fit of the RF model itself.

4.3.4 Performance measures

To compare the performance of RK and RF for spatial data, different metrics can be used to guide the interpretation of the results. For our case study, we assess the accuracy of chosen models, by contrasting the yield prediction against the actual yield.

Let $y(s_j)$ denote the i-th observed value at the site s_j and $\hat{y}(s_j)$ stand for the 10-fold cross validation (CV) prediction for the same site, the root-mean-square prediction error is equal to

$$RMSE = \sqrt{\frac{\sum_{i=1}^{m}(y(s_j) - \hat{y}(s_j))^2}{m}}. \tag{4.8}$$

Given the total number of cross-validation points (m), the average bias in predictions can be quantified through the average ME based on CV, derived as:

$$ME = \frac{1}{m}\sum_{i=1}^{m}(\hat{y}(s_j) - y(s_j)) \tag{4.9}$$

The amount of variation explained by the model is obtained through the R^2 of the k-fold predictions defined as:

$$R^2_{pred} = 1 - \sum_{i=1}^{n}\frac{(\hat{y}(s_j) - y(s_j))^2}{(\hat{y}(s_j) - \overline{y})^2} \tag{4.10}$$

where $\overline{y}$ denotes the average response over all locations.

In addition to R-square, it is possible to quantify how far the observed data deviates from the line of perfect concordance. In this respect, one can derive the Lin's Concordance Correlation Coefficient (Steichen and Cox 2002):

$$\rho_c = \frac{2 \cdot \rho \cdot \sigma_{\hat{y}} \cdot \sigma_y}{\sigma_{\hat{y}}^2 + \sigma_y^2 + (\mu_{\hat{y}} - \mu_y)^2} \tag{4.11}$$

where $\hat{y}$ are the predicted values and y are actual values at CV points, $\mu_{\hat{y}}$ and μ_y are predicted and observed means, and ρ is the correlation coefficient between predicted and observed values.

4.4 Study area and datasets

The study area refers to the Southern Great Plains of the US and spans across three States: Kansas, Oklahoma, and Texas (Figure 4.1).

The Southern Great Plains of the US is an area which covers 7.5 million hectares and accounts for approximately 30% of total US wheat (Lollato et al. 2017). In that region, the winter wheat production achieves 18.8 million metric tons per year.

The winter monitor wheat yield data considered in this study are within the latitudes of 27.47° and 39.72° N and the longitudes of 102.85° and 94.18° W (see Figure 4.2),[1] and are distributed across the counties of Kansas, Oklahoma, and Texas.

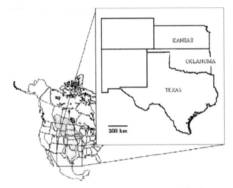

Figure 4.1. Southern Great Plains of US: Kansas, Oklahoma and Texas.

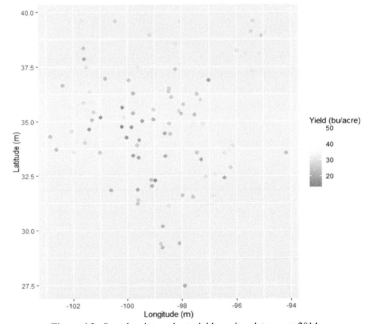

Figure 4.2. Sample winter wheat yield monitor data - year 2014.

---

[1] The coordinate system for this study was based on WGS84.

In our analysis, to improve the spatial prediction of wheat yield data, we also rely on a range of meteorological and environmental predictors. Specifically, the variables listed in Table 4.1 are initially taken into consideration for the experiment.

The original data were downloaded from the enterprise agricultural analytics platform https://aerialintel.blob.core.windows.net/recruiting/datasets/wheat-2014-supervised.csv and then manipulated to create an artificial dataset used for the comparison study.

Table 4.1. Input variables and description.

Variables	Description
Lat	Centroid latitude location of the site
Long	Centroid longitude location of the site
Temp Max	Temperature Max
Temp Min	Temperature Min
Press	Pressure
hum	Humidity
dPoint	dewPoint
Cc	Cloudcover
vis	Visibility
wbearing	Wind Bearing
wspeed	Wind speed
NDVI	Normalized Difference Vegetation Index

4.5 Results

This section evaluates the potentiality of regression kriging and random forest in the spatial estimation of winter wheat yield (Y_w) in the Southern Great Plains of the US. To produce model and predictions for RK and RF, we use the `caret` package (Kuhn 2008) for regression and the `gstat` package (Pebesma 2004) for geostatistical modelling.

The overall R packages required for the analysis are listed below.

```
> library(plyr)
> library(dplyr)
> library(gstat)
> library(raster)
> library(ggplot2)
> library(car)
> library(classInt)
> library(RStoolbox)
> library(caret)
> library(caretEnsemble)
> library(doParallel)
> library(gridExtra)
```

4.5.1 Actual wheat yield and grid locations

The winter wheat yield is the target variable and is expressed in terms of volume per acre (*bu/a*).[2] The following R code loads data, shapefile, and auxiliary variables at grid locations and plots the prediction locations as displayed in Figure 4.3.

```
> #Load data
> train_data<-read.csv("yield_data.csv",header=T,sep=",")
> state<-shapefile("tl_2017_us_state.shp")
> plot(state)
> #Load grid data
> grid<-read.csv("grid_data.csv",header=T,sep=",")
> coord_grid<-grid_data[,c(4:5)]
# Plot points over the US map
>points(coord_grid$Longitude, coord_grid$Latitude, pch=19,
col="red")
```

A total of 119 samples were selected and some statistics for Yw are presented in Table 4.2.

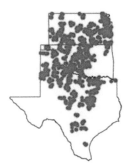

Figure 4.3. Prediction locations.

Table 4.2. Descriptive statistics of wheat yield.

# of samples		119
Actual Winter Wheat Yield (bu/acre)	Min	12.90
	Max	50.10
	Mean	39.10
	St.Dev	9.44

The target variable used in regression and kriging analysis should be normally distributed for obtaining better kriging estimates. In this study, to handle the skewness in the distribution of Y_w a Box-Cox transformation was used.

In R, this is achieved using the following code:

```
> #Power transformation
> powerTransform(train_data$Yield)
```

---

[2] *bu/a* = bushels per acre.

```
Estimated transformation parameters
train_data$Yield
  0.6020114
> # Estimated transformation parameter
> train_data$Yield.bc<-bcPower(train_data$Yield,0.6020114)
```

4.5.2 *Wheat yield prediction models results*

For the kriging regression analysis, we first fit the Generalized Linear Model (GLM) with the meteorological and environmental variables selected through a stepwise regression procedure.

The R code for fitting GLM is:

```
#Set control parameter
> myControl <- trainControl(method="repeatedcv",
+                           number=10,
+                           repeats=5,
+                           allowParallel = FALSE)
> set.seed(1856)
GLM_model<-train(train.x,
            response,
            method = "glm",
            trControl=myControl,
            preProc=c('center', 'scale'))
print(GLM_model)
summary(GLM_model)
119 samples
  5 predictor
Pre-processing: centered (5), scaled (5)
Resampling: Cross-Validated (10 fold, repeated 5 times)
Summary of sample sizes: 107, 107, 107, 107, 108, 107, ...
Resampling results:
  RMSE       Rsquared    MAE
  2.078685   0.29822     1.748143
summary(GLM)
Call:
NULL

Deviance Residuals:
Min      1Q      Median   3Q     Max
-4.837   -1.613  -0.117   1.427  5.192

Coefficients:
              Estimate Std.  Error   t value  Pr(>|t|)
(Intercept)   11.5521         0.1884  61.327   < 2e-16     ***
dpoint        -1.7494         0.3624  -4.827   4.37e-06    ***
hum           1.6220          0.3698  4.386    2.60e-05    ***
vis           0.5063          0.2470  2.049    0.042749    *
```

```
wbearing      0.6519        0.2293 2.844    0.005296  **
wspeed        0.8476        0.2192 3.867    0.000184  ***
---
Signif. codes: 0 `***' 0.001 `**' 0.01 `*' 0.05 `.' 0.1 ` ' 1
(Dispersion parameter for gaussian family taken to be 4.222433)
Null deviance: 683.69 on 118 degrees of freedom
Residual deviance: 477.13 on 113 degrees of freedom
AIC: 516.96
Number of Fisher Scoring iterations: 2
```

Note that in the above code the R object `train.x` is the training `dataframe` containing the predictors while *response* is the vector of the observed values of the target variable.

The further steps in the RK analysis have involved the computation of the variogram of the residuals of the GLM model and the application of simple kriging to the residuals to estimate the spatial prediction of the residuals.

The following code allows to estimate and plot the variogram:

```
# Extract residuals
> train.xy$residuals.glm<-resid(GLM_model)
# Variogram
> var.glm<-variogram(residuals.glm~ 1, data =train.xy)
# Initial parameter set by eye estimation
> mod.glm<-vgm(1,"Exp",5,2)
# least square fit
> mod.fit.glm<-fit.variogram(var.glm, mod.glm)
> mod.fit.glm

windows()
#### Plot variogram and fitted model:
plot(var.glm, pl=F,
     model=mod.fit.glm,
     col="black",
     cex=0.9,
     lwd=0.5,
     lty=1,
     pch=19,
     main="Variogram and Fitted Model\n Residuals of GLM
     model",
     xlab="Distance (m)",
     ylab="Semivariance")
```

The GLM regression predicted results and the simple kriged residuals, to be added to estimate the interpolated winter wheat yield, are achieved in R using the following code:

```
###GLM Prediction at grid location
>grid.xy$GLM_model <- predict(GLM_model, grid.df)
#Simple Kriging Prediction of GLM residuals at grid location
```

```
>SimpleK.GLM<-krige (residuals.glm~ 1,
            loc=train.xy,          # Data frame
            newdata=grid.xy,       # Prediction location
            model = mod.fit.glm,   # fitted varigram model
            beta = 0)              # residuals from a trend;
                                   expected value is 0
## [using simple kriging]

#Kriging prediction (Simple kriging + Regression Prediction)
> grid.xy$SimpleK.GLM<-SimpleK.GLM$var1.pred
# Add RF predicted + SK preedicted residuals
> grid.xy$RK.GLM.bc<-(grid.xy$GLM_model+grid.xy$SimpleK.GLM)
```

The R objects `train.xy` and `grid.xy` are SpatialPointDataFrames containing the observed and prediction locations, respectively. Instead, the object `grid.df` is a dataframe with the auxiliary variables' values at the prediction locations.

The variogram and fitted model for the residuals of GLM model is displayed in Figure 4.4.

For plotting the predicted winter wheat yield, it is necessary to proceed to the back transformation and to use the Box-cox transformation parameters and make the raster conversion. In the subsequent code, there are the correspondent instructions in R and the output:

```
> k1<-1/0.6020114
> grid.xy$RK.GLM <- ((grid.xy$RK.GLM.bc *0.6020114 +1)^k1)
> summary(grid.xy)
Object of class SpatialPointsDataFrame
```

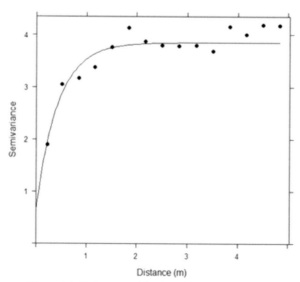

Variogram and Fitted Model
Residuals of GLM model

Figure 4.4. Variogram and fitted model of GLM residuals.

```
Coordinates:
      min       max
x    -103.5    -94.5
y    27.5      39.5
Is projected: NA
proj4string : [NA]
Number of points: 2000
Data attributes:
        ID                 GLM_model          SimpleK.GLM
Min. : 1.0          Min. : 0.4816      Min. :-1.7488
1st Qu.: 500.8      1st Qu.: 9.5122    1st Qu.:-0.9665
Median :1000.5      Median :10.8023    Median :-0.2496
Mean :1000.5        Mean :10.7955      Mean :-0.1941
3rd Qu.:1500.2      3rd Qu.:12.1954    3rd Qu.: 0.5623
Max. :2000.0        Max. :17.2037      Max. : 2.1098
   RK.GLM.bc              RK.GLM
Min. : 0.09261      Min. : 1.094
1st Qu.: 9.02125    1st Qu.:22.010
Median :10.52421    Median :27.389
Mean :10.60131      Mean :28.296
3rd Qu.:12.22090    3rd Qu.:34.012
Max. :18.66659      Max. :64.086
```

```
#Convert to raster
> GLM<-rasterFromXYZ(as.data.frame(grid.xy)[, c("x", "y",
"GLM_model")])
> SK.GLM<-rasterFromXYZ(as.data.frame(grid.xy)[,c("x","y",
"SimpleK.GLM")])
> RK.GLM.bc<-rasterFromXYZ(as.data.frame(grid.xy)[,c("x","y",
"RK.GLM.bc")])
> RK.GLM.YD<-rasterFromXYZ(as.data.frame(grid.xy)[, c("x",
"y", "RK.GLM")])
```

Finally, to plot the predicted winter wheat yield results, we rely on the elegant data visualization available within the ggplot2 package. The R code reported below allows one to arrange in a multiplot, starting from left towards right, the visualization of GLM predicted (Box-Cox), the simple kriging GLM residuals, the RK-predicted (Box-Cox) and RK-GLM predicted values (see Figure 4.5).

```
> #Plot predicted wheat Yield
> glm1<-ggR(GLM, geom_raster = TRUE) +
+ scale_fill_gradientn("", colours = c("orange", "yellow",
"green", "sky blue","blue")) +
+ theme_bw() +
+   theme(axis.title.x=element_blank(),
+       axis.text.x=element_blank(),
+       axis.ticks.x=element_blank(),
+       axis.title.y=element_blank(),
```

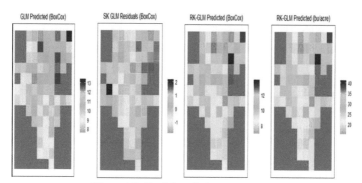

Figure 4.5. Spatial prediction RK results.

```
+       axis.text.y=element_blank(),
+       axis.ticks.y=element_blank())+
+   ggtitle("GLM Predicted (BoxCox)")+
+   theme(plot.title = element_text(hjust = 0.5))
> glm2<-ggR(SK.GLM, geom_raster = TRUE) +
+ scale_fill_gradientn("", colours = c("orange", "yellow",
"green", "sky blue","blue"))+
+ theme_bw()+
+   theme(axis.title.x=element_blank(),
+       axis.text.x=element_blank(),
+       axis.ticks.x=element_blank(),
+       axis.title.y=element_blank(),
+       axis.text.y=element_blank(),
+       axis.ticks.y=element_blank())+
+   ggtitle("OK GLM Residuals (BoxCox)")+
+   theme(plot.title = element_text(hjust = 0.5))
> glm3<-ggR(RK.GLM.bc, geom_raster = TRUE) +
+ scale_fill_gradientn("", colours = c("orange", "yellow",
"green", "sky blue","blue"))+
+ theme_bw()+
+   theme(axis.title.x=element_blank(),
+       axis.text.x=element_blank(),
+       axis.ticks.x=element_blank(),
+       axis.title.y=element_blank(),
+       axis.text.y=element_blank(),
+       axis.ticks.y=element_blank())+
+   ggtitle("RK-GLM Predicted (BoxCox)")+
+   theme(plot.title = element_text(hjust = 0.5))
> glm4<-ggR(RK.GLM.YD, geom_raster = TRUE) +
+ scale_fill_gradientn("", colours = c("orange", "yellow",
"green", "sky blue","blue"))+
+ theme_bw()+
+   theme(axis.title.x=element_blank(),
+       axis.text.x=element_blank(),
```

```
+          axis.ticks.x=element_blank(),
+          axis.title.y=element_blank(),
+          axis.text.y=element_blank(),
+          axis.ticks.y=element_blank())+
+     ggtitle("RK-GLM Predicted (bu/acre)")+
+     theme(plot.title = element_text(hjust = 0.5))
>
> grid.arrange(glm1,glm2,glm3,glm4, ncol = 4) # Multiplot
```

The spatial prediction results of RK are shown in Figure 4.5. The regression kriging predicted winter wheat yield ranges from 1.09 to 64.08 bu/acre, with a mean value of 28.29 and a standard deviation of 9.1. It is worth noting that the predicted winter wheat yields record the highest values in the north-eastern portion of the study area; it is also seen that wheat yield is lower moving towards the southern counties.

In order to understand the worth of RK predicted wheat yield map, it was compared with RF prediction. To fit a RF model, we followed the same steps highlighted above.

Within the `caret` package the switch to RF model is achieved by changing the method argument in the function `train`. The next R code fits an RF model as implemented in the `caret` package:

```
> myControl <- trainControl(method="repeatedcv",
                            number=10,
                            repeats=5,
                            allowParallel = TRUE)
> set.seed(1856)
> mtry <- sqrt(ncol(train.x)) #number of variables randomly
sampled as candidates at each split.
> tunegrid.rf <- expand.grid(.mtry=mtry)

> RF<-train(train.x,
+           response,
+           method = "rf",
+           trControl=myControl,
+           tuneGrid=tunegrid.rf,
+           ntree= 100,
+           preProc=c('center', 'scale'))
> print(RF)
Random Forest

119 samples
  5 predictor
Pre-processing: centered (5), scaled (5)
Resampling: Cross-Validated (10 fold, repeated 5 times)
Summary of sample sizes: 107, 107, 107, 108, 107, 107, ...
Resampling results:
RMSE         Rsquared    MAE
2.031477     0.3253809   1.648224
Tuning parameter 'mtry' was held constant at a value of 2.236068
```

For the variogram modelling of RF residuals, the extraction of RF residuals is required. The RF residuals are obtained through the `resid()` function:

```
>train.xy$residuals.rf<-resid(RF)
```

The following R code is useful for the estimation and plotting of the variogram of RF residuals, which is displayed in Figure 4.6.

```
##variogram
> var.rf<-variogram(residuals.rf~ 1, data = train.xy)
> # Initial parameter set by eye estimation
> mod.rf<-vgm(15,"Exp",5,2)

> # least square fit
> mod.fitt.rf<-fit.variogram(var.rf, mod.rf)
> mod.fitt.rf
   model       psill         range
1    Nug   0.04260472    0.00000000
2    Exp   0.84230380    0.06871389
> windows()
> #### Plot variogram and fitted model:
> plot(var.rf, pl=F,
+         model=mod.fitt.rf,
+         col="black",
+         cex=0.9,
+         lwd=0.5,
+         lty=1,
+         pch=19,
+         main="Variogram and Fitted Model\n Residuals of RF
```

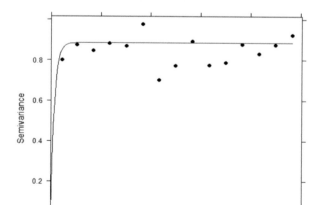

Figure 4.6. Variogram and fitted model of RF residuals.

```
        model",
+         xlab="Distance (m)",
+         ylab="Semivariance")
```

For the simply kriging prediction of RF at grid locations, the R code is:

```
> #Prediction at grid location
> grid.xy$RF <- predict(RF, grid.df)
> #Simpel Kriging Prediction of RF residuals at grid location
> SimpleK.RF<-krige(residuals.rf~ 1,
+               loc=train.xy,        # Data frame
+               newdata=grid.xy,     # Prediction location
+               model = mod.fitt.rf,  # fitted varigram model
+               beta = 0)            # residuals from a trend;
                                     expected value is 0
[using simple kriging]
```

The kriging predictions are obtained as follows:

```
> #Kriging prediction (SK+Regression)
> grid.xy$SimpleK.RF<-SimpleK.RF$var1.pred
> # Add RF predicted + SK preedicted residuals
> grid.xy$RK.RF.bc<-(grid.xy$RF+grid.xy$SimpleK.RF)
```

Hence, for the back transformation, we use the transformation parameters as displayed below:

```
> k1<-1/0.6020114
> grid.xy$RK.RF <-((grid.xy$RK.RF.bc *0.6020114 +1)^k1)
> summary(grid.xy)
Object of class SpatialPointsDataFrame
Coordinates:
      min       max
x   -103.5    -94.5
y    27.5     39.5
Is projected: NA
proj4string : [NA]
Number of points: 2000
Data attributes:
       ID                 GLM                 SK.GLM
Min. : 1.0         Min. : 0.4816       Min. :-1.7488
1st Qu.: 500.8     1st Qu.: 9.5122     1st Qu.:-0.9665
Median :1000.5     Median :10.8023     Median :-0.2496
Mean :1000.5       Mean :10.7955       Mean :-0.1941
3rd Qu.:1500.2     3rd Qu.:12.1954     3rd Qu.: 0.5623
Max. :2000.0       Max. :17.2037       Max. : 2.1098
   RK.GLM.bc            RK.GLM              OK.GLM
Min. : 0.09261     Min. : 1.094        Min. :-1.7488
1st Qu.: 9.02125   1st Qu.:22.010      1st Qu.:-0.9665
Median :10.52421   Median :27.389      Median :-0.2496
```

```
Mean :10.60131      Mean :28.296        Mean :-0.1941
3rd Qu.:12.22090    3rd Qu.:34.012      3rd Qu.: 0.5623
Max. :18.66659      Max. :64.086        Max. : 2.1098
       RF                  OK.RF               RK.RF.bc
Min. : 8.175        Min. :-1.920e-01     Min. : 8.174
1st Qu.:10.627      1st Qu.:-6.044e-03   1st Qu.:10.624
Median :11.343      Median :-2.115e-05   Median :11.335
Mean :11.395        Mean :-7.564e-03     Mean :11.387
3rd Qu.:12.234      3rd Qu.: 3.111e-04   3rd Qu.:12.230
Max. :14.539        Max. : 2.120e-01     Max. :14.572
   RK.RF
Min. :19.19
1st Qu.:27.76
Median :30.48
Mean :30.80
3rd Qu.:34.05
Max. :44.11
```

As before, the raster conversion is achieved by the subsequent code:

```
> #Convert to raster
> RF<-rasterFromXYZ(as.data.frame(grid.xy)[, c("x", "y",
"RF")])
> SK.RF<-rasterFromXYZ(as.data.frame(grid.xy)[,c("x","y",
"SimpleK.RF")])
>RK.RF.bc<-rasterFromXYZ(as.data.frame(grid.xy)[, c("x",
"y", "RK.RF.bc")])
> RK.RF.YD<-rasterFromXYZ(as.data.frame(grid.xy)[, c("x",
"y", "RK.RF")])
```

At the end, to display the predicted winter wheat yield generated by the RK-RF, we adopted the elegant data visualization of the ggplot2 package. The corresponding R code reported below arranges the graphs of interest in a multiplot.

```
#Plot predicted winter wheat yield
> rf1<-ggR(RF, geom_raster = TRUE) +
+ scale_fill_gradientn("", colours = c("orange", "yellow",
"green", "sky blue","blue"))+
+ theme_bw()+
+   theme(axis.title.x=element_blank(),
+       axis.text.x=element_blank(),
+       axis.ticks.x=element_blank(),
+       axis.title.y=element_blank(),
+       axis.text.y=element_blank(),
+       axis.ticks.y=element_blank())+
+   ggtitle("RF Predicted (BoxCox)")+
+   theme(plot.title = element_text(hjust = 0.5))
> rf2<-ggR(SK.RF, geom_raster = TRUE) +
```

```
+ scale_fill_gradientn("", colours = c("orange", "yellow",
"green", "sky blue","blue"))+
+ theme_bw()+
+   theme(axis.title.x=element_blank(),
+        axis.text.x=element_blank(),
+        axis.ticks.x=element_blank(),
+        axis.title.y=element_blank(),
+        axis.text.y=element_blank(),
+        axis.ticks.y=element_blank())+
+   ggtitle("OK RF Residuals (BoxCox)")+
+   theme(plot.title = element_text(hjust = 0.5))
> rf3<-ggR(RK.RF.bc, geom_raster = TRUE) +
+ scale_fill_gradientn("", colours = c("orange", "yellow",
"green", "sky blue","blue"))+
+ theme_bw()+
+   theme(axis.title.x=element_blank(),
+        axis.text.x=element_blank(),
+        axis.ticks.x=element_blank(),
+        axis.title.y=element_blank(),
+        axis.text.y=element_blank(),
+        axis.ticks.y=element_blank())+
+   ggtitle("RK-RF Predicted (BoxCox)")+
+   theme(plot.title = element_text(hjust = 0.5))
> rf4<-ggR(RK.RF.YD, geom_raster = TRUE) +
+ scale_fill_gradientn("", colours = c("orange", "yellow",
"green", "sky blue","blue"))+
+ theme_bw()+
+   theme(axis.title.x=element_blank(),
+        axis.text.x=element_blank(),
+        axis.ticks.x=element_blank(),
+        axis.title.y=element_blank(),
+        axis.text.y=element_blank(),
+        axis.ticks.y=element_blank())+
+   ggtitle("RK-RF Predicted (bu/a)")+
+   theme(plot.title = element_text(hjust = 0.5))
> grid.arrange(rf1,rf2,rf3,rf4, ncol = 4) # Multiplot
```

RF prediction results are displayed in Figure 4.7.

The map of RK-RF in Figure 4.7 predicted values shows a general trend for winter wheat yield to be higher towards the north of the study region. RF coupled with regression kriging generates predictions in the range 19.19-44.11, with a mean of 30.8 and SD equal to 4.2. Overall, the predicted values are lower than those obtained with the previous competitor method, although the resulting maps suggest a similar spatial trend.

To alleviate chance effects in our experiment, the validation is repeated 10 times (10-fold CV). Accordingly, the total dataset was randomly divided into 10 subsets, 9 of which were used for training and 1 subset was used for validation.

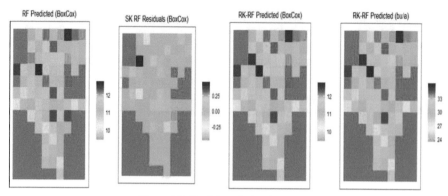

Figure 4.7. Spatial prediction RF results.

To choose the best interpolation method for wheat yield prediction, we contrast the average of MAE, which is a measure for prediction bias, the RMSE, which is a measure for the random prediction error, computed from the predicted and observed values at cross-validation points.

The CV diagnostics summarized in Table 4.3 indicate that better predictions are achieved using the RF model. Specifically, RMSE based on 10-fold cross-validation is about 2.07 for RK and about 2.01 for RF, while we record an MAE value of 1.74 for RK vs 1.61 for RF.

Besides, the amount of variation explained by the RF model is higher (0.33) than RK model (0.29), as revealed by the R-square values.

To summarise, random forest coupled with kriging proved successful in predicting winter wheat yield, outperforming the hybrid interpolation technique based on GLM method, with higher coefficient of determination, and lower MAE and RMSE.

Table 4.3. Cross-validation diagnostics.

Model	RMSE	R-square	MAE
RK-GLM	2.07	0.29	1.74
RK-RF	2.01	0.33	1.61

4.6 Conclusions

It is very important to accurately estimate spatial prediction of agriculture yield. First, this chapter summarizes the employment of DM techniques as promising methods to obtain reliable and accurate crop yield forecasting. Secondly, it focuses on the modelling of the spatial prediction of agriculture yield data using spatial hybrid interpolation techniques: GLM-regression kriging and random forest regression kriging.

The results from the case study showed that there are some quantifiable advantages in performing random forest regression kriging for the dataset at hand, especially in terms of prediction accuracy.

The preference of regression kriging random forest over regression kriging can be explained in that this relatively novel and not entirely worked-out method in the field of spatial predictions has no requirements about the distribution and stationarity of the variable of interest.

Also, RF can fit complex non-linear relationships in the multidimensional space of covariates.

Undoubtedly, the key to success in the RF framework is the training data quality.

However, a limitation of the use of random forests is that the model is usually effective only within the range in covariate values shown by the training data. In gauging the attractive features of the hybrid interpolation techniques employed in this study, one must also note the limitations in using RK to model and map a given spatial phenomenon. Distinctively, practitioners should carefully consider the following issues. First, one prerequisite for RK is to accurately pinpoint auxiliary variables expected to be correlated with the variable of interest. Besides, to avoid artefacts, auxiliary variables should have a constant physical relationship with the target variable: in other terms, they should vary smoothly in space.

For variogram modelling, a reasonable number of point pairs must be available at various spacing. Since RK is completely dependent on the quality of data, it worth noting that the predictions cannot be accurate if data have been sampled using a biased or unrepresentative design.

Finally, a crucial issue of using RK refers to the reliable estimation of covariance/correlation structure: both the regression model parameters and covariance function parameters need to be estimated simultaneously. However, in order to estimate coefficients, we need to know the covariance function of residuals, which can only be estimated after the coefficients. A thorough discussion of RK limitations can be found in Hengl et al. (2007).

References

Arango, R.B., A.M. Campos, E.F. Combarroa, E.R. Canas and I. Díaz. 2016. Mapping cultivable land from satellite imagery with clustering algorithms. *International Journal of Applied Earth Observation* 49: 99–106.

Basso, B., J.T. Ritchie, F.J. Pierce, R.P. Braga and J.W. Jones. 2001. Spatial validation of crop models for precision agriculture. *Agricultural Systems* 68: 97–112.

Bhagawati, R., K. Bhagawati, A.K.K. Singh, R. Nongthombam, R. Sarmah and G. Bhagawati. 2015. Artificial neural network assisted weather based plant disease forecasting system. *International Journal on Recent and Innovation Trends in Computing and Communication* 3: 4168–4173.

Bhargavi, P. and S. Jyothi. 2009. Applying naive Bayes data mining technique for classification of agricultural land soils. *International Journal of Computer Science and Network Security* 9: 117–122.

Bi, C. and G. Chen. 2010. Bayesian networks modeling for crop diseases. *Computer and Computing Technologies in Agriculture* IV: 312–320.

Breiman, L. 2001. Random forests. *Machine Learning* 45: 5–32.

Brisson, N., C. Gary, E. Justes, R. Roche, B. Mary, D. Ripoche, D. Zimmer, J. Sierra, P. Bertuzzi and P. Burger. 2003. An overview of the crop model STICS. *European Journal of Agronomy* 18: 309–332.

Buttafuoco, G. and F. Lucà. 2016. The contribution of geostatistics to precision agriculture. *Annals of Agricultural & Crop Sciences* 1: 1008.

Cressie, N. 2015. *Statistics for Spatial Data.* Wiley Series in Probability and Statistics: New York.

Christensen, R. 2001. *Linear Models for Multivariate Time Series and Spatial Data.* Springer: New York.

Coulston, J.W., C.E. Blinn, V.A. Thomas and R.H. Wynne. 2016. Approximating prediction uncertainty for random forest regression models. *Photogrammetric Engineering & Remote Sensing* 82: 189–197.

Dhanya, C.T. and D. Nagesh Kumar. 2009. Data mining for evolution of association rules for droughts and floods in India using climate inputs. *Journal of Geophysical Research* 114: 1–15.

Everingham, Y., J. Sexton, D. Skocaj and G. Inman-Bamber. 2016. Accurate prediction of sugarcane yield using a random forest algorithm. *Agronomy for Sustainable Development* 36: 27.

Fayyad, U., G. Piatetsky-Shapiro and P. Smyth. 1996. The KDD process for extracting useful knowledge from volumes of data. *Communications of the ACM* 39: 27–34.

Fayad, I., N. Baghdadi, J.S. Bailly, N. Barbier, V. Gond, B. Herault, M. El Hajj, F. Fabre and J. Perrin. 2016. Regional scale rain-forest height mapping using regression-kriging of spaceborne and airborne LiDAR data: application on French Guiana. *Remote Sensing* 8: 1–18.

Franko, U., M. Puhlmann, K. Kuka, F. Böhme and I. Merbach. 2007. Dynamics of water, carbon and nitrogen in an agricultural used Chernozem soil in Central Germany. pp. 245–258. *In*: Kersebaum, K.C., J.M. Hecker, W. Mirschel and M. Wegehenkel (eds.). *Modelling Water and Nutrient Dynamics in Soil Crop Systems*. Springer, Stuttgart, Germany

Gommes, R., M. Bernardi and F. Petrassi. 1996. *Agrometeorological Crop Forecasting*. Environment and Natural Resources Services (SDRN), FAO Research, Extension and Training Division.

Gonzalez-Sanchez, A., J. Frausto-Solis and W. Ojeda-Bustamante. 2014. Predictive ability of machine learning methods for massive crop yield prediction. *Spanish Journal of Agricultural Research* 12: 313–28.

Goovaerts, P. 1997. *Geostatistics for Natural Resources Evaluation* (Applied Geostatistics). New York: Oxford University Press.

Han, J., M. Kamber and J. Pei. 2006. *Data Mining: Concepts and Techniques*. Morgan Kaufmann Publishers.

Hengl, T., G.B.M. Heuvelink and D.G. Rossiter. 2007. About Regression-kriging: from equations to case studies. *Computers & Geosciences* 33: 1301–1315.

Hengl, T., G.B.M. Heuvelink, B. Kempen, J.G. Leenaars, M.G. Walsh, K.D. Shepherd, A. Sila, R.A. MacMillan, J.M. de Jesus, L. Tamene and J.E. Tondoh. 2015. Mapping soil properties of Africa at 250 m resolution: Random forests significantly improve current predictions. *PLOS One* 10: 1–26.

Hengl, T., M. Nussbaum, M.N. Wright, G.B.M. Heuvelink and B. Gräler 2018. Random forest as a generic framework for predictive modeling of spatial and spatio-temporal variables. *Peerj* https://doi.org/10.7287/peerj.preprints.26693v3.

Horie, T., M. Yajima and H. Nakagawa. 1992. Yield forecasting. *Agricultural Systems* 40: 211–236.

Jeong, J.H., J.P. Resop, N.D. Mueller, D.H. Fleisher, K. Yun and E.E. Butler. 2016. Random forests for global and regional crop yield predictions. *PLOS One* 11: e0156571.

Kalpana, R., N. Shanth and S. Arumugam. 2014. A survey on data mining techniques in agriculture. *International Journal on Recent and Innovation Trends in Computing and Communication* 3: 426–431.

Kaul, M., R. Hill and C. Walthall. 2005. Artificial neural network for corn and soybean prediction. *Agricultural Systems* 85: 1–18.

Kitchen, N., K. Sudduth, D. Myers, S. Drummond and S. Hong. 2005. Delineating productivity zones on claypan soil fields using apparent soil electrical conductivity *Computers and Electronics in Agriculture* 46: 285–308.

Klosgen, W. and J.M. Zytkow. 2002. *Handbook of Data Mining and Knowledge Discovery*. Oxford University Press.

Kuhn, M. 2008. *Caret* package. *Journal of Statistical Software* 28(5).

Li, J. and A.D. Heap. 2011. A review of comparative studies of spatial interpolation methods in environmental sciences: Performance and impact factors. *Ecological Informatics* 6: 228–241.

Lobell, D.B. and M.B. Burke. 2009. *Climate Change and Food Security: Adapting Agriculture to a Warmer World*. Springer: Netherlands.

Lobell, D.B. and M.B. Burke. 2010. On the use of statistical models to predict crop yield responses to climate change. *Agricultural and Forest Meteorology* 150: 1443–1452.

Lollato, R.P., J.T. Edwards and T.E. Ochsner. 2017. Meteorological limits to winter wheat productivity in the U.S. southern great plains. *Field Crops Research* 203: 212–226.

Matheron, G. 1969. Le krigeage universel (Universal Kriging). *Cahiers du Centre de Morphologie Mathematique*, Ecole des Mines de Paris, Fontainebleau.

Meinshausen, N. 2006. Quantile regression forests. *Journal of Machine Learning Research* 7: 983–999.

Mentch, L. and G. Hooker. 2016. Quantifying uncertainty in random forests via confidence intervals and hypothesis tests. *Journal of Machine Learning Research* 17: 841–881.

Miller, D., J. McCarthy and A. Zakzeski. 2009. A fresh approach to agricultural statistics: data mining and remote sensing. *Section on Government Statistics in JSM* 2009, 3144–3155.

Nadolnyak, D., D. Vedenov and J. Novak. 2008. Information value of climate-based yield forecasts in selecting optimal crop insurance coverage. *American Journal of Agricultural Economics* 90: 1248–1255.

National Agricultural Statistics Service (NASS). 2006. *The Yield Forecasting Program of NASS by the Statistical Methods Branch, Estimates Division, National Agricultural Statistics Service, U.S. Department of Agriculture*, Washington, D.C., May 2006. NASS Staff Report No. SMB 06-01.

Nendel, C., M. Berg, K. Kersebaum, W. Mirschel, X. Specka, M. Wegehenkel, K. Wenkel and R. Wieland. 2011. The MONICA model: testing predictability for crop growth, soil moisture and nitrogen dynamics. *Ecological Modelling* 222: 1614–1625.

Odeh, I.O.A., A.B. McBratney and D.J. Chittleborough. 1994. Spatial prediction of soil properties from landform attributes derived from a digital elevation model. *Geoderma* 63: 197–214.

Oliver, M.A. and R. Webster. 1990. Kriging: a method of interpolation for geographical information systems. *International Journal of Geographical Information Systems* 4: 313–332.

Pebesma, E.J. 2004. Multivariable geostatistics in S: the gstat package. *Computers & Geosciences* 30: 683–691.

Raorane, A.A. and R.V. Kulkarni. 2013. Review-role of data mining in agriculture. *International Journal of Computer Science and Information Technologies* 4: 270–272.

Schlenker, W. and D.B. Lobell. 2010. Robust negative impacts of climate change on African agriculture. *Environmental Research Letters* 5: 014010.

Scornet, E. 2018. Tuning parameters in random forests. *ESAIM: Proceedings and Surveys* 60: 144–162.

Steichen, T.J. and N.J. Cox. 2002. A note on the concordance correlation coefficient. *Stata Journal* 2: 183–189.

Stöckle, C.O., M. Donatelli and R. Nelson. 2003. CropSyst, a cropping systems simulation model. *European Journal of Agronomy* 18: 289–307.

Supriya, D.M. 2017. Analysis of soil behavior and prediction of crop yield using data mining approach. *Journal of Innovative Research in Computer and Communication Engineering* 5(5).

Tao, F. and Z. Zhang. 2010. Adaptation of maize production to climate change in North China Plain: Quantify the relative contributions of adaptation options. *European Journal of Agronomy* 33(2): 103–116.

Tobler, W.R. 1970. A computer movie simulating urban growth in the Detroit region. *Economic Geography Supplement* 46: 234–240.

Tripathi, S., V. Srinivas and R. Nanjundiah. 2006. Downscaling of precipitation for climate change scenarios: a support vector machine approach. *Journal of Hydrology* 330: 621–640.

van Ittersum, M.K., K.G. Cassman, P. Grassini, J. Wolf, P. Tittonell and Z. Hochman. 2013. Yield gap analysis with local to global relevance—a review. *Field Crops Research* 143: 4–17.

Verheyen, K., D. Adrianens, M. Hermy and S. Deckers. 2001. High resolution continuous soil classification using morphological soil profile descriptions. *Geoderma* 101: 31–48.

Wager, S., T. Hastie and B. Efron. 2014. Confidence intervals for random forests: the jackknife and the infinitesimal jackknife. *Journal of Machine Learning Research* 15: 1625–1651.

Wright, M. and A. Ziegler. 2015. ranger: A fast implementation of random forests for high dimensional data in c++ and r. *Journal of Statistical Software* 77: 1–17.

Wu, X., V. Kumar, J. Ross Quinlan, J. Ghosh, Q. Yang, H. Motoda, G.J. McLachlan, A. Ng, B. Liu, P.S. Yu, Z.-H. Zhou, M. Steinbach, D.J. Hand and D. Steinberg. 2008. Top 10 algorithms in data mining. *Knowledge and Information Systems* 14: 1–37.

Zhang, T., J. Zhu and R. Wassmann. 2010. Responses of rice yields to recent climate change in China: An empirical assessment based on long-term observations at different spatial scales (1981–2005). *Agricultural and Forest Meteorology* 150: 1128–1137.

Yang, C., S. Prasher, S. Sreekanth, N. Patni and L. Masse. 1997. An artificial neural network model for simulating pesticide concentrations in soil. *Transactions of the ASAE* 40: 1285–1294.

Chapter 5
Land Cover/Use Analysis and Modelling

Elisabetta Carfagna[1,]* and *Gianrico Di Fonzo*[2]

5.1 Land pattern recognition

Over the last few decades, several land cover databases have been produced using aerial photos or high, medium and coarse resolution satellite data. These types of databases are an important instrument for knowing and managing the territory at various scales, with particular reference to agriculture food security, natural resources and environment protection. The land cover databases are digital maps created in a Geographic Information System (GIS) whose basic elements are pixels (in a raster approach) or polygons (in a vectorial approach) of specific land cover types and are produced by photo-interpretation of images on the screen or by semiautomatic classification of the set of measures of electromagnetic radiation reflected by the earth's surface corresponding to each area unit. These area units are pixels whose size can range from less than 1 m to 5 km.

Land cover is classified according to classification systems defining the objects or classes and the criteria used to distinguish land cover through a legend that is the expression of the classification system applied at a specific place and defined scale.

Whilst land cover may be observed directly in the field or by remote sensing, the identification of land use and its changes generally requires the integration of natural and social information (expert knowledge, interviews with land managers) to determine which human activities are occurring in the different parts of the landscape, even when land cover appears to be the same. Since land use is the human use of land, it involves the management and modification of natural environment or wilderness into built environment such as fields, pastures, and settlements. Land use is also defined as "the arrangements, activities and inputs to a certain land cover type to produce, change or maintain it". This definition establishes a direct link between land

---

[1] University of Bologna, Department of Statistical Sciences.
[2] Sapienza University Roma and Italian Health Ministry.
  Email: gianrico.difonzo@uniroma1.it
* Corresponding author: elisabetta.carfagna@unibo.it

cover and the actions of socio-economics and livelihoods of communities. The FAO and UNEP have collaborated in the development of a land cover mapping schema that provides a common reference system which is recognized as an ISO standard (ISO 19144-1 and 19144-2) as the Land Cover Classification System (LCCS).

The main steps in the production of a land cover database include (Latham and Rosati (2016)): (1) Selection and acquisition of appropriate satellite image products using crop calendar and Leaf Area Index (LAI)-based on image processing products and vegetation indices (2) Legend generation (3) Photo-keys preparation from the legend to assist the photo-interpretation (4) Pre-processing of the imagery, that depends on the nature of the images acquired. It may include georectification, mosaicking and segmentation of images, that is an object-based processing to generate vector boundaries of consistent parcels (5) Analysis of the images that produces geospatial information and can be performed by automatic classification, that is fast, low-cost and with varying levels of accuracy, depending on the validation points that can be selected, or by visual interpretation, that requires a medium-to-high level of commitment in terms of cost, time and human resources, with generally more reliable results. In some cases, the analysis uses a mix of visual and semi-automatic methods that enhance the possibility to discriminate the land cover features (6) Assessment of the accuracy of the land cover database, through ground data collection or photointerpretation of very high-resolution satellite data on a sample basis.

5.2 Supervised and unsupervised classification

Various kinds of classifiers have been developed for classifying remote sensing data, some of them perform the classification pixel by pixel (e.g., maximum likelihood classifier), some others classify contiguous groups of pixels (e.g., parallelepiped classification), some classifiers perform unsupervised classification, and others supervised (Benedetti et al. 2014).

In unsupervised classification, an image is segmented into unknown classes. It is the aim of the researcher to label afterwards, just in case, which classes. Unsupervised classification aims at grouping pixels with similar spectral reflective characteristics into clusters which are then labeled with a certain class name, without defining the classes a priori.

Supervised classification uses a set of user-defined spectral signatures to classify an image. The spectral signatures are derived from training areas (or sets) that are created by depicting features of interest on an image. Supervised classification, as well as photo-interpretation, is made according to a land cover legend defined in advance, in which each class (or label) represents a land cover type. Often, more classes than the ones foreseen by the legend are used, in order to catch the variability inside the training set; then some classes are aggregated.

Several authors, including Carfagna and Gallego (2005), highlighted that the spectral response and the identification of some land covers, particularly of crops, are not in one to one correspondence. In fact, the radiometric response of the same crop in different conditions can vary across the pixels of an image. A more appropriate approach is to consider the spectral response of a crop as a function of the probability distribution of its spectral reflectance.

5.2.1 Parametric classification

The Maximum Likelihood classifier is the most well-known method for classifying satellite images. It performs supervised classification of remote sensing data through maximum likelihood discriminant analysis. It uses a multivariate probability density function describing the target classes identified on the basis of training data. Each density function represents the probability that the spectral pattern of a class falls within a given region in a multidimensional spectral space. With the Maximum Likelihood classifier, each pixel of a satellite image is assigned to the class it has the maximum probability to belong to.

When the maximum likelihood classifier is used with uniform prior probabilities, large classes tend to be underestimated and small classes tend to be overestimated. On the contrary, if some information is available on the approximate proportion of the different land cover types, a proportional prior probability may be used; but in this case, large classes tend to be overestimated and small classes are underestimated or even disappear. The bias appears also where the theoretical conditions on multivariate Gaussian distribution are true (see Carfagna and Gallego 2005). Moreover, the Gaussian distribution assumption seldom holds when multitemporal data are used; thus, the adoption of other classification methods has become common with the increasing availability of free of charge multitemporal remote sensing data. Gómez et al. (2016) and Defourny (2017) give an overview of strengths and weakness of algorithms used for classification of satellite images data.

Linear methods for classification can be used for classifying remote sensing images. Since the predictor $G(x)$ takes values in a discrete set G, the parametric classification problem can be solved dividing the input space into a collection of regions labeled according to the classification (Friedman et al. 2001). Supposing there are K classes, for convenience labeled $1,2,\ldots,K$, the aim for a class of methods in a Bayesian framework is to estimate the posterior probabilities $Pr(G = k|X = x)$, often using generalized linear models. In particular, the classification is created deriving the linear decision boundaries that divide the observations in the data set. A variety of methods for identifying the linear decision boundaries have been developed, according to the way the linear function fits the training data set.

The logistic regression is a well-known model that estimates the posterior probabilities of the K classes via linear functions in x, ensuring that they sum to one and remain in $[0, 1]$.

The model can be written with the following re-parameterization:

$$\log \frac{Pr(G = 1 \mid X = x)}{Pr(G = K \mid X = x)} = \beta_{10} + \beta_1^T x$$

$$\log \frac{Pr(G = 2 \mid X = x)}{Pr(G = K \mid X = x)} = \beta_{20} + \beta_2^T x$$

$$\ldots..$$

$$\log \frac{Pr(G = K - 1 \mid X = x)}{Pr(G = K \mid X = x)} = \beta_{(K-1)0} + \beta_{K-1}^T x$$

(5.1)

The model is specified in terms of $K - 1$ log-odds or logit transformations and can be expressed as follows:

$$Pr(G = k \mid X = x) = \frac{\exp(\beta_{k0} + b_k^T x)}{1 + \sum_{l=1}^{K-1} \exp(\beta_{l0} + b_l^T x)}, k = 1, \ldots, K-1$$

$$(5.2)$$

$$Pr(G = K \mid X = x) = \frac{1}{1 + \sum_{l=1}^{K-1} \exp(\beta_{l0} + b_l^T x)}$$

These probabilities sum to one. Denoting the entire parameter set $\theta = \{\beta_{10}, \beta_{20}, \ldots, \beta_{(K-1)0}, \beta_{K-1}\}$, the probabilities can be denoted as $Pr(G = k|X = x) = p_k(x; \theta)$.

The fitting in the logistic regression model is usually performed by maximum likelihood, in particular, using the conditional likelihood of G given X. Since $Pr(G|X)$ completely specifies the conditional distribution, the *multinomial* sample distribution is appropriate. The log-likelihood for N observation is:

$$l(\theta) = \sum_{i=1}^{N} \log p_{gi}(x_i; \theta),$$

$$(5.3)$$

where $p_k(x_i; \theta) = Pr(G = k|X = x; \theta)$.

To maximize the log-likelihood, the derivatives are set to zero. The score equations are $p + 1$ equations nonlinear in β:

$$\frac{\delta l(\beta)}{\delta \beta} = \sum_{i=1}^{N} x_i(y_i - p(x_i; \beta)) = 0,$$

$$(5.4)$$

The score equations (5.4) can be solved by the Newton-Raphson algorithm, which requires the second-order partial derivatives or Hessian matrix:

$$\frac{\delta^2 l(\beta)}{\delta \beta \delta \beta^T} = -\sum_{i=1}^{N} x_i x_i^T p(x_i; \beta)(1 - p(x_i; \beta)).$$

$$(5.5)$$

Starting with β^{old}, a single Newton update is:

$$\beta^{new} = \beta^{old} - \left(\frac{\delta^2 l(\beta)}{\delta \beta \delta \beta^T} \right)^{-1} \frac{\delta l(\beta)}{\delta \beta},$$

$$(5.6)$$

where the derivatives are evaluated at β^{old}.

Let y denote the vector of y_i values, X the $N \times (p + 1)$ matrix of x_i values, p the vector of fitted probabilities with the i-th element $p(x_i; \beta^{old})$ and W a $N \times N$ diagonal matrix of weights with i-th diagonal element $p(x_i; \beta^{old} (1 - p(x_i; \beta^{old}))$, then, the scores and Hessian can be rewritten in matrix notation as:

$$\frac{\delta l(\beta)}{\delta \beta} = X^T (y - p)$$

$$\frac{\delta^2 l(\beta)}{\delta \beta \delta \beta^T} = -X^T W X$$

$$(5.7)$$

The Newton step is thus:

$$\beta^{new} = \beta^{old} + (X^T W X)^{-1} X^T (y - p)$$
$$= (X^T W X)^{-1} X^T W (X \beta^{old} + W^{-1}(y - p))$$
$$= (X^T W X)^{-1} X^T W z$$

$$(5.8)$$

In the second and third line, the Newton step is expressed as a weighted least squares step, with the response:

$$z = X\beta^{old} + W^{-1}(y - p) \tag{5.9}$$

sometimes known as the *adjusted response*. These equations are solved repeatedly, since at each iteration p changes, and hence so does W and z.

This algorithm is referred to as *iteratively re-weighted least squares* or IRLS, since each iteration solves the weighted least squares problem:

$$\beta^{new} \leftarrow argmin_\beta (z - X\beta)W(z - X\beta) \tag{5.10}$$

Generally, $\beta = 0$ is a good starting value for the iterative procedure and the algorithm converges, since the log-likelihood is concave, but overshooting can occur.

For the multiclass case ($K \geq 3$) the Newton algorithm can also be expressed as an iteratively re-weighted least squares algorithm, but with a vector of $K - 1$ responses and non-diagonal weight matrix per observation.

The latter precludes any simplified algorithms, and in this case, it is numerically more convenient to work with the expanded vector Θ directly (Friedman et al. 2001). Logistic regression problems, very large both in N and p, can be efficiently fitted using penalized models applied to the logistic regression, such as the Lasso and the Ridge regression (Hoerl and Kennard 1970). The letter method shrinks the regression coefficients by imposing a penalty on their size, minimizing the following penalized residual sum of squares:

$$\hat{\beta}^{ridge} = argmin \left\{ \sum_{i=1}^{N} (y_i - \beta_0 - \sum_{j=1}^{p} x_{ij}\beta_j)^2 + \lambda \sum_{j=1}^{p} \beta_j^2 \right\} \tag{5.11}$$

Here, $\lambda \geq 0$ is a complexity parameter that controls the amount of shrinkage: the larger the value of λ, the greater the amount of shrinkage.

When the categorical response variable G has $K > 2$ levels, the linear logistic regression can be generalized to a multi-logit model. The traditional approach is to extend the binomial case, as shown in expressions (5.1) and (5.2) (Zhu and Hastie 2004; Shen and Tan 2005). As said before, this parameterization is not estimable without constraints, because it gives identical probabilities to any values for the parameters $\{\beta_{0l}, \beta_l, l\}_1^K$, $\{\beta_{0l} - c_0, \beta_l - c\}_1^K$.

The model (5.2) can be fitted by regularized maximum (multinomial) likelihood. Using a similar notation as before, let $p_k(x_i) = Pr(G = k|X = x)$, and let $g_i \in \{1,2,\ldots, K\}$ be the i-th response. Defining the *elastic-net penalty* P_α as:

$$P_\alpha(\beta) = \sum_{j=1}^{p} \left[\frac{1}{2}(1-\alpha)\beta_j^2 + \alpha |\beta_j| \right] \tag{5.12}$$

that is a compromise between the ridge regression penalty ($\alpha = 0$) and the lasso penalty ($\alpha = 1$). The method maximizes the penalized log-likelihood:

$$max_{\{\beta_{0k},\beta_k\}_1 \in \Re^{K(p+1)}} \left[-\sum_{i=1}^{N} \log p_{gi}(x_i) \quad \lambda \sum_{k=1}^{K} P(\beta_k) \right]. \tag{5.13}$$

Denoting by Y the $N \times K$ indicator response matrix, with elements $y_{ik} = I(g_i = k)$, the log-likelihood part of (5.1) can be rewritten in the more explicit form:

$$l(\{\beta_{0k},\beta_k\}_1^K) = \frac{1}{N}\sum_{i=1}^{N}\left[\sum_{k=1}^{K} y_{ik}(\beta_{0k} + b_k^T x_i) - \log\left(\sum_{k=1}^{K} e^{\beta_{0k}+\beta_k^T x_i}\right)\right] \qquad (5.14)$$

The newton algorithm for multinomial can be tedious, because of the vector nature of the response observation (Friedman et al. 2010). However, performing *partial Newton steps* by forming a partial quadratic approximation to the log-likelihood (5.14), the previous complexities can be avoided, allowing only (β_{0k}, β_k) to vary for a single class at a time.

Hence, if the current estimates of the parameters are $(\hat{\beta}_0, \hat{\beta})$, a quadratic approximation to the log-likelihood (Taylor expansion about current estimates), is:

$$l_{Qk}(\beta_{0k},\beta_k) = -\frac{1}{2N}\sum_{i=1}^{N} w_{ik}(z_{ik} - \beta_{0k} - \beta_k^T x_i)^2 + C(\{\hat{\beta}_{0k},\hat{\beta}_k\}_1^K) \qquad (5.15)$$

where:

$$z_{ik} = \hat{\beta}_{0k} + \hat{\beta}_k^T x_i + \frac{y_{ik} - \hat{p}_k(x_i)}{\hat{p}_k(x_i)(1 - \hat{p}_k(x_i))} \qquad (5.16)$$

$$w_{ik} = \hat{p}_k(x_i)(1 - \hat{p}_k(x_i)),$$

For each value of λ, an outer loop which cycles over k is created; it computes the quadratic approximation l_k about the current parameters $(\hat{\beta}_0, \hat{\beta})$, then a coordinate descendent algorithm is used to solve the penalized weighted least-squares problem:

$$\min_{(\beta_{0k},\beta_k)\in\mathcal{R}^{p+1}}\{-l_{Qk}(\beta_{0k},\beta_k) + \lambda P_a(\beta_k)\} \qquad (5.17)$$

This amounts to the sequence of nested loops:

- **outer loop:** Decrement λ
- **middle loop (outer):** cycle over $k \in \{1,2,\dots, K\}$
- **middle loop (inner):** Update the quadratic approximation l_k using the current parameters
- **inner loop:** Run the coordinate descent algorithm on the penalized weighted least squares problem (5.17)

5.2.2 Non-parametric classification

Non-parametric classification models based on a response variable that characterizes the data used in the analysis are known as non-parametric supervised models. The response variable is the inference target that has to be predicted. In land use classification problems, we deal with a qualitative response variable, and the classification algorithms can be divided into two phases:

- Training phase
- Test phase

In the first phase, the classification model is built, while in the second one the model is implemented in order to predict the class each observation belongs to.

5.2.2.1 Ensemble methods

Bagging, Boosting and Random Forest (Louppe 2014) are methods based on decision trees (Watts et al. 2011) belonging to the family of algorithms called Ensemble Methods. They use more than one classifier and finally unify the obtained results. The particularity of these kinds of models is that the various classifiers, defined *weak*, are combined among them, in order to construct a *strong* classifier. These methods are expected to improve the prediction both in terms of variance and bias.

Bagging

The name derives from Bootstrap Aggregation, because it combines the resampling techniques of Bootstrap with a predictive model, such as decision trees. Bagging constructs multiple decision trees, each one with a different sample; thus, with different learners. In particular, the training set is resampled by iteratively extracting a subset of the original data, such that the process creates N random samples with replacement. This way, each sample produces a tree and the final output is averaged with all the intermediate outputs of each tree. Alternatively, the final classification can be obtained by associating a vote based on some weights to each tree. Once the votes are computed, the Bagged classifier assigns the new item to the most voted class; then, the results from the various trees are combined in order to provide a single final output.

Boosting and AdaBoost

Like Bagging, Boosting is a method that allows the construction of a strong classifier starting from weak ones, using predictive models and assigning weights to each classifier. Boosting creates an iterative process in order to reduce the classification errors. Various kinds of algorithms have been developed for Boosting, according to the way each algorithm updates the weights or combines the classifier output. One of the most commonly used is AdaBoost, that generates a sequence of weak classifiers and selects, at each iteration, the best classifier according to the used sample. The samples that are not correctly classified receive higher weights in the next iterations.

The algorithm is made of the following steps:

- *Step* 1: the weights w_n are initialized imposing $w_n^{(1)} = \frac{1}{N}$ for $n = 1,\ldots, N$
- *Step* 2: For $m = 1,2,\ldots, M$, fit a classifier $y_m(x)$ on the training dataset (t_n) in order to minimize the error of the weight function:

$$J_m = \sum_{n=1}^{N} w_n^{(m)} I(y_m(x_n) \neq t_n),\qquad(5.18)$$

where: $I(y_m(x_n) \neq t_n) = \begin{cases} 1 \text{ if } y_m(x_n) \neq t_n \\ 0 \text{ otherwise} \end{cases}$

allows evaluating the quantity:

$$\epsilon_m = \frac{\sum_{n=1}^{N} w_n^{(m)} I(y_m(x_n) \neq t_n)}{\sum_{n=1}^{N} w_n^{(m)}}\qquad(5.19)$$

that is used to compute

$$\alpha_m = \ln\left(\frac{1-\epsilon_m}{\epsilon_m}\right) \tag{5.20}$$

and update the weights:

$$w_n^{(m+1)} = w_n^{(m)} \exp[\alpha_m I(y_m(x_n) \neq t_n)] \tag{5.21}$$

- *Step* 3: the prediction is computed using the final model, that is the following:

$$Y_M = sgn(\sum_{m=1}^M \alpha_m y_m(x)) \tag{5.22}$$

considering the following function:

$$E = \sum_{n=1}^N \exp[-t_n f_m(x_n)] \tag{5.23}$$

where:

$$f_m(x) = \frac{1}{2} \sum_{l=1}^m \alpha_l \tag{5.24}$$

represents a classifier that combines linearly the basic hypothesis; thus, the function E has to be minimized with respect to the coefficients α_l and to the parameters $y_1(x),\ldots,y_m(x)$.

Supposing that the classifiers $y_1(x),\ldots,y_{m-1}(x)$ and the respective coefficients $\alpha_1,\ldots,\alpha_{m-1}$ are fixed, then the function E has to be minimized only with respect to the last classifier and parameter, that are respectively $y_m(x)$ and α_m. So, from equation (5.23), the following expression is derived:

$$E = \sum_{n=1}^N \exp[-t_n f_{m-1}(x_n) - \frac{1}{2} t_n \alpha_m y_m(x_n)]$$

$$= \sum_{n=1}^N w_n^{(m)} \exp[-\frac{1}{2} t_n \alpha_m y_m(x_n)]$$

where $w_n^{(m)} = \exp[-t_n f_{m-1}(x_n)]$ is constant with respect to $y_m(x_n)$ and α_m.

Denoting Γ_m and M_m respectively the whole set of categories correctly and wrongly classified by y_m, it is possible to write the function E as follows:

$$E = e^{-\alpha_m/2} \sum_{n\in\Gamma_m} w_n^{(m)} + e^{\alpha_m/2} \sum_{n\in M_m} w_n^{(m)}$$

$$= (e^{\alpha_m/2} - e^{-\alpha_m/2}) \sum_{n=1}^N w_n^{(m)} I(y_m(x_n) \neq t_n) + e^{-\alpha_m/2} \sum_{n=1}^N w_n^{(m)}$$

Minimizing (5.18) allows one to minimize the function E with respect to $y_m(x_n)$. In fact, E results to be:

$$E = (e^{\alpha_m/2} - e^{-\alpha_m/2}) J_m + c$$

where c represents a constant term.

In the same way, equation (5.20) can be used to minimize E with respect to α_m.

Finally, once optimums $y_m(x_n)$ and α_m have been found, the weights distribution becomes:

$$w_n^{(m+1)} = w_n^{(m)} \exp[-\frac{1}{2} t_n \alpha_m y_m(x_n)] \tag{5.25}$$

and, exploiting $t_n y_m(x_n) = 1 - 2I(y_m(x_n) \neq t_n))$, (5.25) can be written as follows:

$$w_n^{(m+1)} = w_n^{(m)} \exp[-\alpha_m/2] \exp[\alpha_m I(y_m(x_n) \neq t_n)]$$

that is an algorithm update, given that $\exp(-\alpha_m/2)$ is independent of n.

Random Forest

Random Forest (Friedman et al. 2001) combines a group of decision trees with the aim of solving the issue of tree instability. Each tree of the Forest is built using a different *bootstrap* of the original data. Like in *Bagging*, multiple trees are generated starting from the same *training set*, choosing the optimal split in a random set of the possible splits. This way, each tree will result differently, avoiding polarization and ensuring robustness. Due to the Law of big numbers, this model converges and doesn't overfit the data.

The Random Forest algorithm works as follows:

- *Step* 1: A certain number n of *bootstrap* samples are extracted from the set N of original data;
- *Step* 2: For each *bootstrap* sample, the algorithm creates a tree until the maximum extension, with the following variation: given the M classifier, for each split a random number of predictors m such that $m < M$ is randomly sampled. Typically, for classification problems $m \approx M$;
- *Step* 3: In the classification setting, the prediction on the new data is made using the mode of the n tree previsions. An estimate of the error can be obtained as follows:
  - The OOB (*Out-Of-Bag*) data that are not present in the *bootstrap* sample are predicted.
  - The error rate for the prevision is computed.

The model is invariant with respect to monotone transformations, it's robust with respect to *outlier* and allows some strategies for missing data. The accuracy of the *Random Forest* depends on the strength of the trees and on their correlation.

5.3 Classification of Sentinel satellite data using ground data and R

We have compared the behavior of the models described above, one parametric model (Regularized multinomial regression) and two non-parametric ones (Boosting and Random Forest) on Sentinel satellite data from Copernicus project of the European Space Agency and ground data kindly provided by the Italian Ministry of Agriculture (MiPAAF). Particularly, the analysis is based on 574 geo-referenced points in the north of Tuscany Region on which the land use was assessed by the Ministry in 2016 in the framework of AGRIT project. The land use of these points has been grouped into 4 categories, namely grape, olive, sunflower, winter cereals and a class called "other". For these points, the Italian Ministry of Agriculture collected *in situ* information about crops, presence of fences, presence of irrigation and type of vegetation coverage of the field. On the same area, we have acquired the digital elevation model, satellite data from Sentinel 1 and Sentinel 2 (6 images for each satellite), plus 6 vegetation indexes for each of the six images of Sentinel 2; namely, Normalized Difference

Vegetation Index (NDVI), Green Normalized Vegetation Index (GNDVI), Two-band Enhanced Vegetation Index (EVI2), Normalized Difference Water Index (NDWI), Chlorophyll Red-Edge (ClRed-edge) and Soil-Adjusted Vegetation Index (SAVI). We have handled raster data with the help of raster and RGdal packages.

We have used the confusion matrix to summarize the correct classifications and the misclassifications in a contingency table format. In confusion matrices, usually the rows represent the classified labels, and the columns identify the reference labels. The entry of the confusion matrix denotes the proportion of area in a classified land cover class i and reference land-cover class j, for $i, j = 1,..., m$. The row total identifies the proportion of area classified as land-cover class i, and the column total represents the proportion of area attributed to land cover class i in the reference data, for $i = 1,..., m$. These proportions can be derived from pixels or polygon counts. The entries of the confusion matrix can be also reported in terms of counts rather than proportions (see, e.g., Foody 2002), as in our case.

We compare the three models on the basis of the overall accuracy, that is one of the most popular accuracy measures derived from the confusion matrix and represents the overall proportion correctly classified. The overall accuracy expresses the probability that a randomly selected unit is correctly classified and provides a measure of the quality of the classification as a whole.

We compare the three models also on the basis of Cohen's Kappa (Agresti 2002; Banerjee et al. 1999; Cohen 1960) which measures the beyond chance agreement between the classification and the test data; in fact, it discounts the total proportion of agreement (p_o) by the level of agreement expected by chance (p_c):

$$\kappa = \frac{p_o - p_c}{1 - p_c} = \frac{\Sigma_i p_{ii} - \Sigma_i p_{i+} \; p_{+i}}{1 - \Sigma_i p_{i+} \; p_{+i}} \tag{5.26}$$

Cohen's Kappa can be computed only if the same number of categories is used both for classification and test data.

The dataset:

```
>data(AgritSentinel)
>set.seed(681)
```

has been divided into training and test set with the package "splitstackshape"

```
>testandtrain=stratified(data,group=4,0.8,bothSets = TRUE)
>train=testandtrain[[1]]
>table(train$COLTIVATE)
##
##      GRAPE    OLIVE    OTHER    SUNFLOWER    WINTER CEREAL
##       66       88      215        15            74
test=testandtrain[[2]]
table(test$COLTIVATE)
##
##      GRAPE    OLIVE    OTHER    SUNFLOWER    WINTER CEREAL
##       17       22       54         4            19
>dim(train)
## [1] 458 148
```

We have used the R package for Random Forest analysis, called `randomForest` and evaluated the classification performance through the confusion matrix based on the test set.

```
>head(names(train))
## [1] "coords.x1"  "coords.x2"  "ID_PUNTO"  "COLTIVATE"
"IRRIG_PRES"
## [6] "COPERTURA"
>rf.cl<-randomForest(x=train[,c(5:148)], y=train$COLTIVATE,
ntree=1000,      importance=TRUE,data=datset_tes1,      type=
"classification",keep.forest=TRUE)
>predicted=predict(rf.cl,test[,5:148], type="response")
>confusionMatrix(predicted, test$COLTIVATE)
## Confusion Matrix and Statistics
##
##                Reference
##   Prediction  GRAPE  OLIVE  OTHER  SUNFLOWER  WINTER
                                                CEREAL
##       GRAPE     11     1      0        0        0
##       OLIVE      4    21      0        0        0
##       OTHER      2     0     52        2        6
##    SUNFLOWER     0     0      0        2        0
## WINTER CEREAL    0     0      2        0       13
##
## Overall Statistics
##
##              Accuracy  :  0.8534
##               95% CI  :  (0.7758, 0.9122)
##   No Information Rate  :  0.4655
##     P-Value [Acc > NIR]  :  < 2.2e-16
##
##                 Kappa  :  0.7824
##  Mcnemar's Test P-Value  :  NA
##
## Statistics by Class:
##
##                   Class: GRAPE Class: OLIVE Class: OTHER
## Sensitivity            0.64706      0.9545       0.9630
## Specificity            0.98990      0.9574       0.8387
## Pos Pred Value         0.91667      0.8400       0.8387
## Neg Pred Value         0.94231      0.9890       0.9630
## Prevalence             0.14655      0.1897       0.4655
## Detection Rate         0.09483      0.1810       0.4483
## Detection Prevalence 0.10345      0.2155       0.5345
## Balanced Accuracy    0.81848      0.9560       0.9008
```

```
##                        Class: SUNFLOWER  Class: WINTER CEREAL
## Sensitivity                  0.50000                   0.6842
## Specificity                  1.00000                   0.9794
## Pos Pred Value               1.00000                   0.8667
## Neg Pred Value               0.98246                   0.9406
## Prevalence                   0.03448                   0.1638
## Detection Rate               0.01724                   0.1121
## Detection Prevalence         0.01724                   0.1293
## Balanced Accuracy            0.75000                   0.8318
```

We have repeated the analysis with the Boosting model using the adabag package:

```
>adaboost <- boosting(COLTIVATE ~ .,data=train[,4:126],
mfinal=20,boos = T)
>predbosting=predict.boosting(adaboost,newdata=
test[,5:126])
>confusionMatrix(as.factor(predbosting$class),
test$COLTIVATE)
## Confusion Matrix and Statistics
##
##                  Reference
##  Prediction   GRAPE   OLIVE    OTHER   SUNFLOWER   WINTER
##                                                    CEREAL
##  GRAPE          10      3        0        0          0
##  OLIVE           7     19        0        0          0
##  OTHER           0      0       53        4          5
##  SUNFLOWER       0      0        0        0          0
##  WINTER CEREAL   0      0        1        0         14
##
## Overall Statistics
##
##                  Accuracy  :  0.8276
##                     95% CI :  (0.7464, 0.8914)
##     No Information Rate    :  0.4655
##       P-Value [Acc > NIR]  :  8.192e-16
##
##                     Kappa  :  0.7431
##   Mcnemar's Test P-Value   :  NA
##
## Statistics by Class:
##
##                  Class: GRAPE  Class: OLIVE  Class: OTHER
## Sensitivity           0.58824       0.8636        0.9815
## Specificity           0.96970       0.9255        0.8548
## Pos Pred Value        0.76923       0.7308        0.8548
## Neg Pred Value        0.93204       0.9667        0.9815
```

```
## Prevalence              0.14655        0.1897        0.4655
## Detection Rate          0.08621        0.1638        0.4569
## Detection Prevalence 0.11207           0.2241        0.5345
## Balanced Accuracy    0.77897           0.8946        0.9182
##                       Class: SUNFLOWER  Class: WINTER CEREAL
## Sensitivity              0.00000               0.7368
## Specificity              1.00000               0.9897
## Pos Pred Value              NaN                0.9333
## Neg Pred Value           0.96552               0.9505
## Prevalence               0.03448               0.1638
## Detection Rate           0.00000               0.1207
## Detection Prevalence     0.00000               0.1293
## Balanced Accuracy        0.50000               0.8633
```

Finally, we have performed the penalized logistic model, that is available in the `glmnet` package. In order to apply this model, we have standardized the explanatory variables and reparametrized the dummy variables through the `model.matrix` function:

```
>data=data(AgritSentinel.standardize)
>testandtrain=stratified(data,group=4,0.8,bothSets = TRUE)
>train=testandtrain[[1]]
>table(train$COLTIVATE)
##
##    GRAPE    OLIVE    OTHER    SUNFLOWER    WINTER CEREAL
##      66       88      215        15             74
>test=testandtrain[[2]]
>table(test$COLTIVATE)
##
##    GRAPE    OLIVE    OTHER    SUNFLOWER    WINTER CEREAL
##      17       22       54         4             19
>binarycolumns= model.matrix( ~ .-1, train[,5:7])
>binarycolumns2=model.matrix( ~ .-1, test[,5:7])
```

We have used the routine `cv.glmnet` function in the `glmnet` package to estimate the best with the cross-validation method. The Figure 5.1 below shows the results of the process with the best lambda.

```
>cvfit=cv.glmnet(as.matrix(cbind(binarycolumns, train[,8:
126])),train$COLTIVATE,family="multinomial",type.
multinomial="grouped")
>plot(cvfit)
>Predit_lg=predict(cvfit,as.matrix(cbind(binarycolumns2,
test[,8:126])), s = "lambda.min", type = "class")
>confusionMatrix(as.factor(Predit_lg), test$COLTIVATE)
## Confusion Matrix and Statistics
##
##                   Reference
```

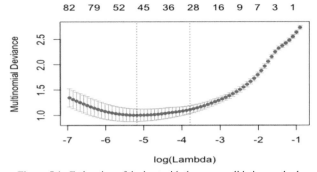

Figure 5.1. Estimation of the best with the cross-validation method.

```
## Prediction     GRAPE   OLIVE   OTHER   SUNFLOWER   WINTER
##                                                    CEREAL
## GRAPE            10       2      0        0          0
## OLIVE             7      19      0        0          0
## OTHER             0       1     51        1          5
## SUNFLOWER         0       0      0        3          0
## WINTER CEREAL     0       0      3        0         14
##
## Overall Statistics
##
##                  Accuracy  :  0.8362
##                    95% CI  :  (0.7561, 0.8984)
##      No Information Rate   :  0.4655
##       P-Value [Acc > NIR]  :  < 2.2e-16
##
##                     Kappa  :  0.7608
##   Mcnemar's Test P-Value   :  NA
##
## Statistics by Class:
##
##                       Class: GRAPE  Class: OLIVE  Class: OTHER
## Sensitivity               0.58824       0.8636        0.9444
## Specificity               0.97980       0.9255        0.8871
## Pos Pred Value            0.83333       0.7308        0.8793
## Neg Pred Value            0.93269       0.9667        0.9483
## Prevalence                0.14655       0.1897        0.4655
## Detection Rate            0.08621       0.1638        0.4397
## Detection Prevalence 0.10345           0.2241        0.5000
## Balanced Accuracy    0.78402           0.8946        0.9158
##                     Class: SUNFLOWER  Class: WINTER CEREAL
## Sensitivity             0.75000              0.7368
## Specificity             1.00000              0.9691
## Pos Pred Value          1.00000              0.8235
## Neg Pred Value          0.99115              0.9495
```

```
## Prevalence              0.03448          0.1638
## Detection Rate          0.02586          0.1207
## Detection Prevalence    0.02586          0.1466
## Balanced Accuracy       0.87500          0.8530
```

Each of the previous models has been repeated for different training and test sets, maintaining the same proportions of the sample (80% training and 20% test set). Random Forest is the best model in terms of accuracy and kappa statistics, followed by the parametric model and then by the boosting one; although the difference among the performances of the various models is limited (Random forest: Accuracy 0.8534, Kappa 0.7824; Regularized multinomial regression: Accuracy 0.8362, Kappa: 0.7608; Boosting: Accuracy 0.8276, Kappa 0.7431). The boxplot in Figure 5.2 shows the distribution of the accuracy of the three models with different splits of the dataset into training and test sets. We can notice that the two nonparametric models generate less dispersed values of the accuracy than the parametric one; particularly, Random Forest generates the least dispersed values of the accuracy.

In general, the level of prevalence of a land use has a strong impact on the capacity of classifiers to generate a correct classification. We obtain very high classification accuracies for grape and olive trees that are the main land uses in our dataset (14% and 19% respectively). Among seasonal crops, sunflower has a very low prevalence in the ground data sample (3%) and the worst performance in terms of prediction, as shown in the confusion matrix.

Winter cereals prevalence is 16% and its accuracy is quite high, although in some cases, winter cereals are confused with the class called "other". Furthermore, a distinction among the various cereals is necessary for agricultural statistics. The classification performance of the class "other" is very good in terms of sensitivity and specificity, due to its prevalence (47%), but not helpful for agricultural statistics.

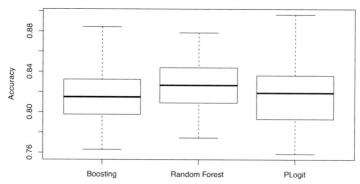

Figure 5.2. Distribution of the accuracy of the models Boosting, Random Forest and Regularized multinomial regression with different splits of the dataset into training and test sets.

5.4 Accuracy assessment of land use databases

A land cover or land use database is useful if its quality is evaluated and is high and it can, thus, be considered reliable. Sometimes, the accuracy is confused with the scale of the product: the more detailed the scale, the higher the accuracy. The scale of remote sensing data represents only the level of detail of the basic material and cannot be considered as the quality of the land cover database. Moreover, the scale of used remote sensing data is strictly linked to the purposes of the project, for example for global projects coarse resolution data have to be used and the accuracy depends on the quality of the production process and on the consequent quality of the product itself, given the scale appropriate for the purposes of the project.

Accuracy assessment should be performed during any phase of the data production process, involving the quality of the classification, as well as the validation of the resulting map, that is the assessment of the degree to which the map agrees with reality (Lunetta et al. 1991; Carfagna and Marzialetti 2009b; Gallego et al. 2010).

Often, a very small amount of resources is devoted to quality control of the photo-interpretation or classification process and to validation of the database through the selection of a probability sample of polygons or points. When polygons are delineated by photo-interpretation, the quality control should be made by repeating the production process for a sample of polygons, with the same basic material and the same procedure. For validating a land cover map, a sample of polygons is compared with the corresponding ground truth, in case the scale of remote sensing data is compatible with ground truth; otherwise, the comparison is generally made with other remote sensing data with a more detailed scale.

When a land cover map is produced by semi-automatic classification of remote sensing data, its quality can be assessed and validated using pixels, blocks of pixels or polygons.

As discussed in Benedetti et al. (2014), the validation of land cover/land use databases involves the assessment of both positional and thematic accuracy (Lunetta et al. 1991; Foody 2002). Positional accuracy is the accuracy of the location of a unit in the map relative to its location in the reference data. The positional deviation of the selected control unit in the map relative to the reference unit has traditionally been measured in terms of root mean square error (Lunetta et al. 1991). The thematic accuracy refers to the accuracy of land cover types depicted in the map compared to the land cover types in the reference data.

Two main types of thematic errors can be identified: the omission and the commission errors. The omission error occurs when a case belonging to a class is not allocated in that class. The commission error occurs when a case belonging to a class is erroneously allocated to another class.

The thematic accuracy is typically assessed through a confusion or error matrix, like the ones showed in previous sections, with the entries estimated on the sampled units in order to obtain estimates of accuracy parameters. The overall accuracy derived from the confusion matrix expresses the probability that a randomly selected unit is correctly attributed to a class by the map, and provides a measure of the quality of the map as a whole.

The accuracy of individual land cover/land use classes may be also assessed. Story and Congalton (1986) distinguished between producer's accuracy and user's accuracy. The producer's accuracy for land cover/land use class i expresses the conditional probability that a randomly selected unit classified as category i by the reference data is classified as category i by the map. It is referred to as producer's accuracy, because the producer of a land cover/land use map is interested in how well a reference category is depicted in the map. The user's accuracy for land cover/ land use class i expresses the conditional probability that a randomly selected unit, classified as category *i* in the map, is classified as category *i* by the reference data. The row and column totals can be also used to quantify the probabilities of omission and commission errors.

In order to collect reference data for map accuracy assessment, the most commonly used probability sample designs are systematic and stratified random sampling. Besides these basic approaches, alternative sampling schemes have been adopted; for example, adaptive procedures which allow efficient quality control. However, they do not allow unbiased estimates of the quality parameters because their efficiency is due to sample selection dependent on previously selected units and stopping rules based on the quality parameter. Carfagna et al. (2008), Carfagna and Marzialetti (2009a) and Carfagna and Marzialetti (2009b) proposed adaptive sample designs for both quality control and validation of land cover databases which allow optimal allocation of sample units when previous information concerning the variability inside the strata is not available.

The parameter taken into consideration is the percentage of area correctly classified or photo-interpreted. The assumption is that the land cover type and the size of the polygons affect the probability of making mistakes in the photo-interpretation as well as in the automatic classification. Thus, the population of the photo-interpreted or classified polygons are stratified according to two variables: the land cover type and the size of polygons.

Carfagna and Marzialetti (2009a) proposed the following adaptive sequential procedure with permanent random numbers. A permanent random number is assigned to each polygon, independently in each stratum. Then, the polygons are organized according to their permanent random numbers. This ordering corresponds to the sample selection. No previous estimate of the variability inside the strata is available; thus, a first stratified random sample of polygons is selected with probability proportional to stratum size. Call *n* the sample size. Neyman's allocation is computed with sample size $n + 1$ and one polygon is selected in the stratum with the maximum positive difference between actual and Neyman's allocation. Then, the quality parameter and its precision are estimated. If the precision is acceptable, the process stops; otherwise, Neyman's allocation is computed with the sample size $n + 2$, and so on, until the precision considered acceptable is reached. This adaptive sequential sample design allows the sample size needed for assessing the quality of the land cover map and its accuracy to be minimized. Moreover, thanks to the use of the permanent random numbers, it allows an unbiased estimate of the accuracy, since the sample size per stratum is dependent on the previously selected units but the sample selection is not and the stopping rule is not based on the estimates of the quality parameter, it is based only on its standard deviation.

This quality control procedure allows for unbiased estimates of the quality parameters, is always more efficient than stratified sampling with proportional allocation and is also more efficient than the procedure proposed by Thompson and Seber (1996, pages 189–191), who suggested stratified random sampling in two or, more generally, *k* phases.

As mentioned above, Cohen's Kappa measures the beyond chance agreement between the classification of remote sensing data and the reference data, since it discounts the total proportion of agreement (p_o) by the level of agreement expected by chance (p_c). Cohen's Kappa is also used for evaluating the quality of databases generated by photo-interpretation of satellite images.

When the sample size is sufficiently large, the standard deviation of Cohen's Kappa can be computed using a large sample variance of the estimate (see Fleiss et al. (1969)):

$$Var(\kappa) = \frac{1}{n(1-p_c)^2}\left(\sum_i p_{ii}\{1-(p_{i+}+p_{+i})(1-\kappa)\}^2 + (1-\kappa)^2\sum_{i\neq j}p_{ij}(p_{i+}+p_{+j})^2 - \{\kappa-p_c(1-\kappa)\}^2\right). \quad (5.27)$$

Carfagna et al. (2008) proposed an adaptive sequential procedure with permanent random numbers also for Cohen's Kappa. Furthermore, since the sequential procedure requires performing a series of operations at each step, a two-phase approach with permanent random numbers was proposed both for the quality measure percentage of area correctly classified (see Carfagna and Marzialetti 2009a) and for the Cohen's Kappa (Carfagna et al. 2008). Although less efficient than the adaptive sequential procedure, the two-step procedure is more efficient than the above-mentioned k phases procedure proposed by Thompson and Seber.

5.5 Land use change

The confusion matrix, traditionally used in map accuracy assessment, could be used for land cover change detection, see Benedetti et al. (2014). One approach is simultaneous analysis of multitemporal data (Serra et al. 2003) through image differencing, principal component analysis or change vector analysis (Serra et al. 2003; Lu et al. 2004). The most common approach is the comparative analysis of independently produced land cover classifications from two different dates (post-classification comparison), due to its applicability to available single date classifications (Foody 2002). It allows one to identify land cover changes between two time periods, and requires high accuracy of the classifications to be compared (see, e.g., Foody 2002). Classifications at two different time periods can be compared through a purposely modified confusion matrix which summarizes land cover changes occurring between the time periods t_1 and t_2.

The matrix rows report the land cover classes at time t_1, and the matrix columns refer to the classes identified at time t_2; the diagonal entries of the matrix, n_{ii}, identify the number of areas classified as land cover class i in both the time periods under investigation, for $i = 1,\ldots, k$. The off-diagonal elements of the matrix, n_{ij}, express the number of units classified as land cover class i at time t_1, and as land cover class j at time t_2, for $i \neq j$, $i,j = 1,\ldots, k$. The row totals express the number of areas classified as land cover class i at time t_1, for $i = 1,\ldots, k$. The column totals express the total number of areas classified as land cover class j at time t_2, for $j = 1,\ldots, k$. The global change index can be computed as 1 minus the proportion of unchanged areas. Change indexes conditional to specific land cover types can also be computed in a similar way.

According to Defourny (2017), current map accuracies, which range from 70 percent to 85 percent for global land cover products, make it impossible to derive any land cover change information from direct comparison of such products, since the annual land cover change rate is significantly lower than the error rate in the land cover maps, particularly at coarse resolutions. This result stresses the importance of quality control and validation of land cover databases. Object-based (Desclée et al. 2006; Ernst et al. 2013) and pixel-based (Hansen et al. 2013) methods have been proposed to identify the land surface change on the basis of the direct comparison of reflectance values or sets of vegetation index values.

References

Agresti, A. 2002. *Categorical Data Analysis*. Wiley Series in Probability and Statistics.

Banerjee, M., M. Capozzoli and L. McSweeney. 1999 Beyond kappa: A review of interrater agreement measures. *Canadian Journal of Statistics* 27: 3–23.

Benedetti, R., P. Postiglione, D. Panzera and D. Filipponi. 2014 Developing more efficient and accurate methods for the use of remote sensing in agricultural statistics. *Technical Report Series GO-05-2014*, http://gsars.org/wp-content/uploads/2014/09/Technical-Report-on-Developing-More-Efficient-and-Accurate-MethodsFINAL.pdf.

Carfagna, E. and F.J. Gallego. 2005. Using remote sensing for agricultural statistics. *International Statistical Review* 73: 389–404.

Carfagna, E., J. Marzialetti and S. Maffei. 2008. Sequential and two Phase sample designs for quality control. Atti della XLIV Riunione Scientifica della Società Italiana di Statistica, Università della Calabria, Arcavacata, 25–27 giugno 2008. http://old.sis-statistica.org/files/pdf/atti/rs08_poster_38. pdf.

Carfagna, E. and J. Marzialetti. 2009a. Sequential design in quality control and validation of land cover databases. *Applied Stochastic Models in Business and Industry* 25: 195–205.

Carfagna, E. and J. Marzialetti. 2009b. Continuous innovation of the quality control of remote sensing data for territory management. pp. 172–188. *In*: Erto, P. (ed.). *Statistics for Innovation, Statistical Design of Continuous Product Innovation*. Springer Verlag.

Cohen, J. 1960. A coefficient of agreement for nominal scales. *Educational and Psychological Measurement* 20: 37–46.

Defourny, P. 2017. Land cover mapping and monitoring. *In*: Delincé, J. (ed.). *Handbook on Remote Sensing for Agricultural Statistics* (Chapter 2). Handbook of the Global Strategy to improve Agricultural and Rural Statistics (GSARS): Rome.

Desclée, B., P. Bogaert and P. Defourny. 2006. Forest change detection by statistical object-based method. *Remote Sensing of Environment* 102: 1–11.

Ernst, C., P. Mayaux, A. Verhegghen, C. Bodart, C. Musampa and P. Defourny. 2013. National forest cover change in Congo Basin: Deforestation, reforestation, degradation and regeneration for the years 1990, 2000 and 2005. *Global Change Biology* 194: 1173–1187.

Fleiss, J.L., J. Cohen and B.S. Everitt. 1969. Large sample standard errors of kappa and weighted kappa. *Psychological Bulletin* 72: 323–327.

Foody, G.M. 2002. Status of land cover classification accuracy. *Remote Sensing of Environment* 80: 185–201.

Friedman, J., T. Hastie and R. Tibshirani. 2001. *The Elements of Statistical Learning*. Volume 1. Springer: New York.

Friedman, J., T. Hastie and R. Tibshirani. 2010. Regularization paths for generalized linear models via coordinate descent. *Journal of Statistical Software* 33: 1.

Gallego, F.J., E. Carfagna and B. Baruth. 2010. Accuracy, objectivity and efficiency of remote sensing for agricultural statistics. pp. 193–211. *In*: Benedetti, R., M. Bee, G. Espa and F. Piersimoni (eds.). *Agricultural Survey Methods*. John Wiley & Sons: Chichester, UK.

Gómez, C., J.C. White and M.A. Wulder. 2016. Optical remotely sensed time series data for land cover classification: A review. *ISPRS Journal of Photogrammetry and Remote Sensing* 116: 55–72.

Hansen, M.C., P.V. Potapov, R. Moore, M. Hancher, S.A. Turubanova, A. Tyukavina, D. Thau, S.V. Stehman, S.J. Goetz, T.R. Loveland, A. Kommareddy, A. Egorov, L. Chini, C.O. Justice and J.R.G.

Townshend. 2013. High-resolution global maps of 21st-century forest cover change. *Science* 342: 850–853.

Hoerl, A.E. and R.W. Kennard. 1970. Ridge regression: Biased estimation for nonorthogonal problems. *Technometrics* 12: 55–67.

Latham, J. and I. Rosati. 2016. Information on land in the context of agricultural statistics. *Technical Report Series GO-15-2016*, Global Strategy to Improve Agricultural and Rural Statistics http://gsars.org/wp-content/uploads/2016/08/TR_Information-on-Land-in-the-Context-of-Ag-Statistics-180816.pdf.

Louppe, G. 2014. Understanding random forests: From theory to practice. arXiv preprint arXiv:1407.7502.

Lu, D., P. Mausel, E. Brondízio and E. Moran. 2004. Change detection techniques. *International Journal of Remote Sensing* 25: 2365–2407.

Lunetta, R.S., R.G. Congalton, L.K Fenstemarker, J.R. Jensen, K.C. McGwire and L.R. Tinney. 1991. Remote sensing and geographic information system data integration: Error sources and research issues. *Photogrammetric Engineering & Remote Sensing* 57: 677–687.

Serra, P., X. Pons and D. Saurí. 2003. Post-classification change detection with data from different sensors: Some accuracy considerations. *International Journal of Remote Sensing* 24: 3311–3340.

Shen, L. and E.C. Tan. 2005. Dimension reduction-based penalized logistic regression for cancer classification using microarray data. *IEEE/ACM Transactions on Computational Biology and Bioinformatics* 2(2): 166–175.

Story, M. and R.G. Congalton. 1986. Accuracy assessment: A user's perspective. *Photogrammetric Engineering and Remote Sensing* 52: 397–399.

Thompson, S.K. and G.A.F. Seber. 1996. *Adaptive Sampling*. Wiley: New York.

Watts, J.D., S.L. Powell, R.L. Lawrence and T. Hilker. 2011. Improved classification of conservation tillage adoption using high temporal and synthetic satellite imagery. *Remote Sensing of Environment* 115: 66–75.

Zhu, J. and T. Hastie. 2004. Classification of gene microarrays by penalized logistic regression. *Biostatistics* 5: 427–443.

CHAPTER 6
Statistical Systems in Agriculture

Cecilia Manzi[1,] and Federica Piersimoni[2]*

6.1 Introduction

Agricultural statistics play an important role in the economic and sustainable development of Countries. Agricultural statistics are pivotal and their area cover a large number of interests. In fact, the FAO portal collecting food and agriculture data (i.e., FAOSTAT) contains 15 domains and 70 data collections for over 245 countries (see Figure 6.1).

FAO approach is not limited to the agricultural sector, in a strict sense, but looks to many other related aspects, such as the environment, the rural development, the food supply chain, and security. This appears as in line with the Global Strategy (GS) to Improve Agricultural and Rural Statistics, launched in 2010 by FAO, World Bank, and United Nations (FAO et al. 2010). The GS represents a comprehensive framework for improving availability and use of agricultural and rural data, necessary for evidence-based decision making. It is based on three pillars:

- produce a minimum set of core data;
- better integrate agriculture into the National Statistical Systems;
- improve governance and statistical capacity building.

The GS tries to address the requirements for agricultural data posed by the Millennium Development Goals and other emerging issues, such as use of crops for biofuels, food security, and the environment. The relevance of those issues emerged gradually during a time of decline in the overall quality and availability of agricultural statistics. In a large number of countries, collection and production of agricultural data is managed outside of national statistical systems. It is often based on availability of administrative data rather than rigorous statistical methodologies.

[1] Italian National Institute of Statistics ISTAT, Agricultural Statistical Service.
[2] Italian National Institute of Statistics ISTAT, Processes Design and Frames Service.
  Email: piersimo@istat.it
* Corresponding author: manzi@istat.it

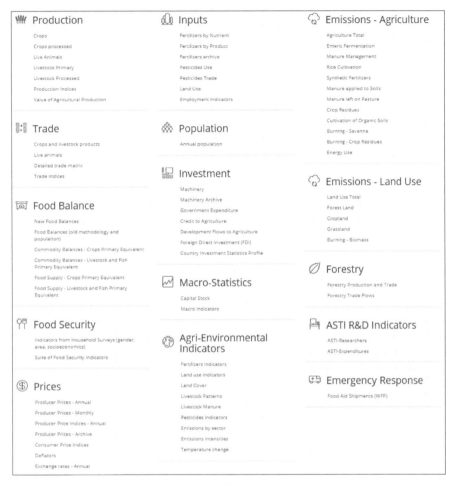

Figure 6.1. Structure of the FAOSTAT database.

The GS addresses this significant challenge, while promoting the integration of agricultural statistics into national systems and official channels.

The GS lays the foundation for a master sampling frame for agricultural and rural statistics. Furthermore, it promotes the use of this frame in integrated survey systems. In fact, the master sampling frame can be adopted as a basis for all data collection through surveys (UDA Consulting 2018). It is used to select samples either for multiple surveys, each with different content (as opposed to building an ad-hoc sampling frame for each survey), or for use in different rounds of a continuing or periodic survey.

The Handbook on Master Sampling Frames for Agricultural Statistics (Vogel et al. 2015) and the Master Sampling Frame for Fisheries and Aquaculture Statistics (FAO 2018b) are considered as two important guidelines on master sampling frames for agriculture.

Generally, national agricultural statistical systems across the world focus on the three divisions categorized by the International Standard Industrial Classification (ISIC). Those categories are the following:

- 01 (Crop and Animal production, Hunting and related service activities);
- 02 (Forestry and Logging);
- 03 (Fishing and Aquaculture).

For example, this approach is followed into the European Union (Eurostat 2020; see Figure 6.2).

In the agricultural statistics, data are collected by different sources: direct surveys, registers and administrative data, estimations made by experts, and so on. However, new data sources, such as research projects in the inter-linked fields and big data, are becoming more readily available. Some key requirements for these new sources are that unique identifiers have to be introduced and that GIS information is extensively included in data collection.

Administrative data offer the possibility to reduce the costs and the statistical burden, but their use depends on their availability and quality (Eurostat 2013).

Figure 6.2. Structure of Eurostat database.

Eurostat has estimated that only 24% of the variables were coming from administrative sources in the Farm Structure Survey 2013 (see Figure 6.3). Information and Communication Technologies (ICT) are among the new forces that are changing data collection methods (Eurostat 2015). The use of big data in the agriculture statistics framework is still developing and best practices are rapidly growing. Among big data techniques, the ones derived from the use of remote sensing are helpful in agriculture (Zhang et al. 2010), as they can supply new high-quality standards, especially in estimating coverage and detail.

From a more theoretical point of view, the implementation of more robust methodologies and best practices may help to improve quality and timeliness of agriculture statistics. Nevertheless, new methodologies in data collection and preparation introduce new concerns, mainly related to data access, privacy issues, actual costs, and IT requirements. Additionally, the impacts on consolidated methods and data processing have to be carefully considered (Alleva 2016).

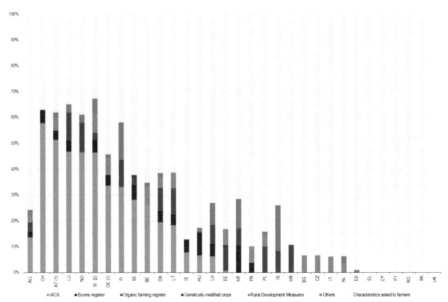

Figure 6.3. Percentage of variables collected from administrative registers and statistical questionnaires in FSS 2013, by country.

Although National Statistical Offices are increasingly integrating well-known procedures and new technologies, direct surveys still represent the main source of statistical data in agriculture. A large number of methods are used in the agricultural sector to directly interview respondents. Among the different methods, the most common are:

- Paper and Pencil interviewing (PAPI);
- Computer-assisted personal interviewing (CAPI);
- Computer-Assisted Self-Interviewing (CASI) also known as Computer-assisted web interviewing (CAWI);

- Computer-Assisted Telephone Interviewing (CATI);
- Mail-out/Mail-back and Drop-off/Pick-up or the similar drop-off/mail-back and drop-off/pick-up.

Often, in surveys based on a large population, a combination of methods is used (mixed mode survey) in order to capitalize on greater potential response rates and/or improve coverage of the population of interest, reducing survey non-response error.

In order to deliver statistics on a certain population, the statistical offices differ between two different strategies: census and sample surveys. Therefore, the same distinction applies to agricultural statistics and will be developed in the following section.

6.2 The census of agriculture

The Census of Agriculture is a key pillar for all national agricultural statistical systems. In many developing countries, this is often the only way to produce information on the structure and other main aspects of the agricultural sector. Additionally, agricultural censuses are the only instruments able provide statistics on farms at lower geographical levels. Hence, it is an essential source of information for governments and decision-makers (FAO 2015; 2018a).

During the ninth round of FAO World Programmes for the Census of Agriculture (2006–2015), 127 countries and territories have carried out an agricultural census (FAO 2019). Historically, the Census of Agriculture has been aimed at providing data on the structure of agricultural holdings, with particular attention to small administrative units. Agricultural censuses have also been used to provide benchmarks to improve current crop and livestock statistics. Their relevance stands also in building useful frames for follow-up agricultural sample surveys.

Since agricultural censuses are usually undertaken every ten years, it is straightforward to associate them with variables that change relatively slowly over time. However, some countries have conducted five-year interval censuses to bring more up-to-date data for agricultural policy purposes.

Agricultural censuses are concerned with data on the basic structure of agricultural holdings. Thus, they mainly cover aspects such as size of holding, land tenure, land use, crop area, irrigation, livestock numbers, labour, use of machinery, and other agricultural inputs. Further, agricultural censuses do not involve variables that show great yearly dynamics, such as agricultural outputs or prices.

In recent years, increasing efforts have been made to improve the informative potential of agricultural censuses. On this line, achieving better integration of different statistical activities remains a key issue. Integration, in a statistical sense, means that each statistical collection is not carried out in isolation, but as a component of a dynamic and comprehensive national statistics system.

The census of agriculture can be conducted in many different ways. FAO distinguishes four modalities:

- classical approach: single one-off operation in which all the census information is recorded. It also includes the short-long questionnaire concept, even though the long questionnaire may be completed at a second visit;

- modular approach: clearly distinguishable core module and supplementary module(s), using information collected in the core module as the frame for the supplementary module(s). The supplementary modules use the frame generated by the core module to target specific populations, which can be all holdings, all holdings above a certain size or subsets of agricultural holdings, such as livestock producers or crop producers–again, with or without size considerations;
- use of registers and administrative records as a source of census data. In principle, when greater amounts of information can be obtained from administrative sources, the production of census-type statistics will be faster, cheaper and more complete. The most complete use of registers will be when all the essential census items can be based on administrative sources;
- register-based census combined with full enumeration/surveys.

The latter may be less work-intensive and result in a cheaper collection of items from the whole population. In fact, in a register-based census:

- the burden on enumerators and respondents can be reduced;
- the non-response rate in cases where information is driven from registers is lower;
- survey data for differing levels of non-response in different population groups can be more easily corrected.

The use of a unique identifier is essential in order to link information successfully in agricultural censuses. Unique identifiers also assist in the detection and correction of identical statistical units (i.e., duplicates).

Agricultural censuses are conducted using different approaches adapted to resource constraints and national conditions.

In order to increase efficiency, some countries carry out their agricultural censuses using a sample enumeration, where the target population is usually the total of agricultural holdings in the country.

The main advantage of sample enumeration is the reduction in costs derived from the reduction in field work. To enumerate only a sample of holdings implies fewer field personnel (enumerators and supervisors), fewer training venues and trainers, fewer forms and questionnaires to deliver and collect, and less information to check and process. This strategy is also selected when some areas of the country can be difficulty accessed and contribute for only a small proportion. If those conditions are satisfied, sample enumeration may still provide reliable statistics.

Some weaknesses of sample enumeration are:

- the amount of subnational and other finely classified data that can be produced is limited. Usually, sample sizes should be large enough to retain many of the attributes of a complete enumeration census providing data at the subnational level, even if finer level data, such as for villages, cannot be provided;
- cross-tabulations that were not planned at the stage of sampling design may not be possible or may have high sampling errors;

- in the case of rare events, such as areas of minor crops or the number of rare types of livestock, few of these events would be picked in the sample and the data would, therefore, be subject to high sampling errors;
- its application requires personnel who are well trained in sampling methods and analysis, which is not always the situation in many developing countries;
- it may not provide an adequate or complete frame for agricultural surveys.

FAO suggests a list of items for census in agriculture. It distinguishes between "essential" items and "frame" items. The "essential" items are those that every country should collect, regardless of the methodological approach used for the agricultural censuses, as they are needed for national purposes and international comparisons. Conversely, the "frame" items are used primarily for building the frame for supplementary modules or subsequent surveys and relate specifically to censuses using the modular approach. Other items in the census program are additional, with no distinction regarding their particular suitability for the classical or modular approach. The items are presented according to 15 themes (see Figure 6.4), corresponding to areas of interest for the census program.

Fisheries are normally outside the scope of the agricultural census, but item "Engagement of household members in fishing activity" is included in the list of items as it is suitable for wider scopes.

In the European Union, a different approach has been adopted with the aim of reducing costs. Regulation on Integrated Farm Statistics (EU) no. 2018/1091 foresees that the census shall cover 98% of the total Utilized Agricultural Area (UAA) and 98% of the livestock units of each Member State. It means the Member States may exclude the smallest units from the coverage of the survey until the threshold is reached. This solution does not come without concerns. For example, according to the results of 2016 Farm Structure Survey, Hungary, could reach 98% of UAA including in the population only 121.980 units on a total of 430.000.

Identification and general characteristics
Land
Irrigation
Crops
Livestock
Agricultural practices
Services for agriculture
Demographic and social characteristics
Work on the holding
Intra-household distribution of managerial decisions and ownership on the holding
Household food security
Aquaculture
Forestry
Fisheries
Environment / GHG emissions

Figure 6.4. Themes suggested by FAO for agricultural census.

The agricultural Census is the most important structural survey in agriculture. In the European Union it is embodied in the framework of the Regulation (EU) no. 2018/1091 on Integrated Farm Statistics, which includes also two sample surveys in 2023 and 2026, while the census is referred to 2020. The same Regulation covers statistical releases on the structure of orchards and vineyards and other agri-environmental indicators.

A second milestone of the EU agricultural statistical system is the SAIO regulation, which will be adopted before 2022. The SAIO regulation contains aggregated crop and animal production statistics, agri-environmental statistics on fertilizers, nutrient balances and pesticides, and potentially agricultural price statistics. SAIO deals with variables of agricultural inputs (prices of feed, pesticides, etc.) and outputs (crop and animal production and prices). All this data will provide aggregated statistics, while no micro-data will be sent to Eurostat. This regulation also allows data to be collected from farms, administrative sources, intermediaries, wholesale entities, market organizations and other sources, and can include a certain number of expert estimates (Lazar et al. 2016).

6.3 The farm accountancy data network

In the EU statistical system, the Farm Accountancy Data Network (FADN), launched for the first time in 1965 (European Commission 2020), represents a source of agricultural data worth mentioning. The aim of the network is to gather accountancy data from farms in order to determine incomes and business analysis of agricultural holdings. The network was launched to give a more insightful evaluation of the Common Agricultural Policy (CAP).

The FADN consists of an annual survey carried out by the Member States, collecting both physical/structural data (location, crop areas, livestock numbers, labour force) and economic/financial data (value of production of the different crops, stocks, sales and purchases, production costs, assets, liabilities, production quotas and subsidies, including those connected with the application of CAP measures). The survey does not cover all the agricultural holdings in the EU, but only those considered as "commercial" (i.e., approximately 80.000 holdings on a population of about 5.000.000 farms). This group covers approximately 90% of the total utilized agricultural area (UAA) and accounts for about 90% of the total agricultural production. The methodology applied helps to provide representative data along three important dimensions: region, economic size, and type of farming.

6.4 Agricultural statistics in other countries: some examples

In the United States, Department of Agriculture National Agricultural Statistics Service (NASS) maintains an annual program for a vast array of commodities. The program covers crops, economics, and livestock sectors of agriculture.

Every five years, following the completion of the Census of Agriculture, NASS conducts a review of the program. Its primary purpose is to ensure that the NASS annual estimating program targets commodities and states the most relevant and latest available information. For this reason, after the review is announced, stakeholder inputs are requested.

The primary source of information for the program review is the Census of Agriculture, as it is the most comprehensive source of data. However, estimates from the current annual program and administrative data are also taken into consideration. In the special case of 2019, the program included:

- citrus;
- crop progress and condition;
- field crops;
- floriculture;
- livestock;
- mushrooms;
- non-citrus fruit and tree nut;
- vegetables.

Estimates are subject to revision, as more reliable data become available from commercial or government sources.

The Census of Agriculture in the USA is conducted every five years to collect information on the number of farms, land use, production expenses, value of land, buildings, farm products, farm size, characteristics of farm operators, market value of agricultural production sold, acreage of major crops, inventory of livestock and poultry, and farm irrigation practices.

The Census of Agriculture in the USA is a complete count of farms and ranches and the people who operate them. Even small plots of land—whether rural or urban—growing fruit, vegetables or some food animals count if $1,000 or more of such products were raised and sold, or normally would have been sold, during the census year. Related to the Census, NASS have carried out other surveys to go into in-depth details. They concern:

- census of aquaculture;
- census of horticultural specialities;
- irrigation and water Management;
- local food marketing practices;
- on-farm energy production;
- organics;
- tenure, ownership and transition of agricultural land.

The Natural Resource Conservation Service (NRCS) and the National Agricultural Statistics Service (NASS) have each developed area sampling frames. The NASS also includes a list sampling frame, which is used in multiple frame contexts. The NASS area frame covers all land in the US.

The land is stratified by land characteristics. Segments of approximately equal size and identifiable boundaries are delineated within each stratum and designated on aerial photographs. The main survey using this frame, the June Area Survey, is conducted in June each year and obtains data on crop areas, livestock inventories and economic statistics. A probability sample of nearly 11,000 segments is selected within each stratum. All farm operators operating within the boundaries of the selected segments are interviewed. In a given year, approximately 85,000 agricultural and non-agricultural land use tracts are identified within the sampled segments. From that identification, over 35,000 detailed personal interviews are conducted with farmers operating farms inside the segment boundaries or who have the potential to qualify as a farm.

A large effort involves drawing the detailed boundaries of every field in each sample segment. In major producing areas, subsamples of corn, soybean, cotton and wheat fields are selected for objective measurement surveys for yield forecasts. The cropland and land use boundaries drawn on the photos are digitized and used as "ground truth" for the crop land data layers prepared using remote sensing data. The frame is used to measure the incompleteness of the list of frames also used by NASS.

The use of segments with identifiable boundaries creates issues through all stages. Satellite imagery and other topographic mapping materials are used to determine the stratum and PSU (Primary Sampling Units) boundaries. The most permanent boundaries, such as paved roads, railroads, canals, rivers, etc., are used. The area frames are expected to remain in use for 15–20 years. Hence, its determination represents a major investment.

NASS has developed a list frame with large emphasis on large farms, rare items suppliers, and farm characteristics that are used as key variables for stratification on other sampling methods. For the five-year census of agriculture, NASS attempts to build a list that is as complete as possible. The area frame is then used to account for farms that are not on the list frame (NASS-USDA 2020).

Statistics Canada carries out 72 surveys and statistical programs in the agricultural and food sector (Statistics Canada 2020). They are grouped as follows:

- animal production (18);
- crop production (24);
- farm business (17);
- farm characteristics (3);
- food (1);
- land use (6);
- other content related to Agriculture and food (3).

The last item includes the Census of Agriculture, which takes place every five years. The census is performed on a population including all farms producing crops, livestock or other products for sale. Census forms are delivered to farms by Post. Once completed, the questionnaire is mailed back to the Data Operations Centre for processing. Respondents also have the option to complete and submit a questionnaire on-line.

The 2021 Census of Agriculture involves electronic questionnaire collection. Invitation letters are delivered to farmers by post. Farm operators are asked to complete the online form by using an access code provided in the invitation. When a questionnaire is considered as not been received or missing, a follow-up is conducted by telephone. Furthermore, the Canadian Census of Agriculture will reduce the statistical burden by replacing questionnaire answers with administrative data where possible.

In Brazil, the agricultural statistical system is based on different sources. The agricultural census collects information on agricultural establishments, forests and/ or aquaculture of all municipalities of a country. The goal of this research is to update previous census data and to provide information about economic, social, and environmental farming. It usually occurs every 10 years and it is carried out by the Brazilian Institute of Geography and Statistics (IBGE).

The information collected consolidates structural data about the agricultural sector, comprising statistics on the number of establishments, land use, number of tractors, implements, machinery and vehicles, characteristics of the establishment and of the producer, employed persons, livestock heads, vegetable and animal productions, among other aspects.

The Systematic Survey of Agricultural Production (LSPA) provides monthly information about the forecast and monitoring of agricultural harvests, with estimates for production, average yield and planted and harvested areas, having the municipal districts as collection units. A mixed survey technique is adopted: PAPI (Paper and pencil interview), expert estimates and administrative data (IBGE 2020).

The Municipal Agricultural Production survey (PAM) provides information on planted area, area to be harvested, harvested area, amount produced, average yield and average price paid to the producer in the reference year for 64 agricultural products (31 of temporary crops and 33 of permanent crops). PAM is integrated with the LSPA so that all data on agricultural products investigated by the LSPA during the crop cycle automatically migrate to PAM on December 31. This enables an annual consolidation based on the monthly data to be done. PAM is carried out annually and covers the whole national territory, with results disclosed down to the level of Municipalities. For this survey, a technique similar to that of LSPA is adopted.

The survey of stocks provides conjunctural data every semester on the volume and spatial stocks distribution of the main farm products and on the units where they are stored, having as collection units the establishments dedicated to storage and dry storage services or those which store farm products or their derivatives. IBGE disseminates the results of the Municipal Livestock Survey together with information on herd inventories, quantity and value of animal products, as well as the quantity and value of animal products, including aquaculture.

In Australia, in 2015 The Australian Bureau of Statistics (ABS) and the Australian Bureau of Agricultural and Resource Economics and Sciences (ABARES) had undertaken the National Agricultural Statistics Review (NASR) to consider all aspects of the current agricultural statistical system and assess its capacity to inform decision making. ABARES is the research arm of the Australian Government Department of Agriculture, Water and the Environment and its activity covers the agriculture and water resource portfolio, from agricultural commodities and forecasting to land use, fishery and forestry. The Australian Bureau of Statistics (ABS) is in charge of the agricultural Census, every five years, and agricultural surveys, conducted annually.

The NASR has provided an opportunity for the ABS and ABARES, as the principal producers of agricultural statistics, to hear directly from other statistical users and producers across government, industry and the research sector about these issues, and to identify opportunities to address them.

Through the NASR, the two organizations have identified a pathway to establish a contemporary, best practice Australian agricultural statistical system for the future. The NASR has also enhanced the relationship between the two agencies and paved the way for further collaboration to provide stronger and more effective leadership across the system.

The Agricultural Census in Australia collects data from businesses across a variety of agricultural industries in Australia. It is currently the key vehicle through which the ABS produces agricultural data for regional geographies and a wide range of agricultural industries. It provides benchmark data to support the agricultural statistical programs of both ABS and the ABARES, including through the maintenance of a high-quality frame of agricultural businesses. Data are collected from businesses ranging above a minimum threshold and cover land use, crop and horticultural area and production, livestock numbers, farm management and demographic information.

The Agricultural Survey (AS) has the objective to act as a source of basic agricultural statistics about a wide variety of commodities. The frame population of the AS is all establishments with an Estimated Value of Agricultural Operations (EVAO) of $5,000 or more. AS collects area and production data for a wide range of agricultural commodities, mainly cereals and other broad acre crops, vegetables, fruit, vineyards, livestock. This commodity data is used to produce data on the Value of Agricultural Commodities Produced (VACP). Information on inputs to the production process is also collected and disseminated. Two complementary collections are run in conjunction with the Agricultural Survey, namely Apples and Pears (user-funded) and Vineyards (user-funded).

In India, the statistics are under the control of the Ministry of Statistics and Programme Implementation that has two wings, one related to Statistics and the other concerned about Programme Implementation.

The Statistics Wing, called the National Statistical Office (NSO), consists of the Central Statistical Office (CSO), the Computer center and the National Sample Survey Office (NSSO). Besides these two wings, a National Statistical Commission was created by a Resolution of Government of India (MOSPI) and one autonomous Institute, the Indian Statistical Institute, was declared as an institute of National importance by the Parliament (Government of India 2020).

The agricultural statistical system in India includes a huge number of projects:

- crop area statistics;
- crop production;
- crop forecasts;
- production of horticultural crops;
- land use;
- irrigation statistics;
- land holdings and agricultural census;
- agricultural prices;
- agricultural market intelligence;
- cost of cultivation of principal crops;
- livestock numbers;
- fisheries statistics;
- forestry statistics;

- marketable surplus and post-harvest losses;
- market research surveys;
- index numbers in agriculture;
- recording of area under mixed crops;
- input statistics.

Information is obtained using classical statistical approaches, such as sample surveys in crop area statistics, or alternative sources such as arrivals, exports and growers' associations. The Ministry of Agriculture and the National Crop Forecasting Centre (NCFC) is also studying an innovative method using space, agro-meteorology, and land-based observations (FASAL), which uses Remote Sensing to estimate the area under principal crops that should be actively pursued.

The Agricultural Census in the country is conducted every five years because the structure of holdings in the country is changing fast due to fragmentation of holdings and urbanization/industrialization dynamics. The census provides detailed statistics on the structure of operational holdings and their main characteristics, like number and area, land use, irrigation, tenancy and cropping pattern. It is conducted on census-cum-sample survey basis.

The Census operations are completed in three phases. In the first phase, a list of holdings with their operated area and social characteristics is prepared on census basis in land record States/UTs (covering about 86% of reported area) and on 20 percent sample villages in non-land record States/UTs. In the second, detailed data on agricultural characteristics is collected on a sample basis from 20 percent selected villages, both in Land Record and Non-Land Record States, and the parameters are then estimated at the Tehsil/District/State level. In the third phase, data on input use pattern are collected on a sample basis from selected holdings from selected 7 percent villages and the parameters are estimated at the District/State level (Dhar 2012).

6.5 Conclusion

The sources of information in agricultural statistics are many: censuses, sample surveys, administrative data, reports submitted by regional offices, cooperatives, and agricultural enterprises, data obtained from bookkeeping and production accounting at such farms, new data sources such as big data.

Data collection, analysing and publishing of agriculture sector differs country by country. In many cases, data are collected by sector using different sampling frames and surveys. This division of data by sector does not allow for cross-sector analysis or the ability to measure the impact of actions in one sector on other sectors.

Surveys on crop production are often carried out separately from livestock production surveys, using different sampling frames. This does not allow the analysis of holdings characteristics that produce both crops and livestock or for comparing them to holdings that specialize in either crops or livestock.

Both households and holdings are often included in the target population of the same surveys.

In many countries, the responsibility to collect agriculture, fishery and forestry data is shared by different organizations, mainly National Statistical Offices and Ministry of Agriculture. This entails difficulties in coordination. Data are kept separate and often produce conflicting results, which confuses issues and data users. All these aspects make the agricultural statistical systems quite complex, hence, a strategy of sources and methods integration is needed.

An important point is that countries should develop a master sampling frame for agriculture, in order to improve data collection systems and the quality of statistics. Even taking into account the differences in terms of agricultural and economic structures, this should be a common element to develop a reliable and quality agricultural statistical system.

References

Alleva, G. 2016. Modernization of agricultural statistics to respond to new multidimensional demands. *Proceedings ICAS VII Seventh International Conference on Agricultural Statistics* I Rome 24–26 October 2016. DOI: 10.1481/icasVII.2016.pl1.

Dhar, V. 2012. *Agriculture Census in India*, http://www.fao.org/fileadmin/templates/ess/ess_test_folder/Workshops_Events/APCAS_24/PPT_after/APCAS-12-31-Agri_Census_India_APCAS24.pdf.

European Commission. 2020. Farm accountancy data network, https://ec.europa.eu/info/food-farming-fisheries/farming/facts-and-figures/farms-farming-and-innovation/structures-and-economics/economics/fadn_en.

Eurostat. 2013. *Task Force on Linkage Administrative Data with Statistics*. Standing Committee for Agricultural Statistics (CPSA).

Eurostat. 2015. *Strategy for Agricultural Statistics for 2020 and Beyond*. https://ec.europa.eu/eurostat/documents/749240/749310/Strategy+on+agricultural+statistics+Final+version+for+publication.pdf/9c7787ca-0e00-f676-7a64-7f56e74ec813.

Eurostat. 2020. https://ec.europa.eu/eurostat/data/database.

FAO. 2015. *World Programme for the Census of Agriculture 2020 Volume 1: Programme, Concepts and Definitions*. FAO Statistical development Series 15. Rome (http://www.fao.org/3/a-i4913e.pdf).

FAO. 2018a. *World Programme for the Census of Agriculture 2020 Volume 2: Operational Guidelines*. FAO Statistical Development Series 16 (http://www.fao.org/3/CA1963EN/ca1963en.pdf).

FAO. 2018b. *Master Sampling Frames (MSF) for Fishery and Aquaculture Statistics* (http://www.fao.org/3/ca6433en/ca6433en.pdf).

FAO. 2019. *Main Results and Metadata by Country (2006–2015)*. World programme for the Census of Agriculture 2010. FAO Statistical Development Series no. 17. Rome.

FAO. 2020. http://www.fao.org/faostat/en/#home.

FAO, World Bank and UN. 2010. *Global Strategy to Improve Agricultural and Rural Statistics*. September 2010 Report No. 56719-GLB.

Government of India. 2020. http://mospi.nic.in/agriculture-statistics.

IBGE. 2020. https://www.ibge.gov.br/en/statistics/economic/agriculture-forestry-and-fishing/17174-systematic-survey-of-agricultural-production.html?=&t=o-que-e.

Lazar, A.C., J. Selenius and M. Jortay. 2016. *Strategy for Agricultural Statistics 2020 and Beyond: For the Future European Agricultural Statistics System (EASS)*. European Commission, Eurostat 2920 Luxembourg, DOI: 10.1481/icasVII.2016.f37c.

NASS-USDA. 2020. https://www.nass.usda.gov/index.php.

Statistics Canada. 2020. https://www150.statcan.gc.ca/n1/en/type/surveys.

UDA Consulting. 2018. *An Overview of Agricultural Statistical Systems in the Selected Countries*, http://www.udaconsulting.com/sites/default/files/2018-03/AGRI_2017_UDA.pdf.

Vogel, F., N. Keita, M. Galmés, F.J. Gallego Pinilla and C. Ferraz. 2015. *Handbook on Master Sampling Frames for Agricultural Statistics: Frame Development, Sample Design and Estimation*. Publication Prepared in the Framework of the Global Strategy to Improve Agricultural and Rural Statistics, FAO. http://www.fao.org/3/ca6398en/ca6398en.pdf.

Zhang, F., Z. Zhu, Y. Pan, T. Hu and J. Zhang. 2010. Application of remote sensing technology in crop acreage and yield statistical survey in China. *Meeting on the Management of Statistical Information Systems (MSIS)*, Daejeon, Republic of Korea, 26–29 April 2010.

CHAPTER 7
Exploring Spatial Point Patterns in Agriculture

*M Simona Andreano** and *Andrea Mazzitelli*

7.1 Introduction

Point pattern analysis is a part of spatial statistics that analyses the spatial distribution of points in two-dimensional space (Schabenberger and Gotway 2004). This framework considers methods to explain the spatial patterns of statistical units using microdata and by characterizing each unit by its geographic coordinates (Diggle 2003).

The analysis of point patterns has recently interested the agricultural economists, the geographers, and the statisticians.

In the following, we give an recent review of this topic. Kelly et al. (2004), using spatial point pattern statistics, show that farmers' practices can influence girth distribution as well as the spatial pattern shea butter tree. The degree of this influence may vary according to the site. Two different sites in southern Mali and three treatments (cultivated field, fallow and forest) per site were involved in the mentioned study.

Murwira and Skidmore (2005) investigate whether spatial heterogeneity of a normalised difference vegetation index is related to the probability of African elephant (presence in different parts of an agricultural landscape in North-Western Zimbabwe) based on data from the early 1980s and early 1990s. In other words, the aim is the study of the coexistence of arable cultivation and wildlife management outside the protected areas. To this end, the authors consider the elephant distribution map as a spatial point pattern.

Bamière et al. (2011) use a spatially explicit mathematical programming farm-based model, which accounts for three spatial levels (field, farm, and landscape) and a spatial pattern index (i.e., the L-function, see Section 7.3) to analyse the scheme and the implementation of an agri-environmental programme aimed to preserve the *Tetrax tetrax* (i.e., the little bustard) in the Plaine de Niort (France).

Universitas Mercatorum, Economic Faculty.
Email: a.mazzitelli@unimercatorum.it
* Corresponding author: s.andreano@unimercatorum.it

Long et al. (2014) aim at investigating changes in specific cropping sequences in northeast Montana (USA) using a spatio-temporal framework. In particular, the authors used point pattern analysis to assess the spatial dependence among fields by cropping sequence and year to determine if the locations were randomly distributed across the area.

Chaiban et al. (2019) investigate the spatial point pattern distribution of extensive and intensive chicken farms in Thailand, founding their analysis on detailed 2010 census data. The results evidence that both the level of clustering and location of clusters are well described with reasonable accuracy through point pattern methodologies.

The main research questions that we may address using spatial point pattern analysis in agriculture are, for example:

- Why does land concentration occur in a given area?
- Why do the points spread uniformly in particular areas of the study region?
- Is the distribution of farm sizes (from the smallest to the largest) concentrated or sparse?

To this end, this chapter describes the basic steps in the explorative analysis of spatial point patterns, offering a comprehensive description of point processes and applications on agricultural farms, providing in particular, the introductive statistical methods to analyse spatial clustering. The empirical exercises are performed using the R-package `spatstat` (Baddeley and Turner 2005).

The layout of the paper is as follows: in Section 7.2, we introduce the main statistical concepts of a spatial point process. Section 7.3 is devoted to the description of some basic statistics for the analysis of spatial point pattern. Section 7.4 contains an empirical illustration of the theory using the R-package `spatstat`. Finally, Section 7.5 concludes the chapter.

7.2 The spatial point process

In spatial statistics, we can recognize different typologies of spatial data: geostatistical, lattice, and spatial point patterns (see Chapter 1 for further details). Generally, spatial data are classified by the nature of the spatial domain (Cressie 1993; Schabenberger and Gotway 2004). Let us define a generic spatial process in d dimensions as:

$$\{X(\mathbf{z}_i): \mathbf{z}_i \in D \subset \mathbb{R}^d\} \tag{7.1}$$

where X denotes the attribute we may observe, $\mathbf{z}_i$ is the i-th location, defined using a $d \times 1$ vector of coordinates. If only two coordinates are considered (i.e., $d = 2$), the two-dimensional spatial process is defined. The observed point pattern x is the agricultural variable under study and represents a realisation of the random point process X.

Diggle (2003) defines a spatial point pattern as "*a set of locations, irregularly distributed within a designated region, and presumed to have been generated by some form of stochastic mechanism*".

In agricultural studies, the points, for instance, could represent trees, farms, animal nests, and so on.

Point locations may represent all possible events (mapped point pattern) or only a subset (sampled point pattern). As evidenced by Gatrell et al. (1996), the researchers should be mindful of the presence of the edge effects in the analysis that tend to distort the estimation and inference close to the boundary.

In the simplest case, the data set includes only the event locations. In other cases, the points may have some additional information describing the analysis, denoted as marks. The marks could be categorical (for example, type of agricultural crop) or quantitative (for example, tree diameter) or even multivariate. In this case, the pattern is defined as marked spatial point pattern, while conversely is referred to as unmarked. In this Chapter, we review the main techniques for explanatory analysis only for unmarked spatial point patterns.

Consider a stochastic process $\{X(\mathbf{z}_i): \mathbf{z}_i \in D \subset \mathbb{R}^d\}$, where $\mathbf{z}_i = (z_{i1}, z_{i2}, ..., z_{id})$, with $i = 1, ..., n$, are random points that represent the locations of the events of interest.

The properties of a spatial point process can be formalised through the intensity function $\lambda(\mathbf{z}_i)$, which is the average number of events per unit area in a small area around $\mathbf{z}_i$. The intensity $\lambda(\mathbf{z}_i)$ may be constant (i.e., homogeneous) or different across different locations (i.e., inhomogeneous).

The approach assumes that the realisation of the point process X is observed only inside a region W that is denoted as sampling window. The window W is fixed and known. The study region can be represented by a rectangular or more complex polygonal region. Our aim is to make inference on the parameters of the process X.

The investigation of the intensity should represent the first step of an exploratory analysis for spatial point patterns. If the point process $X(\mathbf{z}_i)$ is homogeneous, the expected numbers of points for any area A (i.e., $N(A)$) is proportional to the area of A, where A is a sub-region of a larger region on which the process is defined, i.e.,

$$E(N(A)) = \lambda |A| \tag{7.2}$$

where $|A|$ is the area of A and the constant of proportionality λ is the intensity. Under the homogeneity of the process, the empirical intensity of points is:

$$\hat{\lambda} = \frac{n(x)}{|W|} \tag{7.3}$$

where $n(x)$ is the total count of points and $|W|$ is the area of the sampling window W. The (7.3) is an unbiased estimator of the true intensity λ (Diggle 2003).

However, the intensity of a spatial point process may differ across the locations (inhomogeneous process). In this case, the intensity function can be estimated through non-parametric techniques, such as kernel smoothing. See Diggle (2003) for further details.

The simplest theoretical model for modelling a spatial point pattern is called the complete spatial randomness (CSR) process that is described through the homogeneous Poisson process. The CSR presents two main properties (Diggle 2003):

1) The number of events in any region, A, is distributed according to a Poisson distribution with a mean of $\lambda |A|$, where $|A|$ is the area of A.

2) The $n(x)$ points in A represent an independent sample from the uniform distribution on A.

The above properties ensure that, according to (1), a homogeneous Poisson distribution is being followed, and, according to (2), there are no attractive or inhibitory interactions between the points. In essence, item (2) states that the location of one point in space does not affect the probabilities of nearby points.

The homogeneous Poisson point process plays a central role in many statistical tests, as reference or null model. More complicated models are built starting from CSR, and many mathematical concepts are defined relative to CSR (Cressie 1993). One important research question that derives from the properties of CSR is analysing if there is a tendency of events to display a pattern over an area that is opposed to randomness. The main two violations from CSR pattern are: the regular (inhibition) and clustered patterns. In the uniform pattern, every point is regularly distanced from all of its neighbours, underlining inhibition between events. In the clustered pattern, there is *contagion* (agglomeration) among points.

Although Diggle (2003) denoted the homogenous Poisson process as *"an unattainable standard"*, the empirical analysis usually begins with a hypothesis test on CSR validity. The null hypothesis (H_0) is that the points follow the CSR distribution; while the alternative hypothesis (H_1) is that the events are spatially clustered or dispersed. If the CSR assumption is rejected, the researcher can perform further analysis to understand the appropriate nature of spatial point pattern under investigation. See Diggle (2003) for excellent details about these methods.

The methods for testing CSR can be broadly divided into two groups: quadrat methods and distance methods. These topics will be analysed in depth in the next section.

7.3 Explorative analysis for point patterns

7.3.1 The quadrat counts method

The quadrat analysis concerns a class of methods of spatial analysis which is relatively easy to implement and largely used by plant and animal ecologists (Greig-Smith 1964).

Quadrat count analysis is a typology of variance analysis. The procedure can be formalized in the following steps:

1. we partition the area under study W into m quadrats or sub-regions of equal area;
2. we count the number of events that occur in each quadrat;
3. we use the distribution of quadrat counts as an indicator of spatial pattern.

The techniques can be implemented using either an exhaustive census of quadrats or considering a sample of quadrats randomly located across the area of interest. There are no special constraints for the shape and the size of a quadrat. However, the size should be reasonable compared to the area of the region.

The choice of quadrat size can greatly influence our analysis. Large quadrats obviously produce a coarse picture of the pattern. Conversely, if the quadrat is too small, many quadrats may include only one point, or none at all. For a particular analysis, however, the shape of size of quadrats are fixed.

To verify the hypothesis of CSR, we can use the simple test statistic defined as:

$$X^2 = \frac{\sum_{i=1}^{m}(n_i - \bar{n})^2}{\bar{n}},$$

(7.4)

where $n_1, \ldots, n_m$ are the number of events in m quadrats, and $\bar{n} = \sum_{i=1}^{m} n_i / m$ is the expected number of events under CSR. The null distribution of (7.4) is χ^2_{m-1}, provided that $m > 6$ and the expected number of events per quadrat is greater than 1 (Diggle 2003). This test statistic is a standard Chi-square goodness of fit test and evaluates if the n points are distributed uniformly and independently or, in other terms, whether the quadrat counts are independent Poisson variates with common mean (Schabenberger and Gotway 2005).

The statistic (7.4) is $m - 1$ times the sample variance to mean ratio, i.e., *Index of*

$$Dispersion = \left((m-1) \frac{s^2}{\bar{n}} \right)$$ of the observed counts that was used by Fisher et al. (1922)

to test the hypothesis that the counts are distributed according a Poisson distribution. Under a CSR pattern, the sample variance to mean ratio is approximately equal to 1. This evidence is based on the circumstance that in a Poisson process (i.e., a CSR pattern) mean and variance are equal.

Note that failing of the null hypothesis of this test may indicate a non-uniform distribution of the points in the area under investigation or a dependence between events. In particular, significantly large values of statistic (7.4) indicate aggregation, while significantly small values suggest regularity.

7.3.2 The distance-based methods

A valued alternative to the quadrat counts method for testing CSR is represented by a distance-based approach. This framework is largely recommended by scientists and practitioners. In fact, as observed by Cressie (1993) "*...the reduction of complex point patterns to counts of the number of events in random quadrats and to one-dimensional indices results in a considerable loss of information*".

Distance methods are based on the distances between neighbouring events, and between sample points and neighbouring events. Specifically, we may consider:

1. inter-event distances (i.e., pairwise distances) $d_{ij} = \|x_i - x_j\|$ between all different pairs of events x_i and x_j $(i \neq j)$ in the pattern;
2. nearest neighbour distances $t_i = \min_{j \neq i} d_{ij}$, the distance from each point x_i to its nearest neighbour;
3. point-to-nearest-event distances (i.e., empty space distances) $d(u) = \min_i \|u - x_i\|$, the distance from a fixed reference location u in the window to the nearest point.

To define a formal statistical analysis, we usually use the empirical distribution function (EDF) of these distances.

Let us start our description from point-to-nearest event distances. The distance $d(u) = \min_i \|u - x_i\|$ from a fixed location $u \in \mathbb{R}^2$ to the nearest point x_i is denoted as empty space distance and can be computed for all u locations. In this case, the EDF of the observed empty space distances on a grid of possible locations u_j, $j = 1,..., m$ is defined as (Diggle 2003):

$$\hat{F}(r) = m^{-1}\#(d(u_j) \leq r) \tag{7.5}$$

where # means "the number of". The F function is called the empty space function, since it is a measure of the average space left between events (Bivand et al. 2013).

The estimation of F is prevented by the presence of edge effects resulting from the non-observability of events outside the window observed that usually is a part of a larger region on which the process works. For this reason, an edge correction is needed in order to reduce bias (Ripley 1988). Under CSR, the F function is:

$$F_{CSR}(r) = 1 - exp(-\lambda\pi r^2). \tag{7.6}$$

To test the CSR hypothesis, we compare $\hat{F}(r)$ with the value of $F_{CSR}(r)$ obtained by plugging in the estimated intensity $\hat{\lambda} = \dfrac{n(x)}{|W|}$. Values $\hat{F}(r) > F_{CSR}(r)$ suggest a regularly spatial point pattern. Conversely, values $\hat{F}(r) < F_{CSR}(r)$ suggest a clustered pattern.

More formally, it is possible to compute a plot where $F_{CSR}(r)$ (i.e., theoretical Poisson distribution under CSR) is the abscissa and $\hat{F}(r)$ (i.e., the empirical distribution) the ordinate. If the data are compatible with CSR, the plot should be near linear. To assess the significance of any departure from linearity, we can use lower and upper simulation envelopes under CSR (Diggle 2003, p. 13) and check if the empirical function is contained inside.

Alternatively, a formal Monte Carlo test can be defined using the l-th largest and l-th smallest values of k simulations, where l is a chosen rank. See Diggle (2003) for further details.

In analogy to point-to-nearest distances, it is possible to define the EDF for nearest neighbour distances from i-th event to the nearest event in A, $t_i = \min_{j \neq i}\|x_i - x_j\|$, as (i.e., the G function):

$$\hat{G}(r) = n(x)^{-1}\#(t_i \leq r) \tag{7.7}$$

where $i = 1,..., n(x)$ are the events in W. For a homogeneous Poisson spatial point process (CSR) of intensity λ, the nearest neighbour distance distribution function is:

$$G_{CSR}(r) = 1 - exp(-\lambda\pi r^2). \tag{7.8}$$

The (7.8) is identical to $F_{CSR}(r)$ function for the Poisson process. This circumstance is intuitively justified from the fact that points of the Poisson process are independent of each other, the information that u is a point of the process does not influence any other points of the process, hence, G is equivalent to F.

Nevertheless, the interpretation of $\hat{G}(r)$ is the reverse of $\hat{F}(r)$ (Baddeley et al. 2014a). Values $\hat{G}(r) > G_{CSR}(r)$ suggest a clustered spatial point pattern. Conversely, values $\hat{G}(r) < G_{CSR}(r)$ suggest a regular pattern.

Also, in this case, the $\hat{G}(r)$ can be compared with upper and lower simulation envelopes exactly as for $\hat{F}(r)$.

The most popular nearest neighbour technique is, however, based on the Clark–Evans statistic (Clark and Evans 1954). This statistic is based on the mean of the nearest neighbour event-to-event distance, $\bar{V} = \sum V_i/n$. Under the assumption of independent nearest-event-to-event distances, it can be shown that the mean and standard deviation of $\overline{V}$ is:

$$E(\overline{V}) = \frac{1}{2\sqrt{\lambda}}; \sigma_{\overline{V}} = \sqrt{\frac{4-\pi}{4m\lambda\pi}} \tag{7.9}$$

where $\lambda = m/|W|$ for m distances sampled in some region A. Using the central limit theorem with CSR, the test statistic is:

$$CE = \frac{\overline{V} - E(\overline{V})}{\sigma_{\overline{V}}} \propto N(0;1). \tag{7.10}$$

The Clark–Evans statistic can be used to test the randomness of a point pattern. Significantly large values indicate regularity, whereas small values indicate a clustered pattern.

Finally, also inter-event distances $d_{ij} = \|x_i - x_j\|$ may be summarised through the EDF:

$$\hat{H}(r) = \left\{ \frac{1}{2}n(n-1) \right\}^{-1} \#(d_{ij} \leq r) \tag{7.11}$$

The procedure to test for CSR is very similar to the previous definitions of distance and depends on the definition of the theoretical homogeneous Poisson distribution function $H_{CSR}(r)$. For further details about this approach see Diggle (2003).

However, the observed inter-event distances usually represent a biased sample of inter-event distances in the spatial point process. In fact, there is a bias in favour of the smaller distance since it is impossible to observe inter-event distances greater than the diameter of the window W.

Ripley's K function (Ripley 1977) is an appropriate alternative that can be used to summarize a point pattern, to test hypotheses, to estimate parameters, and to fit models. Ripley (1977) defined the K function for a stationary point process so that $\lambda K(r)$ is the expected number of other points of the process within a distance r of a typical point of the process. Ripley's K function is defined as:

$$K(r) = (1/\lambda)E(n(r)) \tag{7.12}$$

where $n(r)$ is the number of additional events within a distance r of a randomly chosen point, and λ is the intensity of the process.

The naive estimator of $K(r)$ is its empirical average (Diggle 2003):

$$\hat{K}(r) = \frac{\sum_{i=1}^{n} \sum_{j=1}^{n} I_{ij}(r)}{\hat{\lambda}n} \tag{7.13}$$

where $\hat{\lambda} = n/|W|$ is the estimated intensity in terms of the observed number of events per unit area and:

$$I_{ij}(r) = \begin{cases} 1 \text{ se } d_{ij} < r \\ 0 \text{ se } d_{ij} \geq r \end{cases} \tag{7.14}$$

and d_{ij} is the Euclidean distance between i and j.

Under the CSR hypothesis $K_{CSR}(r) = \pi r^2$. $\hat{K}(r) > \pi r^2$ indicates that a typical point is a part of a cluster and, as result of this, it has more neighbours than it would be expected if the pattern were completely random. Conversely, $\hat{K}(r) < \pi r^2$ displays a regular point process because a typical point in this pattern has fewer neighbours than it would be expected if the pattern were completely random.

However, $L(r) = [K(r)/\pi]^{1/2}$ and its estimator $\hat{L}(r) = [\hat{K}(r)/\pi]^{1/2}$ are a commonly employed transformation. Under CSR, $L_{CSR}(r) = r$. In practice, the value $L_{CSR}(r) = r$ is used as a benchmark. In fact, if $\hat{L}(r) > r$, the points are aggregated. Conversely, if $\hat{L}(r) < r$ a regular pattern is suggested.

As evidenced before, the boundaries of the area under investigation are usually subjective and, for this reason, the edge effects can affect the results obtained. If we ignore the edge effects, we introduce a bias into the estimator $\hat{K}(r)$, especially for large values of r.

Edge effects arise because points outside the boundary are not counted in the numerator, even if they are within a distance r of a point in the study area. Many authors proposed corrections to the K function to account for edge effects. Several methods have been proposed in the literature to account for edge effects (e.g., Cressie 1993; Diggle 2003).

7.4 Empirical application

The first step involves loading the R package `spatstat`, which enables us to perform the empirical illustration (Baddeley and Turner 2005):

```
>library(spatstat) #loading the spatstat package
```

Once we upload this R package, three different processes are generated. The first is obtained starting from a Poisson distribution. Hence, coordinates are drawn independently and uniformly over the study area. Before generating coordinates, a generic 1×1 window is defined for our analysis. Function `as.owin` is adopted with this aim in mind, by concatenating coordinates for the vertical and horizontal axes.

```
>set.seed(24)
>w <- as.owin(c(0,1,0,1)) #defines the window for the analyses
```

A point pattern (A) is generated following a homogeneous Poisson process:

```
>A<-rpoispp(lambda=100, win=w, nsim=1) #returns a pattern
following a homogeneous Poisson process
```

where `lambda=100` is the parameter of the intensity for the Poisson distribution and `w` is the pre-selected window. The point pattern A will have a number of about 100 simulated points in the selected window (Baddeley and Turner 2005). The parameter `nsim` provides the number of realisations to be generated and it is set to one. The function returns a `ppp` object containing, among other values, the coordinates for the point process represented in Figure 7.1(A) and the intensity on the area.

Complete spatial randomness is a starting point for our analysis. However, this assumption may be difficult to verify in the real case. Different vegetations often tend to cluster because of very local condition of terrain and humidity (e.g., Picard et al. 2009). Additionally, in agriculture, a farmer may tend to create regular patterns,

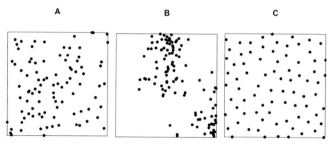

Figure 7.1. Point process generated following the homogenous Poisson process in (A), a Thomas clustered process realization as in (B), and a Simple Sequential Inhibition process pattern in (C).

as for the case of vineyards (Li et al. 2017). For these reasons, in this empirical application, we also provide other different spatial distributions.

A possible way to simulate a clustered pattern is to consider a Poisson cluster process. A specific case is the Thomas' process (Thomas 1949), in which clusters consist of a Poisson number of random points, each having an isotropic Gaussian displacement from its parent (Vinatier et al. 2011). For this reason, an example of clustered process is obtained using the command rThomas in this application.

```
>B<-rThomas(kappa=3,scale=0.1, mu=40) #returns a pattern
following a Thomas process
```

where kappa defines the number of centres regulated by a Poisson process, the scale defines the displacement of a point from cluster centre, and mu is a parameter of the function for the mean of units for each cluster. This pattern is shown in Figure 7.1(B).

As aforementioned, coordinates of points may emerge on a more regular pattern, which also results in a violation of the CSR. This sort of repulsion between different points may be simulated according, for example, to the Simple Sequential Inhibition (SSI) process. This algorithm starts with an empty window, adding new points one-by-one. Each new point is simulated uniformly and independently of preceding points. If the new point is closer than r from an existing point, then it is rejected and another random point is generated.

```
> C<-rSSI(r=0.09, n=80, win = w, nsim=1) #returns a pattern
following a SSI process
```

where r is the inhibition distance and n is the maximum number of points allowed.

Plots for A, B, and C may be obtained by the conventional plot function.

```
>plot(A,pch=20,cex=1.2, main="A") #plot the pattern A
>plot(B,pch=20,cex=1.2, main="B") #plot the pattern B
>plot(C,pch=20,cex=1.2, main="C") #plot the pattern C
```

The three processes included into the ppp can be also summarized:

```
>summary(A) #summary for pattern A
Planar point pattern: 94 points
Average intensity 94 points per square unit
Coordinates are given to 8 decimal places
```

```
Window: rectangle = [0, 1] x [0, 1] units
Window area = 1 square unit
>summary(B) #summary for pattern B
Planar point pattern: 111 points
Average intensity 111 points per square unit
Coordinates are given to 8 decimal places
Window: rectangle = [0, 1] x [0, 1] units
Window area = 1 square unit
>summary(C) #summary for pattern C
Planar point pattern: 80 points
Average intensity 80 points per square unit
Coordinates are given to 8 decimal places
Window: rectangle = [0, 1] x [0, 1] units
Window area = 1 square unit
```

The average intensity is shown, corresponding to the number of points as we use square unit windows.

To obtain a picture of local spatial variations in intensity, we can plot a kernel estimate of intensity. The function can be written as:

```
>plot(density(A), main="A") #plot for density from the pattern A
>plot(density(B), main="B") #plot for density from the pattern B
>plot(density(C), main="C") #plot for density from the pattern C
```

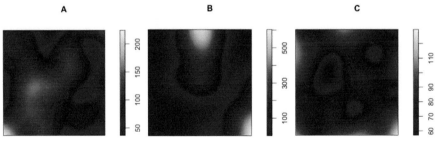

Figure 7.2. Kernel density for patterns A, B, and C.

In Figure 7.2, colors close to the red indicate higher concentration of point, while blue zones suggest lower number of realizations in the surroundings.

From Figure 7.2(B), it is quite clear that the second pattern is very clustered around two zones.

By especially focusing on the B pattern, we can assign the kernel density to a new object to have a 3D representation. This sort of graphics gives us a clearer representation on the distribution of point across study area. The function persp is run, while specifying the level of theta and phi (i.e., the azimuthal and the colatitude for the plot).

```
>den_B <- density(B) #kernel density from the pattern B
>persp(den_B, theta = 40, phi = 10, main="B kernel density")
#3D plot for the pattern B kernel density
```

B kernel density

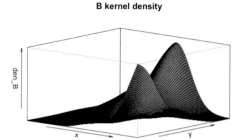

Figure 7.3. Perspective plot of density for a clustered pattern B.

An explorative analysis used to verify the assumption of spatial randomness is based on the quadrat count. For the three processes, the identification of the number of quadrants is quite easy as the window involved is a 1 × 1 square. Maps of quadrat can be easily derived using the spatstat package. First, quadrat counts for each pattern are computed using the quadratcount function.

```
>Qc_A<-quadratcount(A, nx=4, ny=4) #quadrat count for the A case
>Qc_A
```

	x			
y	[0,0.25)	[0.25,0.5)	[0.5,0.75)	[0.75,1]
[0.75,1]	5	2	7	4
[0.5,0.75)	6	7	10	4
[0.25,0.5)	5	10	4	4
[0,0.25)	9	9	5	3

```
>Qc_B<-quadratcount(B, nx=4, ny=4) #quadrat count for the B case
>Qc_B
```

	x			
y	[0,0.25)	[0.25,0.5)	[0.5,0.75)	[0.75,1]
[0.75,1]	0	16	24	0
[0.5,0.75)	1	9	14	2
[0.25,0.5)	1	5	4	7
[0,0.25)	0	0	3	25

```
>Qc_C<-quadratcount(C, nx=4, ny=4) #quadrat count for the C case
> Qc_C
```

	x			
y	[0,0.25)	[0.25,0.5)	[0.5,0.75)	[0.75,1]
[0.75,1]	5	7	5	4
[0.5,0.75)	6	2	6	5
[0.25,0.5)	4	5	4	4
[0,0.25)	6	5	6	6

where nx and ny define the numbers of rectangular quadrats in the *x* and *y* directions to divide the area in 16 portions. Then, we plot each process by also including the grids for the three quadrat counts (add=TRUE).

```
>plot(A,xlab='x',ylab='y')
>plot(Qc_A, add=TRUE, cex = 2) #add quadrats on pattern A plot
```

```
>plot(B,xlab='x',ylab='y')
>plot(Qc_B, add=TRUE, cex = 2) #add quadrats on pattern B plot
>plot(C,xlab='x',ylab='y')
>plot(Qc_C, add=TRUE, cex = 2) #add quadrats on pattern C plot
```

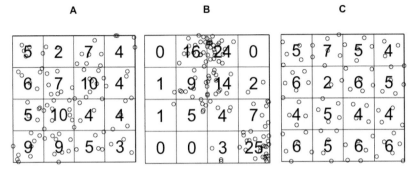

Figure 7.4. Quadrat count plots for patterns A, B, and C.

To test the hypothesis of spatial randomness, we can use the following codes:

```
>quadrat.test(A, 4, 4) #quadrat test for pattern A
    Chi-squared test of CSR using quadrat counts
data: A
X2 = 16.298, df = 15, p-value = 0.7251
alternative hypothesis: two.sided

Quadrats: 4 by 4 grid of tiles
>quadrat.test(B, 4, 4,alternative="clustered")  #quadrat
test for pattern B
    Chi-squared test of CSR using quadrat counts
data: B
X2 = 154.08, df = 15, p-value < 2.2e-16
alternative hypothesis: clustered

Quadrats: 4 by 4 grid of tiles
>quadrat.test(C, 4, 4,alternative="two.sided")  #quadrat
test for pattern C
    Chi-squared test of CSR using quadrat counts
data: C
X2 = 4.4, df = 15, p-value = 0.007794
alternative hypothesis: two.sided

Quadrats: 4 by 4 grid of tiles
```

As expected, the test is not significant for pattern A (generated from the homogeneous Poisson distribution), while it is very significant for pattern B (generated from the Thomas' process), and C (inhibition pattern). As it can be noted, the B pattern is tested against the hypothesis of clustered (alternative= "clustered"), while the C point pattern is tested against the two.sided option.

Moving a step ahead, we may apply some distance-based functions to the point patterns under investigation. The `spatstat` package allows us to implement different functions within the R environment. We start from the F function, which considers the distance from points to the nearest event.

To obtain interpretable results, we have to build upper and lower envelopes. Those envelopes are characterized by the intensity of the `ppp` object, built under a CSR assumption. As in the F function, different distances are considered, a set of different values for r has to be defined. The function `seq` is used, including starting point (`from`), maximum (`to`), and the extent of the interval between the elements of the vector (`by`).

```
>r<-seq(from=0,to=sqrt(2)/6,by=0.00001) #defines distances
at which the different F,G, and K functions will be computed
```

While building envelopes, practitioners have to specify the selected `ppp` object, the kind of function to compute (`fun=Fest` for the case of the F-function), and the `nrank` indicating the rank of the envelope value amongst the simulated values (in our case `nsim=99`).

By default, upper and lower critical envelopes are computed for each value of the distance r (i.e., pointwise), by ordering the `nsim` values, and considering the m-th lowest and m-th highest values, where m = `nrank`. The pointwise envelopes specify the critical points for a Monte Carlo test (Ripley 1981). The test is performed by choosing a *fixed* value of r, and rejecting the null hypothesis (i.e., CSR pattern) if the observed function value is place outside the envelope for this value of r. This test has a significance level $\alpha = 2 \cdot$ `nrank`/(1+`nsim`) (Baddeley et al. 2014b). The code is:

```
>F_A <- envelope(A, fun = Fest, r = r, nrank = 2, nsim = 99)
#simulate envelopes in A for the F function
>F_B <- envelope(B, fun = Fest, r = r, nrank = 2, nsim = 99)
#simulate envelopes in B for the F function
>F_C <- envelope(C, fun = Fest, r = r, nrank = 2, nsim = 99)
#simulate envelopes in C for the F function

>plot(F_A)
>plot(F_B)
>plot(F_C)
```

Figure 7.5 shows the empirical function $\hat{F}(r)$, the function $F_{CSR}(r)$, and the 96% pointwise envelopes. The plot clearly shows that the first pattern (black line in Figure 7.5(A)) respects the theoretical F-function for the CSR hypothesis (red line). Differently, Figure 7.5(B) indicates lower levels, due to the fact that this pattern is largely clustered. Moreover, the third case (Figure 7.5(C)) presents an empirical F-function above the grey bands that limit spatial randomness hypothesis.

Furthermore, the G function can be computed. The 96% pointwise envelopes are obtained by using the same `envelope` function but setting the parameter `fun` to `Gest`. Plots are obtained accordingly.

```
>G_A <- envelope(A, fun = Gest, r = r, nrank = 2, nsim = 99)
#simulate envelopes in A for the G function
>G_B <- envelope(B, fun = Gest, r = r, nrank = 2, nsim = 99)
#simulate envelopes in B for the G function
```

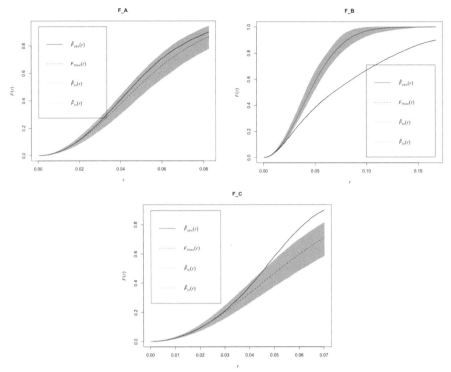

Figure 7.5. Plots of the *F*-functions for patterns A, B, and C.

```
>G_C <- envelope(C, fun = Gest, r = r, nrank = 2, nsim = 99)
#simulate envelopes in C for the G function
>plot(G_A)
>plot(G_B)
>plot(G_C)
```

In Figure 7.6(B) and 7.6(C), it can be noted that the two patterns differ from CSR configuration.

On the one side, Figure 7.6(B) indicates a level of the empirical values above the envelope intervals (i.e., a clustered pattern). On the other side, the third pattern (Figure 7.6(C)) shows an empirical *G* function lower than the grey band (i.e., an inhibition pattern). Those results were expected considering previous values for the *F* function.

The Clark–Evans test is another important evidence that can be retrieved by using `clarkevans.test` function. We perform the test for the B and C configurations. Attention has to be paid to the definition of the alternative hypotheses. For B, we can specify the `alternative = "clustered"`.

```
>ce_B<-clarkevans.test(B, alternative="clustered") #Clark-
Evans test for the process B
>ce_B
    Clark-Evans test
```

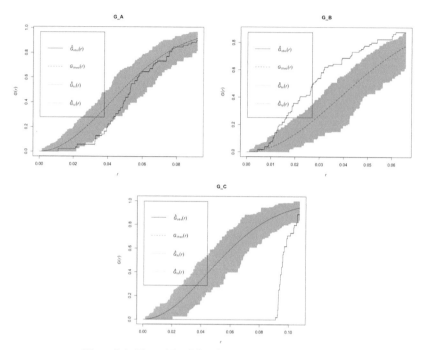

Figure 7.6. Plots of the *G*-functions for patterns A, B, and C.

```
   No edge correction
   Z-test
data: B
R = 0.73267, p-value = 3.56e-08
alternative hypothesis: clustered (R < 1)
```

The test is highly significant against the null hypothesis of spatial randomness. The same test can be applied to the third process C. This time, the alternative is set as "regular", to verify the significance of a regular pattern.

```
> ce_C<-clarkevans.test(C, alternative = "regular")
> ce_C
   Clark-Evans test
   No edge correction
   Z-test
data: C
R = 1.7682, p-value < 2.2e-16
alternative hypothesis: regular (R > 1)
```

Hence, we accept the alternative and we reject the null hypothesis of complete randomness. For both tests, Kaplan-Meier type edge correction may be applied by turning `correction=` "cdf". However, as data are generated on purpose within the selected window, corrections are not applied throughout the chapter.

Lastly, Ripley's *K*-function is calculated to summarize different patterns. The estimated values of *K* from the empirical distribution are compared to the theoretical values for different values of *r*. Higher values are typical of a clustered process. Conversely, values lower than the theoretical envelope indicate a regular pattern. As for other functions, envelopes are derived by the envelope function and setting fun="Kest".

```
>K_A <- envelope(A, fun = Kest, r = r, nrank = 2, nsim = 99)
#simulate theoretical envelopes in A for the K function
>K_B <- envelope(B, fun = Kest, r = r, nrank = 2, nsim = 99)
#simulate theoretical envelopes in B for the K function
>K_C <- envelope(C, fun = Kest, r = r, nrank = 2, nsim = 99)
#simulate theoretical envelopes in C for the K function
```

A comparison between the 96% pointwise envelopes (grey band) and the observed functions (black line) is plotted for the three cases under study.

```
>plot(K_A)
>plot(K_B)
>plot(K_C)
```

Figure 7.7(B) shows how the pattern has higher values of theoretical function: this situation corresponds to the clustered case. Finally, in Figure 7.7(C) a regular pattern can be recognized. In fact, the observed function lies under the grey band, suggesting competition in the point locations.

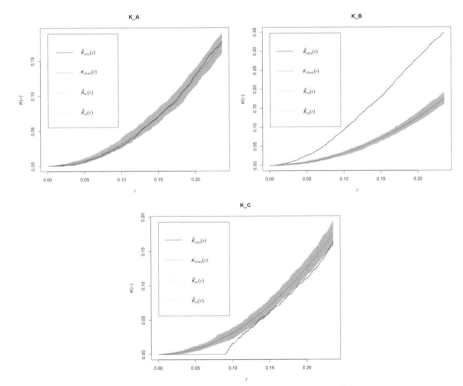

Figure 7.7. Plots of *K*-functions for patterns A, B, and C.

7.5 Concluding remarks

In this chapter, we provided some examples in the field of point pattern analysis by using the R package `spatstat`. To make its appealing more evident, we applied a wide range of different techniques to analyse spatial patterns and assess the deviation from the complete spatial randomness. Particularly, three different patterns corresponding to potential real case situations are examined. We observed how the analysis of point pattern can be performed by adopting a set of graphical and parametrical techniques, as for quadrat counts and the Clark-Evans test. Additionally, the estimation of different functions has been carried out applying inter-event distances, neighbour to neighbour distance, and point-to-nearest-event distance in order to recover information from event localization.

The methodologies presented in this chapter have to be greatly addressed to agricultural applications; considering, for example, a set of farms as a realisation of a spatial point process. Using these tools we can derive useful information on spatial distribution of agricultural activities that would be hidden if analysing the phenomena with standard instead of spatial methods.

References

Baddeley, A. and R. Turner. 2005. spatstat: An R package for analyzing spatial point patterns. *Journal of Statistical Software* 12: 6.

Baddeley, A., P.J. Diggle, A. Hardegen, T. Lawrence, R.K. Milne and G. Nair. 2014a. On tests of spatial pattern based on simulation envelopes. *Ecological Monographs* 84: 477–489.

Baddeley, A., E. Rubak and R. Turner. 2014b. *Spatial Point Patterns: Methodology and Applications with R.* Chapman & Hall/CRC, London.

Bamière, L., P. Havlík, F. Jacquet, M. Lherm, G. Millet and V. Bretagnolle. 2011. Farming system modelling for agri-environmental policy design: The case of a spatially non-aggregated allocation of conservation measures. *Ecological Economics* 70: 891–899.

Bivand, R.S., E. Pebesma and V. Gómez-Rubio. 2013. *Applied Spatial Data Analysis with R.* Springer-Verlag, New York.

Chaiban, C., C. Biscio, W. Thanapongtharm, M. Tildesley, X. Xiao, T.P. Robinson, S.O. Vanwambeke and M. Gilbert. 2019. Point pattern simulation modelling of extensive and intensive chicken farming in Thailand: Accounting for clustering and landscape characteristics. *Agricultural Systems* 173: 335–344.

Clark, P.J. and F.C. Evans. 1954. Distance to nearest neighbor as a measure of spatial relationships in populations. *Ecology* 35: 445–453.

Cressie, N.A.C. 1993. *Statistics for Spatial Data.* John Wiley and Sons, New York, second edition.

Diggle, P.J. 2003. *Statistical Analysis of Spatial Point Patterns.* Hodder Arnold, London, second edition.

Fisher, R.A., H.G. Thornton and W.A. Mackenzie. 1922. The accuracy of the plating method of estimating the density of bacterial populations: with particular reference to the use of Thornton's agar medium with soil samples. *Annals of Applied Biology* 9: 325–359.

Gatrell, A.C., T.C. Bailey, P.J. Diggle and B.S. Rowlingson. 1996. Spatial point pattern analysis and its application in geographical epidemiology. *Transactions of the Institute of British Geographers* 21: 256–274.

Greig Smith, P. 1964. *Quantitative Plant Ecology.* Butterworth, Washington, D.C.

Kelly, B.A., J. Bouvet and N. Picard. 2004. Size class distribution and spatial pattern of *Vitellaria paradoxa* in relation to farmers' practices in Mali. *Agroforestry Systems* 60: 3–11.

Li, S., F. Bonneu, J. Chadoeuf, D. Picart, A. Gégout-Petit and L. Guerin-Dubrana. 2017. Spatial and temporal pattern analyses of Esca grapevine disease in vineyards in France. *Phytopathology* 107: 59–69.

Long, J.A., R.L. Lawrence, P.R. Miller, L.A. Marshall and M.C. Greenwood. 2014. Adoption of cropping sequences in northeast Montana: A spatio-temporal analysis. *Agriculture, Ecosystems & Environment* 197: 77–87.

Murwira, A. and A.K. Skidmore. 2005. The response of elephants to the spatial heterogeneity of vegetation in a Southern African agricultural landscape. *Landscape Ecology* 20: 217–234.

Picard, N., A. Bar-Hen, F. Mortier and J. Chadœuf. 2009. The multi-scale marked area-interaction point process: a model for the spatial pattern of trees. *Scandinavian Journal of Statistics* 36: 23–41.

Ripley, B.D. 1977. Modelling spatial patterns. *Journal of the Royal Statistical Society Series B* 39: 172–212.

Ripley, B.D. 1981. *Spatial Statistics*. John Wiley and Sons.

Ripley, B.D. 1988. *Statistical Inference for Spatial Processes*. Cambridge University Press.

Schabenberger, O. and C.A. Gotway. 2005. *Statistical Methods for Spatial Data Analysis*. Chapman and Hall/CRC, Boca Raton.

Thomas, M. 1949. A generalization of Poisson's binomial limit for use in ecology. *Biometrika* 36: 18–25.

Vinatier, F., P. Tixier, P.F. Duyck and F. Lescourret. 2011. Factors and mechanisms explaining spatial heterogeneity: a review of methods for insect populations. *Methods in Ecology and Evolution* 2: 11–22.

CHAPTER 8
Spatial Analysis of Farm Data

Alfredo Cartone* and Domenica Panzera

8.1 Introduction

The agricultural sector includes a wide range of farming activities that are driven by distinctive climate and geographical conditions (Sun et al. 2016). In addition to the absolute location, the relative location of farming activities could matter. This means that the agricultural activities could be affected by location-specific characteristics as well as by certain features of agricultural activities at neighbouring sites. These aspects emphasize the importance of performing a spatial analysis of agricultural production to identify the degree of spatial interaction among farms and local instabilities.

Spatial econometric techniques for modelling spatial effects have been applied in some contributions in the field of agricultural economics. Spatial factors have been explicitly modelled, for example, in the study of crop yields in Anselin et al. (2004). Spatial as well as temporal effects have been introduced into analysing the determinants of land values and rent levels by Huang et al. (2006), so that spatial techniques have been applied to the analysis of crop insurance program in Woodard et al. (2011). Spatial features are also of great interest for developing precision agriculture (Lamour et al. 2021).

The concepts of spatial dependence and spatial heterogeneity are pivotal for spatial analysis. Both these spatial effects are likely to affect outputs from georeferenced data (Anselin 1988). Spatial dependence refers to the extent of similarity (or dissimilarity) of observed data in space, and it is often referred to as spatial autocorrelation (Anselin 1988, 1990; Haining 2003). The analysis of spatial dependence is essential to the full comprehension of some interdependences, spillover effects, and to the understanding of misspecification effects in the presence of a model. The spatial heterogeneity effect refers to the instability over the space of relations and behaviours. This may be evidenced in spatial regimes for variables, functional forms or model coefficients (Anselin 1988, 1990; Haining 2003).

"G. d'Annunzio" University of Chieti-Pescara, Department of Economic Studies.
Email: domenica.panzera@unich.it
* Corresponding author: alfredo.cartone@unich.it

Considering spatial heterogeneity is useful, again, to avoid model misspecification as well as to identify spatial clusters.

Model specification facilitates the introduction of spatial effects in regression models, which express the relation between variables of interest. The introduction of spatial dependence typically requires including in the model spatially lagged variables, that are weighted averages of the observations collected for neighbour units of a given location (Anselin 2010). The specification of spatial heterogeneity requires distinguishing between discrete heterogeneity and continuous heterogeneity (Anselin 2010). Discrete heterogeneity concerns a pre-specified set of spatial units or spatial regimes, among which the regression coefficients are allowed to vary. Continuous heterogeneity relates to how the regression coefficients may differ across the geographical space. Model parameters can be allowed to vary randomly (Swamy 1971) or according to a pre-determined functional form, as in the spatial expansion methods (Casetti 1997). Alternatively, spatial heterogeneity can be accommodated by computing local relationships that vary across space by considering how they behave nearby, thus considering spatial locations explicitly (Brunsdon et al. 1998).

This chapter is focused on detecting and modelling spatial effects in the analysis of the agricultural production at fixed locations, corresponding to farms.

The agricultural output can be represented as following a standard Cobb-Douglas production function with three factors as:

$$P = f(T, L, K) = AT^{\beta_1} L^{\beta_2} K^{\beta_3} \tag{8.1}$$

where P denotes the output, A is the total factor productivity, T denotes the land, L is the labour, and K denotes the capital. The parameters β_1, β_2, and β_3 express the elasticities of output with respect to land, labour and capital, respectively. The function (8.1) can be linearized using logarithms, such as:

$$\log(P) = \log(A) + \beta_1 \log(T) + \beta_2 \log(L) + \beta_3 \log(K) \tag{8.2}$$

The detection of spatial effects in the data is implemented by using exploratory spatial data analysis tools, and a geographical weighted regression approach that is employed to develop local models.

The chapter is organised as follows: In Section 8.2, the main techniques of exploratory spatial data analysis are presented, illustrating their application through the R Software. In Section 8.3, the geographically weighted regression approach and its application with the R Software is illustrated. In Section 8.4, some concluding remarks are outlined.

8.2 Exploratory spatial data analysis

Exploratory spatial data analysis (ESDA) can be defined as a collection of techniques to describe and visualise the spatial distribution of a certain phenomenon, discover patterns of spatial association, or suggest forms of spatial instability (Good 1983; Anselin and Getis 1992). Patterns of spatial dependence or spatial heterogeneity could be investigated using the ESDA tools.

To detect spatial effects, ESDA includes a set of indicators, among which the most popular is the Moran's I for spatial dependence (Moran 1950). The Moran's I statistic can be applied to variables observed at points or areal units and compares the value of the variable at any one location with the value at all other locations. Given the observations on a variable of interest **z**, collected at n sites, z_i, $i = 1,2,.., n$, the Moran's I is specified as:

$$I = \frac{n}{\Sigma_i \Sigma_j d_{ij}} \frac{\Sigma_i \Sigma_j d_{ij}(z_i - \overline{z})(z_j - \overline{z})}{\Sigma_i(z_i - \overline{z})^2} \qquad j \neq i \tag{8.3}$$

where d_{ij} is an element of the matrix **D**, expressing the proximity relations between locations i and j. The spatial weight matrix **D** is a square matrix of dimension n, where n is the number of spatial units. Typically, the diagonal elements of **D** are zero, while, for $i \neq j$, $d_{ij} = 0$, if locations i and j are not neighbours and $d_{ij} = 1$ if i and j are neighbours, according to a specified proximity criterion. Furthermore, most applications in spatial econometrics scale the individual rows (or columns) of **D** by the row totals, so that rows of **D** sum to 1. Different proximity criteria could be used in the definition of **D**, such as: geographical contiguity, physical distance or travel time distance. Social and economic distances, such as the race and ethnicity distance or the occupational distance, could be also considered, as well as any combination of other criteria (Conley and Topa 2002).

Under the null hypothesis of no spatial correlation, the expected value of the Moran's I is $E(I) = -\frac{1}{n-1}$. Values of I above the expected value indicate positive spatial autocorrelation, while values of I below $E(I)$ indicate negative spatial autocorrelation. Whether the value of I is statistically significant depends on its statistical distribution. Cliff and Ord (1981) derived the expression of the sample variance of the Moran's I, and showed that the index has an asymptotic normal distribution. Inference can be based also on the permutation approach, by empirically generating a reference distribution for I.

An approach towards visualising the spatial association between the value observed at one location and the observations at neighbouring locations is offered by the Moran scatterplot (Anselin 1996). The Moran scatterplot displays the value pairs for the observation at each location and its spatial lag (i.e., a weighted average of the observations at neighbour locations), with the x-axis being the original value of the variable under study, and the y-axis its spatial lag. The Moran scatterplot is based on the interpretation of the Moran's I as the slope of the linear relationship between the variables depicted on the axes, and it is useful for identifying clusters of units or possible spatial outliers.

Another indicator of global spatial association is represented by the Getis-Ord global G statistic (Getis and Ord 1992). The global G statistic calculates the spatial interaction of the value observed at one location with the observations at neighbouring locations, but unlike the Moran's I, it allows one to distinguish whether spatial autocorrelation is linked to the effect of higher or lower values of the variable under consideration. The Global G indicator is specified as:

$$G = \frac{\Sigma_i \Sigma_j d_{ij} z_i z_j}{\Sigma_i \Sigma_j z_i z_j} \qquad j \neq i \tag{8.4}$$

Under the null hypothesis of no spatial correlation, its expected value is $E(G) = \frac{S}{n(n-1)}$, with S denoting the sum of linkages in the matrix **D**. Values of

G above $E(G)$ indicates that the spatial autocorrelation is mainly due to higher values of the variable under consideration (i.e., hot spots), values of G above $E(G)$ indicate that the spatial autocorrelation is mainly due to lower value of the variable (i.e., cold spots). Any inference on G can be based upon its asymptotically Gaussian distribution (Getis and Ord 1992).

All the aforementioned indicators measure global spatial association. Moreover, identifying patterns of spatial heterogeneity requires spatial association to be analysed locally. Local indicators of spatial association (LISA), like the local Moran (Anselin 1995), are useful in identifying the presence of local spatial effects, and to spread more light on the presence of spatial heterogeneity. The local Moran for unit i is specified as:

$$I(i) = \frac{n(z_i - \overline{z})\Sigma_j d_{ij}(z_j - \overline{z})}{\Sigma_i(z_i - \overline{z})^2} \qquad j \neq i \tag{8.5}$$

and its expected value is given by $E(I(i)) = -\dfrac{\eta_i}{n-1}$ with η_i denoting the sum of local neighbours $\eta_i = \Sigma_i d_{ij}$. A positive value for $I(i)$ reveals groups of similar values (high or low), a negative value for $I(i)$ denotes clustering of dissimilar values. Local Moran's statistics can be used as indicators of local spatial clusters and to recognise outliers of the spatial distribution (Anselin 1995). The local clusters identified by the LISA can be pictured (i.e., the LISA cluster map) that highlights situations of local spatial associations between lower and higher values of the variable under investigation. Also, the Moran scatterplot can be used in the same fashion of the local Moran's statistics for the identification of spatial clusters.

Local patterns of spatial association can also emerge by using the local statistics G_i and G_i^*, introduced by Getis and Ord (1992). The statistic G_i consists of a ratio of the weighted average of the values observed in the locations that are neighbours of location i, to the sum of all values, not including the value at the location i, that is:

$$G_i = \frac{\Sigma_j d_{ij} z_j}{\Sigma_j z_j} \qquad j \neq i \tag{8.6}$$

Conversely, the G_i^* statistic includes the value y_i at both numerator and denominator, by considering diagonal elements of **D** different from zero and leading to the following expression:

$$G_i^* = \frac{\Sigma_j d_{ij} z_j}{\Sigma_j z_j} \tag{8.7}$$

The expected values for the statistics G_i and G_i^* are defined, respectively, as $E(G_i) = \dfrac{S_i}{n-1}$, with $S_i = \Sigma_j d_{ij}$ for $j \neq i$, and $E(G_i^*) = \dfrac{S_i^*}{n-1}$, with S_i^* denoting the sum of the weights d_{ij}, including the spatial unit i. Values of the statistics above the expected values suggest a high-high cluster, or hot spot, values below the expected values suggest a low-low cluster or cold spot.

8.2.1 *Example of the application of ESDA*

In this section, an application of ESDA applied to simulated farm survey data is presented, showing the results that can be obtained using the R software. In our example, farm survey data are generated for the log of the agricultural output $z_i = \ln(p_i)$ for 500 farms according to the following equation:

$$z_i = \gamma + \gamma_1 t_i + \gamma_2 l_i + \gamma_3 k_i + \delta_i \qquad\qquad i = 1,2,\dots, n \qquad\qquad (8.8)$$

where the intercept is specified as $\gamma = 5 \ln(\eta + v)$, and the parameters corresponding to three inputs are set to $\gamma_1 = 0.5 \ln(\eta + v)$, and $\gamma_2 = \gamma_3 = 0.2 \ln(\eta + v)$. The values η and v represent the coordinates selected from a uniform distribution $U(1.1; 1.3)$, so that the level of the parameters of each farm tend to change according to both latitude and longitude across the studied area. Values for the linear predictors, the log of the three agricultural inputs (i.e., t_i, l_i, k_i), are independently drawn from a uniform distribution $U(1.5; 4)$. Lastly, δ_i is an element of the vector of disturbs $\delta \sim N(0, 1)$.

The code for obtaining our data set is the following.

```
> set.seed(3)
> n<-500
> eps <- rnorm(n = n, mean = 0, sd = 1)
> X <- cbind(constant = 1,
           t = runif(n,1.5,4),      # Land Used
           l = runif(n,1.5,4),      # Labour
           k = runif(n,1.5,4))      # Capital
> coords <- cbind(long = runif(n,1.1,1.3), lat = runif
(n,1.1,1.3))
#coordinates for the farms

> beta_gwr <- cbind(5*log(coords[,1]+coords[,2]), 0.5*log
(coords[,1]+coords[,2]), 0.20*log(coords[,1]+coords[,2]),
0.2*log(coords[,2]+coords[,2]))
#the set of parameters for the different generated variables
changing according to the coordinates
> z<-matrix(0,nrow(coords),1)
> kkk<-0
> for (iii in 1:nrow(coords))
  {
    kkk<-kkk+1
    z[kkk]<-t(X[iii,])%*%beta_gwr[iii,]+0.2*eps[iii]
  }
#creates a vector of agricultural output according to the
parameters generated
```

All variables of the production function, included the geographical coordinates, can be stored, in the file "data.txt" and uploaded as follows:

```
> data<-as.data.frame(cbind(z,X))
> colnames(data)<-c("z","const","t","l","k")
#creates a dataset for the three covariates generated, the
Intercept, and the agricultural output
```

The variables included in the dataset `data` can be accessed by their names, by attaching the dataset as follows:

```
> attach(data)
```

As a first step, values of the dependent variable z_i can be plotted to obtain a map of the agricultural output of 500 farms. We can use the following R code to get this figure.

```
> library(ggplot2) #we use the library ggplot2 for a nice
graphic
>   sp_z<-ggplot(data,   aes(x=coords[,1],   y=coords[,2],
color=z)) + geom_point()
>   sp_z+scale_color_gradient(low="white",   high="black")
+ggtitle("")+xlab("x")+ylab("y")  #return  a  map  for  the
dependent variable
```

In Figure 8.1, darker points represent a higher level of production mainly situated in the high-right corner and in the higher part of the plot.

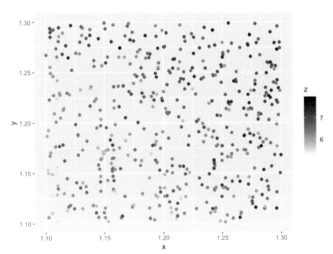

Figure 8.1. Geographical map agricultural output for 500 farms.

The Moran's *I* statistic is computed for the dependent variable of the production function z_i. The computation of this index requires the spatial weight matrix to be defined. Using the library `spdep` (Bivand et al. 2013) a binary proximity matrix based upon the *k* nearest neighbour criterion, with $k = 10$, can be constructed as:

```
> library(spdep)
> knn_b<-knearneigh(coords,k=10)
#The function returns a matrix with the indices of points
belonging to the set of the k nearest neighbours of each other
> nb_b<-knn2nb(knn_b)
#The function converts a knn object returned by knearneigh
into a neighbours list of class nb with a list of integer
vectors containing neighbour units number
```

```
> listw_b<-nb2listw(nb_b,style="B")
#This function supplements a neighbours list with spatial
weights with values given by the coding scheme style chosen.
Style = "B" returns binary weights.
```

The Moran's *I* is then computed with the following R code as:

```
> z<-as.vector(z)
> moran.test(z,listw=listw_b)
# This function prints a list containing among its components
the value of the standard deviate of Moran's I, the p-value of
the test, the value of the observed Moran's I, its expectation
and variance.
```

The output of the Moran's *I* is

```
Moran I test under randomisation
data: z
weights: listw_b
Moran I statistic standard deviate = 13.544, p-value <2.2e-16
alternative hypothesis: greater
sample estimates:
Moran    I statistic     Expectation        Variance
         0.2551317265    -0.0020040080    0.0003604622
```

A Moran scatterplot for the data of interest can be obtained using the following R code:

```
> moran.plot(z,listw=listw_b)
# A plot of spatial data against its spatially lagged values,
augmented by reporting the summary of influence measures for
the linear relationship between the data and the lag.
```

The R output is the following:

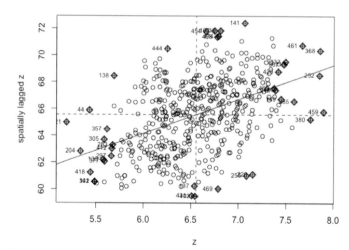

Figure 8.2. Moran scatterplot of the agricultural output of 500 farms. Implied Moran's *I* = 0.255.

The global *G* statistic can be calculated with R as

```
> globalG.test(z, listw=listw_b, alternative="greater")
# This function returns a list that contains among its
components, the statistic, the p-value, the value of the
observed statistic, its expectation and variance. Note that
weights should be binary
    Getis-Ord global G statistic
data: z
weights: listw_b
standard deviate = 2.092, p-value = 0.01822
alternative hypothesis: greater
sample estimates:
  Global G statistic      Expectation          Variance
        2.007578e-02     2.004008e-02      2.912652e-10
```

The level of Moran's *I* is equal to 0.255, higher than the expected value of the statistic. Furthermore, the value of the Global *G* can be computed to evidence the presence of cold spots and hot spots at a global level. Both statistics are computed adopting a $k = 10$ nearest neighbour binary contiguity matrix. While Moran's *I* is clearly higher than its expected value, the Global *G* displays a statistic slightly higher than the expected value. However, in the explorative analysis, the researcher should consider also local indicators of spatial association in order to improve knowledge of local spatial patterns.

In this application, local indicators are particularly adopted to verify and display local instabilities.

Using the R Software, the local Moran index can be computed for the variable of interest as:

```
> locmor<-localmoran(z,listw_b)
> locmor
# This function returns the local spatial statistic Moran's I
calculated for each unit based on the spatial weights object
used. First 6 rows of the output are shown. Ii is the local Moran
statistic, E.Ii is the expectation of local Moran statistic,
Var.Ii is the variance of local Moran statistic, Z.Ii is the
standard deviate of local Moran statistic, Pr() is the p-value
of local Moran statistic
            Ii        E.Ii    Var.Ii          Z.Ii      Pr(z > 0)
1 -3.191308942 -0.02004008 9.785974 -1.013750383 8.446491e-01
2 -0.660885160 -0.02004008 9.785974 -0.204857101 5.811581e-01
3 -6.332609063 -0.02004008 9.785974 -2.017920746 9.782002e-01
4  3.216978806 -0.02004008 9.785974  1.034768504 1.503885e-01
5 -0.071735277 -0.02004008 9.785974 -0.016525255 5.065923e-01
6  8.564390900 -0.02004008 9.785974  2.744160327 3.033295e-03
...
```

In Figure 8.3, values of the local Moran's *I* are reported for the agricultural output of 500 farms. This graph is obtained with the following R code.

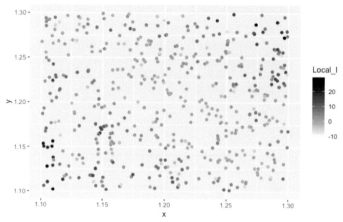

Figure 8.3. Values of local Moran's *I* for 500 farms calculated adopting a *k* = 10 nearest neighbour proximity matrix.

```
> Local_I<-locmor[,1]
#values for the local Moran are extrapolated as a vector in the
first column of the previous output
> sp_lm<-ggplot(as.data.frame(Local_I), aes(x=coords[,1],
y=coords[,2], color=Local_I)) + geom_point()
> sp_lm+scale_color_gradient(low="white", high="black") +
ggtitle("") +xlab("x")+ylab("y")
#returns a plot displaying local Moran for each unit
```

As shown by Figure 8.3, plotting the values of local Moran's *I* for the 500 farms clearly indicates the presence of positive spatial autocorrelation in the opposite corners of the map, where both high values and low levels of agricultural output are present. Particularly, the level of the local indicator of spatial association is positive when each unit has neighbours characterized by similar level of the variable, a circumstance that is indicated in the low left corner and in the high right corner of the map and that suggests potential instabilities.

In order to analyse the source of instabilities more thoroughly, we also perform an analysis based on local *G* statistic. This statistic gives an insight into the eventual sources of instabilities, pointing potential local cluster between low-values or high values.

For the sake of simplicity, we calculate only the statistics G_i with R as:

```
> lg<-localG(z,listw_b)
> lg
# The function returns the value of local G calculated for each
spatial unit based on the spatial weights object used. First
18 statistics in the overall output are shown
[1]   0.669155849   1.034865295   1.703170779   -2.198935310
0.342406993 -2.676776569 0.330761115
[8]   -0.078254576   1.805237793   1.203713509   1.684790916
-3.502730692 -2.094192143 1.163197645
```

```
[15] -0.834736349 -1.607459221 -0.208284316 -0.691131462
4.260862214 2.404553769 -0.410188958 ...
```

The code for obtaining Figure 8.4 is the following.

```
> z_values<-lg[1:500]
> sp_z_glG<-ggplot(as.data.frame(data), aes(x=coords[,1],
y=coords[,2], color=z_values))+geom_point()
> sp_z_glG+scale_color_gradient(low="white", high="black")
+ggtitle("")+xlab("x")+ylab("y")
```

Figure 8.4 shows *z*-values of the local *G* statistic. We spot the presence of a local clusters in the high part of the map, characterised by aggregation of a higher level of agricultural output. On the other side, white points in the low left corner of the map are an evident signal of a spatial cluster due to a cold spot. Therefore, the use of local indicators of spatial association are able to return to the analyst a picture of potential heterogeneity underlying the variables under study and, at the same time, offers to the researcher an insight into the potential causes of spatial instabilities in the agricultural output.

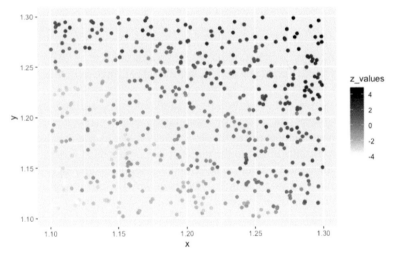

Figure 8.4. Map of the z-values obtained from the local *G* spatial statistic of 500 farms calculated adopting a *k* = 10 nearest neighbour binary contiguity matrix.

8.3 Local models for spatial data: the geographically weighted regression

The use of geographically distributed data imposes great attention in the definition of statistical models. As highlighted in the first section, this is mainly due to the presence of spatial heterogeneity and local instabilities. Agricultural, environmental, as well as economic and ecological phenomena may particularly display a large amount of spatial heterogeneity (Panzera and Postiglione 2014). This issue is strictly connected to the presence of structural instabilities and to the problem of heteroscedasticity, and it is deeply linked to some relevant feature detectable in an explorative analysis

(Anselin 1988). For example, the presence of hot-spots and cold-spots in the explorative analysis of a model residuals may be considered as a sign of instability.

In some cases, the use of spatially augmented models considering spatial dependence may not be, by itself, a suitable choice to consider the presence of spatial effects. As noted by Bivand and Brunstad (2003), a poor increase in the representativeness obtained using spatial models considering spatial dependence offers a reason to explore the presence of spatial instabilities. Considering global coefficients in regression models may present some potential withdrawals in terms of model representativeness. Spatial heterogeneity may cause model misspecification, potentially turning into biased and inefficient estimates.

Geographically weighted regression (GWR) is a technique for taking into account potential variability in regression models coefficients across different geographical units represented by points (Brunsdon et al. 1998; Fotheringham et al. 2002). GWR is a common tool, adopted in a regression frame, to consider significant geographical differences that may depend on latitude, altitude, and other geographical factors.

In the field of estimating production function for agricultural farm data, we can consider log-linearization of a standard production function for the three classical factors—land, labour, and capital. The log-linearization may take the form of a standard linear regression, such as:

$$z_i = \alpha + \beta_1 t_i + \beta_2 l_i + \beta_3 k_i + \varepsilon_i \qquad \varepsilon_i \sim N(0, \sigma^2) \qquad i = 1,2,\dots, n \qquad (8.9)$$

where n is the number of spatial units, and, for each unit i, $z_i = \ln(p_i)$ agricultural outputs, α is the model intercept, t_i, l_i, k_i are the log of three agricultural inputs as aforementioned, and $\beta_1, \beta_2, \beta_3$ are the other parameters to be estimated.

The use of a standard linear regression model is based on the hypothesis that spatial variability does not lead to non-stationarity in the parameter estimates. As mentioned, in many applications the analyst may be interested in assessing the variability of coefficients linked to the presence of geographical factors. In fact, the relationship between independent variables and agricultural output may differ across different farms and a relevant feature may be represented by the presence of potential heterogeneity between different units. Technological differences, specialization, and natural conditions as well as ecological and environmental factors are many examples that may justify this sort of differences (Cho et al. 2007).

In the analysis of spatial data, GWR has often been used for modelling in the presence of spatial heterogeneity, allowing for smooth estimation of the regression parameters. GWR is a locally linear semi-parametric estimation method, which is aimed at capturing, for each observation, the spatial variations of the regression coefficients. For this purpose, a different set of parameters is estimated for each unit considering the characteristics of neighbours. Modelling non-stationarity at unit level adopting GWR leads us to the definition of a local model of the log-linear production function that may be expressed as:

$$z_i = \alpha_i + \beta_{1i} t_i + \beta_{2i} l_i + \beta_{3i} k_i + \varepsilon_i \qquad (8.10)$$

where α_i is the local intercept, t_i, l_i, k_i are the unit observations for the covariates, β_{pi} represent local parameters of the production function, $p = 1,\dots,3$, and ε_i is the error component. The estimation of the model in (8.10) is pursued adopting a geographical version of weighted least squares estimator.

In matrix form, we have:

$$\hat{\theta}_i = (\mathbf{X}'\,\mathbf{W}_i\,\mathbf{X})^{-1}\,\mathbf{X}'\,\mathbf{W}_i\mathbf{y} \tag{8.11}$$

where $\mathbf{X}$ is the matrix containing the log of the agricultural inputs t, l, k and a first column of ones, $\mathbf{W}_i$ is the diagonal matrix of weights, with w_{ij} that express the level of proximity of other units from i, z is the log of agricultural output, and $\hat{\theta}_i$ is the estimate, for each unit i, of the parameters vector $\theta' = (\alpha_i\ \beta_{1i}\ \beta_{2i}\ \beta_{3i})$.

An important aspect in the context of GWR is the setting of the geographical weights contained in the diagonal matrix $\mathbf{W}_i$. The weights w_{ij}s (with $w_{ij} \in \mathbf{W}_i$) may be expressed according to a desirable kernel function of the geographical distance b_{ij} between locations i and j (McMillen 1996; McMillen and McDonald 1997). One of the most commonly used weighting functions is the Gaussian kernel that, for a given location i, is defined as:

$$w_{ij} = \exp\left(-\frac{1}{2}\frac{b_{ij}^2}{h^2}\right) \qquad j = 1,\ldots,n \tag{8.12}$$

where h is the bandwidth that represents a measure of the distance-decay in the weighting function and indicates the extent to which the local calibration results are smoothed (Fotheringham et al. 2002). The distances b_{ij} are generally Euclidean distances. As the bandwidth gets larger, the weights approach unity and the local GWR model approaches the global OLS model.

Another option could be the truncated kernel which sets weight to zero outside a certain range b_{ij}, and is obtained, for instance, according to a bi-square kernel as:

$$w_{ij} = \begin{cases} \left[1 - \left(\dfrac{b_{ij}}{h}\right)^2\right]^2, & \text{if } b_{ij} < h \\[2ex] 0, & \text{if } b_{ij} > h \end{cases} \tag{8.13}$$

In addition to Gaussian and bi-square, many other kernel functions may be considered depending on the nature of the application (e.g., tri-cube and exponential).

Moreover, kernel may be set as fixed or adaptive, where in adaptive kernel bandwidth is referred to a certain number of neighbours. Adaptive kernel is often preferred to a fixed one as it guarantees necessary degrees of freedom for estimation of local parameters in the presence of non-regular spatial configurations.

Bandwidth is a relevant parameter for the estimation of GWR and its selection should be carefully carried out. Several criteria may be adopted (Fotheringham et al. 2002). Many studies identify the level of bandwidth by using the Akaike Information Criterion (AIC, Harris et al. 2010). Instead of the AIC criterion, the researcher can compute cross validation (CV) scores to select bandwidth. However, the use of CV has been considered as sometimes returning too small bandwidth leading to very extreme coefficients (Farber and Paez 2007).

In our example, we will refer to a GWR estimation adopting an adaptive bi-square kernel on which bandwidth is selected according to a corrected AIC (AICc) criterion (Hurvich et al. 1998).

8.3.1 *Example of the application of GWR*

As observed in the explorative analysis, the use of LISA indicators suggests the presence of spatial instabilities in the agricultural output. In this example, we start from the estimation of the log-linearized version of a production function adopting basic linear regression specification. Therefore, this strategy does not consider the presence of spatial effects and the underlying spatial non-stationarity.

Using the R software, the model parameters can be estimated through OLS as:

```
> ols<-lm(z~t+l+k)
# This function is used to fit linear model, and returns a list
containing among its components the coefficient estimates
> summary(ols)
# This function returns several information about the estimated
model including estimated coefficients, parameters inference,
and fit.
Call:
lm(formula = z ~ t + l + k)
Residuals:
     Min            1Q      Median          3Q         Max
-0.91191     -0.20084    -0.00606     0.22809     0.88771
Coefficients:
               Estimate   Std. Error    t value    Pr(>|t|)
(Intercept)    4.41622      0.10265     43.023     < 2e-16    ***
t              0.42850      0.02124     20.170     < 2e-16    ***
l              0.19189      0.02022      9.489     < 2e-16    ***
k              0.16020      0.02081      7.698     7.56e-14   ***
---
Signif. codes: 0 '***' 0.001 '**' 0.01 '*' 0.05 '.' 0.1 ' ' 1
Residual standard error: 0.326 on 496 degrees of freedom
Multiple R-squared: 0.5247,      Adjusted R-squared: 0.5218
F-statistic: 182.5 on 3 and 496 DF, p-value: < 2.2e-16
```

Results from the OLS estimation returns significative estimates for all the three agricultural inputs and for the intercept. The F-statistic confirms the significance of the complete model for three inputs. However, the global production function presents a level of representativeness around 50%. Hence, the analyst should consider spatial heterogeneity that may lead to a poor representativeness of the global model.

The residuals of OLS estimation can be plotted using the following R code:

```
> Residuals<-ols$residuals
> sp_res_ols<-ggplot(as.data.frame(data), aes(x=coords[,1],
y=coords[,2], color=Residuals)) + geom_point()
>    sp_res_ols+scale_color_gradient(low="white",    high=
"black") +ggtitle("")+xlab("x")+ylab("y") #returns plot of
residuals from OLS
```

Figure 8.5 shows the clear presence of spatial patterns in the structure of the residuals, where the higher and lower levels of residuals tend to be clustered. Hence, the spatial structure of the residuals is far from being random. Additionally, the researcher may repeat the explorative analysis carried on in the previous section on the residuals and verify the significance of the LISA indicators. In this case, the linear estimation highlights some of the difficulties of a-spatial models when values of the parameters are strictly linked to geographical factors.

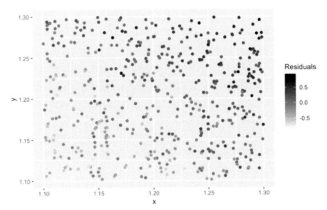

Figure 8.5. Plot of residuals from OLS estimation of a production function for 500 farms.

In the presence of spatial non-linearities, the use of GWR allows for the consideration of spatial features, due to its structure. GWR returns local estimates for the log-linearized production function.

The library GWmodel (Gollini et al. 2015) is used for the implementation of GWR model in R. In the preparation of the GWR estimation, both the coordinates and the data will be conveyed in a format SpatialPointsDataFrame-Class by the special function available in the library sp.

```
> library(sp)
> sp.data<-SpatialPointsDataFrame (coords=coords, data = data)
# This function conveys the dataset and the vector coordinates "coords" into a SpatialPointsDataFrame object.
```

The bandwidth selection will be carried on automatically by relying on the command bw.gwr, selecting as parameters of the R function, the AIC criterion, and the specific adaptive kernel. Defining optimal bandwidth allows us to compute weights matrix to perform the estimation of local models by GWR.

```
> library (GWmodel)
> bwl<-bw.gwr(z~t+l+k, data=sp.data, approach="AIC", adaptive=TRUE,kernel="bisquare", p=2)
#This function computes optimal bandwidth for the production function at farm level using an adaptive bi-square kernel. The power of the Minkowski distance is specified at the default level: 2 (Euclidean distance). Last six rows of the output are shown
```

...
```
Adaptive bandwidth (number of nearest neighbours): 125 AICc
value: -98.60629
Adaptive bandwidth (number of nearest neighbours): 121 AICc
value: -98.27793
Adaptive bandwidth (number of nearest neighbours): 127 AICc
value: -98.71336
Adaptive bandwidth (number of nearest neighbours): 129 AICc
value: -98.27467
Adaptive bandwidth (number of nearest neighbours): 126 AICc
value: -98.59563
Adaptive bandwidth (number of nearest neighbours): 127 AICc
value: -98.71336
```

The object `bwl` includes the level of the optimal bandwidth. At this stage, once the level of the bandwidth is computed, we can proceed with model estimation by adopting the function `gwr.basic`. This function allows one to estimate a local model for the considered production function by geographically weighted regression adopting an adaptive bi-square kernel as:

```
> gwr<- gwr.basic(z~t+l+k, data=sp.data, bw=bwl,adaptive=
TRUE, kernel="bisquare", p=2)
#The function returns an object of class gwr containing
several values, including diagnostic and local coefficients.
Print of the created gwr object displays both ols and a summary
of GWR estimates.
> gwr
***********************************************************
*                    Package GWmodel                      *
***********************************************************
Call:
gwr.basic(formula = z ~ t + l + k, data = sp.data, bw = bwl,
kernel = "bisquare", adaptive = TRUE, p = 2)
Dependent (y) variable: z
Independent variables: t l k
Number of data points: 500
  ...
***********************************************************
*       Results of Geographically Weighted Regression     *
***********************************************************
*****************Model calibration information*****************
Kernel function: bisquare
Adaptive bandwidth: 127 (number of nearest neighbours)
Regression points: the same locations as observations are
used.
Distance metric: Euclidean distance metric is used.
*************Summary of GWR coefficient estimates:*************
```

	Min.	1st Qu.	Median	3rd Qu.	Max.
Intercept	3.93034	4.28101	4.38314	4.61638	4.8111
t	0.35372	0.41386	0.43711	0.45211	0.4981
l	0.13227	0.16580	0.17748	0.19705	0.2775
k	0.10526	0.13428	0.16162	0.18458	0.2432

```
*******************Diagnostic information********************
Number of data points: 500
Effective number of parameters (2trace(S) - trace(S'S)): 56.07803
Effective degrees of freedom (n-2trace(S) + trace(S'S)): 443.922
AICc (GWR book, Fotheringham, et al. 2002, p. 61, eq 2.33):
-98.71336
AIC (GWR book, Fotheringham, et al. 2002, GWR p. 96, eq. 4.22):
-150.2809
Residual sum of squares: 19.94972
R-square value: 0.8201522
Adjusted R-square value: 0.7973818
****************************************************************
```

Output indicates the extent of spatial differences for the coefficients related to all the three agricultural inputs and for the intercept. Furthermore, in the table we observe the resulting increase in the model fit, due to the ability of the GWR to model spatial heterogeneity. Another way to summarize spatial differences in local estimates is by mapping coefficients.

The comparison offered, by printing the created gwr object, between measures of representativeness of linear regression and GWR, is especially interesting. In fact, the object displays the level of adjusted R-squared, AIC, and AICc for the GWR model, useful to assess the improvement in the representativeness due to the local specification (e.g., the R-squared for the OLS model is around 52%).

In Figure 8.6, the spatial patterns of the estimated coefficients are reported, showing the differences across the area for the parameters of the production function. The code for obtaining the plots included in Figure 8.6 is:

```
> est_pars<-gwr$SDF
>Intercept<-est_pars$Intercept #extrapolate parameters for
the Intercept
>sp_int<-ggplot(as.data.frame(sp.data), aes(x=coords[,1],
y=coords[,2], color=Intercept)) + geom_point()
>sp_int+scale_color_gradient(low="white",
high="black")+ggtitle("")+xlab("x")+ylab("y")  #map   the
parameters for the Intercept
>t<-est_pars$t #extrapolate parameters for the Land
>sp_t<-ggplot(as.data.frame(est_pars), aes(x=coords[,1],
y=coords[,2], color=t)) + geom_point()
>sp_t+scale_color_gradient(low="white",
high="black")+ggtitle("")+xlab("x")+ylab("y")  #map   the
parameters for the Land
>l<-est_pars$l #extrapolate parameters for the Labour
```

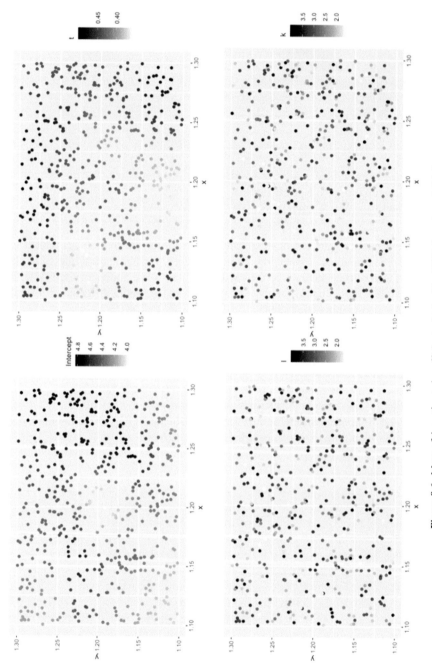

Figure 8.6. Maps of the estimated coefficients obtained by GWR from data of 500 farms.

```
>sp_l<-ggplot(as.data.frame(sp.data),    aes(x=coords[,1],
y=coords[,2], color=l)) +geom_point()
>sp_l+scale_color_gradient(low="white",
high="black")+ggtitle("")+xlab("x")+ylab("y")   #map  the
parameters for the Labour
>k<-est_pars$k #extrapolate parameters for the Capital
>sp_k<-ggplot(as.data.frame(sp.data),    aes(x=coords[,1],
y=coords[,2], color=k)) + geom_point()
>sp_k+scale_color_gradient(low="white",
high="black")+ggtitle("")+xlab("x")+ylab("y")   #map  the
parameters for the Capital
```

As expected, the coefficients corresponding to the intercept and to the land used (*t*) vary greatly with the coordinates. On the other hand, the level of the two other agricultural inputs indicate a less evident trend. The obtained GWR estimates highlight the ability of this technique to recover spatial patterns when the level of the parameters is linked to geographical factors and in case of spatial non-stationarity.

From Figure 8.7, we observe how the residuals from GWR estimation can be mapped. The spatial distribution appears different from the OLS residuals, and does not present any visible cluster due to the fact that the spatial structure of the data is captured by the local models estimated through GWR.

```
> Residuals_GWR<-gwr$SDF$residual #obtain residuals from
GWR estimation
> sp_resGWR<-ggplot(as.data.frame(sp.data), aes(x=coords
[,1], y=coords[,2], color=Residuals_GWR)) + geom_point()
> sp_resGWR+scale_color_gradient(low="white", high="black")
+ggtitle("")+xlab("x")+ylab("y") #map of the GWR residuals
```

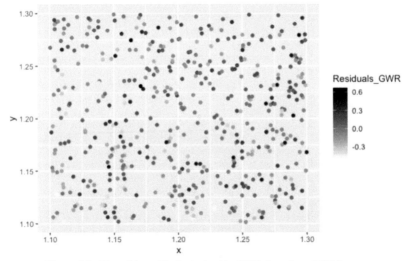

Figure 8.7. Map of the residuals obtained by GWR from data of 500 farms.

8.4 Concluding remarks

In this chapter, we provided some examples in the field of explorative spatial data analysis with R. Hence, we applied this set of techniques to the problem of spatial dependence and spatial instabilities in agricultural output. The presence of both spatial effects has been highlighted as key features in the analysis. Additionally, the estimation of the production function has been carried out applying local models and estimating GWR parameters to study how the influence of agricultural inputs may depend on space. This circumstance has been studied in the presence of local differences that affect agricultural production.

References

Anselin, L. 1988. *Spatial Econometrics: Methods and Models*. Dordrecht: Kluwer Academic Publishers.

Anselin, L. 1990. Spatial dependence and spatial structural instability in applied regression analysis. *Journal of Regional Science* 30: 185–207.

Anselin, L. and A. Getis. 1992. Spatial statistical analysis and geographic information systems. *The Annals of Regional Science* 26: 19–33.

Anselin, L. 1995. Local indicators of spatial association-LISA. *Geographical Analysis* 27: 93–115.

Anselin, L. 1996. The Moran scatterplot as an ESDA tool to assess local instability in spatial association. pp. 111–126. *In*: Fischer, M., H. Scholten and D. Unwin (eds.). *Spatial Analytical Perspectives on GIS*. London: Taylor and Francis.

Anselin, L., R. Bongiovanni and J. Lowenberg-DeBoer. 2004. Spatial econometric approach to the economics of site-specific nitrogen management in corn production. *American Journal of Agricultural Economics* 86: 675–687.

Anselin, L. 2010. Thirty years of spatial econometrics. *Papers in Regional Science* 89: 3–25.

Bivand, R.S. and R.J. Brunstad. 2003. Regional growth in Western Europe: an empirical exploration of interactions with agriculture and agricultural policy. pp. 351–373. *In*: Fingleton, B. (ed.). *European Regional Growth*. Berlin, Heidelberg: Springer.

Bivand, R.S., E. Pebesma and V. Gómez-Rubio. 2013. *Applied Spatial Data Analysis with R. Second Edition*. New York: Springer.

Brunsdon, C., S. Fotheringham and M. Charlton. 1998. Geographically weighted regression. *Journal of the Royal Statistical Society: Series D (The Statistician)* 47: 431–443.

Casetti, E. 1997. The expansion method, mathematical modeling and spatial econometrics. *International Regional Science Review* 20: 9–33.

Conley, T.G. and G. Topa. 2002. Socio-economic distance and spatial patterns in unemployment. *Journal of Applied Econometrics* 17: 303–327.

Cho, S.H., Z. Chen, S.T. Yen and B.C. English. 2007. Spatial variation of output-input elasticities: Evidence from Chinese county-level agricultural production data. *Papers in Regional Science* 86: 139–157.

Cliff, A.D. and J.K. Ord. 1981. *Spatial Processes: Models and Applications*. London: Pion Limited.

Farber, S. and A. Páez. 2007. A systematic investigation of cross-validation in GWR model estimation: empirical analysis and Monte Carlo simulations. *Journal of Geographical Systems* 9: 371–396.

Fotheringham, A.S., C. Brunsdon and M. Charlton. 2002. *Geographically Weighted Regression – The Analysis of Spatially Varying Relationships*. Chichester, UK: Wiley.

Getis, A. and J.K. Ord. 1992. The analysis of spatial association by use of distance statistics. *Geographical Analysis* 24: 189–206.

Gollini, I., B. Lu, M. Charlton, C. Brunsdon and P. Harris. 2015. GWmodel: an R package for exploring spatial heterogeneity using geographically weighted models. *Journal of Statistical Software* 63: 1–50.

Good, I.J. 1983. The philosophy of exploratory data analysis. *Philosophy of Science* 50: 283–295.

Haining, R. 2003. *Spatial Data Analysis: Theory and Practice*. Cambridge: Cambridge University Press.

Harris, P., A.S. Fotheringham, R. Crespo and M. Charlton. 2010. The use of geographically weighted regression for spatial prediction: an evaluation of models using simulated data sets. *Mathematical Geosciences* 42: 657–680.

Huang, H., G.Y. Miller, B.J. Sherrick and M.I. Gómez. 2006. Factors influencing Illinois farmland values. *American Journal of Agricultural Economics* 88: 458–470.

Hurvich, C.M., J.S Simonoff and C.L. Tsai. 1998. Smoothing parameter selection in nonparametric regression using an improved Akaike information criterion. *Journal of the Royal Statistical Society: Series B (Statistical Methodology)* 60: 271–293.

Lamour, J., G. Le Moguédec, O. Naud, M. Lechaudel, J. Taylor and B. Tisseyre. 2021. Evaluating the drivers of banana flowering cycle duration using a stochastic model and on farm production data. *Precision Agriculture* 22: 873–896.

McMillen, D.P. 1996. One hundred fifty years of land values in Chicago: A nonparametric approach. *Journal of Urban Economics* 40: 100–124.

McMillen, D.P. and J.F. McDonald. 1997. A nonparametric analysis of employment density in a polycentric city. *Journal of Regional Science* 37: 591–612.

Moran, P.A.P. 1950. Notes on continuous stochastic phenomena. *Biometrika* 37: 17–23.

Panzera, D. and P. Postiglione. 2014. Economic growth in Italian NUTS 3 provinces. *The Annals of Regional Science* 53: 273–293.

Sun, S., J. Liu, P. Wu, Y. Wang, X. Zhao and X. Zhang. 2016. Comprehensive evaluation of water use in agricultural production: a case study in Hetao Irrigation District, China. *Journal of Cleaner Production* 112: 4569–4575.

Swamy, P.A.V. 1971. *Statistical Inference in Random Coefficient Regression Models.* Berlin: Springer.

Woodard, J.D., G.D. Schnitkey, B.J. Sherrick, N. Lozano-Garcia and L. Anselin. 2011. A spatial econometric analysis of loss experience in the U.S. crop insurance program. *The Journal of Risk and Insurance* 79: 261–286.

CHAPTER 9
Spatial Econometric Modelling of Farm Data

Anna Gloria Billé,[1,]* Cristina Salvioni[2] and Francesco Vidoli[3]

9.1 Spatial modelling of farm data

Since the publication in 2002 of the Special Issue in Agricultural Economics entitled *"Spatial analysis for agricultural economists: Concepts, topics, tools and examples"* the number of applications of spatial econometric methods in agricultural economics has steadily increased over time.

Spatial explicit models have been largely used to analyse the drivers of technology adoption (Case 1992; Florax et al. 2002; Anselin et al. 2004; Abdulai and Huffman 2005; Krishna and Qaim 2012; Krishnan and Patnam 2013; Lapple and Kelley 2015; Fang and Richards 2018; Lapple et al. 2017), with specific attention to organic farming (Lapple and Kelley 2015; Schmidtner et al. 2012; Wollni and Andersson 2014), and spatial regimes in technologies (Billé et al. 2018).

Another relevant stream of literature estimates spatial models that address both spatial dependence and spatial heterogeneity to explain variation in farmland values (Nickerson and Lynch 2001; Patton and McErlean 2003; Cavailhès and Wavresky 2003; Livanis et al. 2006; Kostov 2009; Maddison 2009; Wang 2018; Yang et al. 2019).

The importance of neighbourhood effects has been recognized in studies explaining market participation (Holloway and Lapar 2007), farm household diversification (Corral and Radchenko 2017), farm survival (Saint-Cyr et al. 2018), land rental intention (Skevas et al. 2018), pesticide use (Aida 2018) and in the analysis of policy intervention in agriculture (Grogan and Goodhue 2012; Storm et al. 2014; Marconi et al. 2015; Feichtinger and Salhofer 2016; Fruh-Muller et al. 2019).

Finally, an increasing group of studies analyses the issue of spatial dependence in the analysis of technical efficiency in the context of stochastic frontier (Druska and

[1] University of Padova, Department of Statistical Sciences.
[2] "G. d'Annunzio" University of Chieti-Pescara, Department of Economic Studies.
[3] University of Roma Tre, Department of Political Science.
  Emails: cristina.salvioni@unich.it; francesco.vidoli@uniroma3.it
* Corresponding author: annagloria.bille@unipd.it

Horrace 2004; Areal et al. 2012; Fusco and Vidoli 2013; Tsionas and Michaelides 2016; Pede et al. 2018).

In the rest of the chapter, we first describe the general formulation of the production function (Section 9.1.1) and then we present the R codes to fit and test spatial production and spatial frontier functions (Section 9.2). In Section 9.3, we explain how to fit and test spatial production models when farm data exhibit unobserved spatial heterogeneity, focusing on a possible way to identify unobserved spatial regimes (Subsection 9.3.1) and to identify contiguous spatial clusters (Subsection 9.3.2).

9.1.1 Production function and the frontier production function

In agriculture, there is a long tradition of statistics-gathering and it is one of the economic sectors with the deepest supply of micro data set on input and output from the production process. This explains the large number of studies that estimate agricultural production functions, analyse productivity, and measure the efficiency of farms.

A production function describes the technical relationship that transforms inputs into output. For each level of input use, the function assigns a unique output level. A general way of writing a production function is

$$y = f(x_1, \ldots, x_n)$$

where y is an output and x_i are the productive inputs that can include labour, capital, knowledge (human capital), energy consumption, raw materials, natural resources (land, water, minerals), and others. It is usually assumed that production functions fulfil some properties: essentiality of inputs, positive returns, diminishing returns and/or proportional returns to scale (see Chambers 1998, 9). Many mathematical specifications can be used to estimate the production function (Griffin et al. 1987). The choice of functional form brings a series of implications with respect to the shape of the isoquants and the values of elasticities of factor demand and factor substitution. The simplest specification is the linear production function:

$$y = \alpha + \sum_{i=1}^{N} \beta_i x_i$$

Despite its mathematical simplicity, this linear form is rarely used since it violates the property of essentiality.[1]

A widely used form is the Cobb Douglas production technology

$$y = A \prod_{i=1}^{N} x_i^{\alpha_i}$$

where A represents the Hicksian neutral efficiency level of firm i, which is unobserved by the econometrician. The natural logs transformation of the previous equation can be expressed as a linear equation in the form:

$$\ln(y) = \alpha_0 + \sum_{i=1}^{N} \alpha_i \ln(x_i)$$

---

[1] Essentiality of inputs: If at least one $x_i = 0$, then $y = 0$, i.e., production is not possible without any of the inputs.

The Cobb Douglas function is often chosen because it has economic properties (diminishing returns to each input and constant returns to scale) superior to the simple linear function and because its parameters are easy to obtain from real data. Another largely used functional form is the translog for one output and K inputs

$$\ln(y) = \alpha_0 + \sum_i^N \alpha_i \ln(x_i) + \frac{1}{2} \sum_i \sum_j \alpha_{ij} \ln(x_i) \ln(x_j)$$

This specification is a more flexible extension of the Cobb-Douglas function, it fulfils a set of desirable characteristics and it is easy to derive and allows the imposition of homogeneity.

Once a functional form is chosen, a key issue in the estimation of production functions is that direct OLS estimation of the production function is problematic because of endogeneity (Marschak and Andrews 1944; Ackerberg et al. 2015).

Endogeneity can arise when observed inputs are correlated with unobserved shock. Under such circumstances, OLS will yield biased and inconsistent estimates. In a linear framework, the standard approach for addressing the potential endogeneity bias is to use instrumental variables or fixed effects.

The estimated model of production is the means to the objective of measuring inefficiency. This is because the production function represents the maximum output attainable given a set of inputs. Measurement of (in)efficiency is, then, the empirical estimation of the extent to which observed agents (fail to) achieve the production frontier as originally argued by Debreu (1951) and Farrell (1957).

One of the main approaches to study productivity and efficiency is the Stochastic Frontier Model (SFM), independently proposed by Aigner et al. (1977) and Meeusen and Broeck (1977). The SFM is motivated by the theoretical idea that no economic agent can exceed the frontier and the deviations from this extreme represent the individual inefficiencies. For the description, implementation and testing of the non-spatial and spatial Stochastic Frontier Analysis models see Section 9.2.

For a review of some of the most important developments in the econometric estimation of the stochastic frontier models (endogeneity issues, recent advances in generalized panel data stochastic frontier models, etc.) see for example Greene (2008) and Kumbhakar et al. (2017).

In the stochastic production frontier estimation, the endogeneity of input problem has been neglected until recently. In such environments, the endogeneity issue can be solved for example by using the semi-parametric approach proposed by Olley and Pakes (1996) and Levinsohn and Petrin (2003) (see for example Shee and Stefanou (2014), and Latruffe et al. (2017) for an application to agriculture).

9.2 Fitting and testing spatial stochastic frontier models

In this section, we explain how to fit and test spatial models with a specific focus on farm data. In particular, we define a spatial frontier model by making use of the stochastic frontier approach. Log-log transformations of the Cobb-Douglas production function can be used to implement the well-known spatial linear model specifications. For details on these types of models in the context of regional data, the reader is referred to Chapter 12. Note that in agricultural production function/ stochastic frontier models a problem of potential endogeneity of some inputs (regressors) is often present. In this Section, we assume the exogeneity of all the regressors in our model specifications. For details on the use of the spatial linear production function and potential endogeneity (Billé et al. 2018).

In the following subsection, we consider the simulation setup for the definition of the true values of the parameters, the sample size, the generation of the spatial coordinates and the spatial weighting matrix used to assume a particular spatial process. The assumed weighting matrix/matrices is/are based on a k-nearest neighbour approach (k-nn), i.e., we define a Boolean matrix with the same number $k \in \mathbb{N}$ of nearest neighbours for each random variable in space. Let $\mathbf{W} = \{w_{ij}\}$ be the spatial weighting matrix with elements equal to the weights among pairs of random variables (y_i, y_j) for $i, j = 1,..., n$, with n the sample size, then

$$\begin{cases} w_{ij} = 1 & \Leftrightarrow \quad y_j \in \mathcal{N}_k \\ w_{ij} = 0 & otherwise \end{cases}$$

where $\mathcal{N}_k$ is the set of nearest random variables y_j to y_i defined by k. Finally, $\mathbf{W}$ is row-normalized such that $\sum_j w_{ij} = 1$, $\forall i$. Discussions on the definition on different spatial weighting matrices as well as model specifications can now be found in several spatial book references (Anselin 1988; LeSage and Pace 2009; Elhorst 2014; Arbia 2014; Kelejian and Piras 2017).

To set the simulation setup, consider the following codes. We first load the following package

```
>library(spdep)
```

in order to use some useful spatial functions inside, see Bivand and Piras (2015). Then we set the seed, the number of nearest neighbours k, the true values of the vector of parameters $\boldsymbol{\beta} = (\beta_0, \beta_1, \beta_2, \beta_3)'$ for the constant, Land, Labour and Capital inputs, the true values of the vector of parameters $\boldsymbol{\theta} = (\theta_1, \theta_2, \theta_3)'$ for spatially-lagged Land, Labour and Capital inputs, and the true values of the autoregressive coefficients, ρ and λ, in the dependent variables and among the error terms (or the inefficiencies), respectively, as follows

```
>set.seed(3)
>k          <- 30
>beta       <- c(10, 0.5, 0.3, 0.2)
>theta      <- c(0.6,0.2,0.2)
>rho        <- 0.6
>lambda     <- 0.4
```

We set the sample size n and we generate the longitude and latitude coordinates by using two Uniform distributions from 0 to 50 and from -70 to 20, i.e., $\mathcal{U}(0, 50)$ and $\mathcal{U}(-70, 20)$, respectively,

```
>n<- 500
>coords <- cbind(long = runif(n,0,50), lat = runif(n,-70,20))
>head(coords)
##              long           lat
## [1,]    8.402076    -43.33272
```

```
##  [2,]    40.375820   -49.23735
##  [3,]    19.247118    12.29245
##  [4,]    16.386716   -40.08144
##  [5,]    30.105034   -19.92414
##  [6,]    30.219703   -17.64876
```

and we generate an *n*-dimensional Identity matrix and the weighting matrix **W** by using some functions into the spdep package as follows

```
>I_n<- as(diag(n), "CsparseMatrix")
>nb <- knn2nb(knearneigh(coords, k = k, longlat=TRUE))
>W <- as(nb2mat(nb, style="W"), "CsparseMatrix")
```

In particular, the function knearneigh provides a list of class *knn* with the information into the first member of the region number ids to define the nearest neighbours for each random variable. The argument longlat=TRUE selects the Great Circle geographical distances among pairs of units in space. The knn2nb function transforms the object of class knn into an object of class nb (neighbour list), while the nb2mat function transforms an object of class nb into an *n*-dimensional weighting matrix. The argument style = "W" directly row-normalizes the weights. Both the Identity matrix **I**$_n$ and the weighting matrix **W** are sparse by using the function as (,"CsparseMatrix").

Finally, we set the matrix of regressors which include the constant, Land, Labour and Capital inputs by drawing numbers from $\mathcal{U}(1.5, 4)$

```
>X<-cbind(constant=1, A=runif(n,1.5,4),L=runif(n,1.5,4),K
=runif(n,1.5,4))
```

This section is devoted to the description, implementation and testing of the Stochastic Frontier Analysis models (SFA) and of a recent extension of SFA, called Spatial Stochastic Frontier Analysis (SSFA).

A caveat, however, is mandatory: the number of different techniques proposed in the literature for estimating production (or cost) efficiency is wide, differentiating among parametric (as SFA, Aigner et al. (1977) and Meeusen and Broeck (1977), R packages Benchmarking, frontier), non-parametric (as Data Envelopment Analysis (DEA), Farrell (1957) and Charnes, Cooper, and Rhodes (1978), R packages Benchmarking, nonparaeff or FEAR Wilson (2008)) or semi-parametric techniques (Park and Simar (1994), Kuosmanen and Kortelainen (2012) and Ferrara and Vidoli (2017), R package semsfa).

If, at first glance, non-parametric techniques seem to be particularly flexible and generalizable, the main disadvantage lies precisely in their deterministic nature, since it is not even possible to recognize if the difference in terms of efficiency among units is caused by technical inefficiency or by exogenous/accidental effects (Fried and Lovell 2008). The parametric model of stochastic frontier overcomes the main limits associated with deterministic models, providing a detailed analysis of the inefficiency sources that are not directly associated to farm policy and/or random disturbances, too. On the other hand, the most significant disadvantage associated with the SFA approach is the lack of flexibility associated with the specification of a given functional form.

The SFA approach implies the construction of the stochastic optimum frontier, based on an underlying production/cost function, identified through the relative comparison of the firm performance in a set economic system. The observed deviations from the optimum frontier may be split into the combination of two effects: the effect caused by the random noise and the technical/cost inefficiency. More formally, the stochastic frontier model can be written as:

$$y_i = f(x_i; \boldsymbol{\beta}) + v_i - u_i, \qquad i = 1,\dots, n$$

where $Y_i \in R_+$ is the single output of unit i, $X_i \in R_+^p$ is the vector of p inputs, $f(\cdot)$ defines a production (frontier) relationship between inputs X and the single output Y, v_i is a symmetric two-sided error representing random effects, usually assumed Normal $v \sim N(0, \sigma_v^2)$, and $u_i > 0$ is one-sided error term which represents technical inefficiency, usually assumed Half-Normal ($u \sim N^+(0, \sigma_u^2)$). Please note that (i) the inefficiency and the error terms must both be orthogonal to the input, output, or to the other variables used in the functional specification, and (ii) it is usually assumed that v and u are each identically independently distributed (*i.i.d.*).

Starting from the Data Generating Process (DGP) designed in the previous chapter (i.e., same simulation setup, sample size, spatial coordinates and spatial weighting matrix), the next step is merely to set up the inefficiency (positive) term u using the `rsn` function (`sn` package) that generates a random sample of n units from a skew-normal distribution with 1 as location parameter and 4 as the scale parameter.

```
>u <- abs(rsn(n, 0, 1, 4))
```

This hypothesis, namely the independence among production units, can often be violated, especially in the regional applications where exogenous conditional factors or the different resiliencies of specific territories may lead to comparative advantages or disadvantages for units within the same region. For this reason, and to test different models of stochastic frontier, the inefficiency term u has been simulated as dependent by two distinct effects (which in practical applications can be substitutive or complementary): (i) a global spatial spillover effect and (ii) a drift effect.

From an economic point of view, therefore, u_2 may catch the spillover effect from which neighbouring units may benefit, influencing each other and benefiting from the agglomeration or Marshallian atmospheric externalities or may grasp the different resilience of the territories as a consequence of an economic shock. More formally, u_2 can be defined as: $\mathbf{u}_2 = (\mathbf{I}_n - \lambda\mathbf{W})^{-1}\mathbf{u}$.

Practically in R, $\mathbf{u}_2$ can be expressed as:

```
>u2 <- solve(I_n - lambda * W) %*% u
```

The second effect, which can be found in some regional or agricultural applications, is the drift effect. This is linked not so much to the neighbourhood as to the physical position, in terms of latitude and longitude, of the units in the analysed region. A different impact on land productivity, for example, may be the result of a different spatial distribution of temperatures or rainfall. Obviously, the presence of the two spatial effects may or may not overlap depending on the application problem.

Given these premises, we modify $\mathbf{u}_2$ in such a way that the inefficiency of the units is a function of their absolute position in the region, too, multiplying it with drift normalized value function of latitude and longitude.

```
>drift = 1-0.5*(coords[,1]+mean(coords[,1]))*(coords[,2]+
mean(coords[,2]))
>u2 <- u2 * drift/mean(drift)
```

We have, therefore, all the elements to calculate a simulated *y* (named *y_sfa*) that takes into account the covariates **X** (Land, Labour and Capital inputs as defined in the previous section), the random error term, but also an inefficiency term that depends on the neighbourhood and the relative position of the single unit in the region.

```
>y_sfa <- as.matrix(X%*%beta + eps - u2)
```

It is, therefore, possible to assess how the standard SFA model can help to estimate the simulated DGP and whether the basic assumptions regarding errors (*i.i.d.*) are respected even in the presence of a strong component linked to the economic space within which farms produce. Different packages are available on CRAN; in this application exercise, Benchmarking package has been chosen for the large number of options and functions complementary to the main function sfa. Please note that, in this package, *X* (first parameter) and *y* (second parameter) must be passed as matrix.

```
>library(Benchmarking)
>x_sfa = as.matrix(cbind(df$A, df$L, df$K))
>y_sfa = as.matrix(y_sfa)
>fit.sfa<- sfa(x_sfa,y_sfa)
>summary(fit.sfa)
##                 Parameters   Std.err   t-value   Pr(>|t|)
## (Intercept)        10.7183   0.43005    24.924      0.000
## x1                  0.5646   0.08844     6.384      0.000
## x2                  0.1849   0.08545     2.164      0.030
## x3                  0.1223   0.08579     1.426      0.154
## lambda              2.9148   0.43275     6.736      0.000
## sigma2 5.1373
## sigma2v = 0.5409906 ; sigma2u = 4.596338
## log likelihood = -887.1153
## Convergence = 4 ; number of evaluations of likelihood
function 24
## Max value of gradien: 7.733437e-06
## Length of last step: 0
## Final maximal allowed step length: 0.33075
```

Standard diagnostics of the SFA function (obtained by the usual function summary) reports a good adaptation of the model to the data (the estimated β are very similar to the simulated ones, the intercept is higher—as it must be for the estimation of the productive frontier); different from other non-frontier models two key values of this analysis are reported here: σ_v^2 (equal to 0.719) and σ_u^2 (equal to 3.367). They represent, respectively, the estimated variance of the random component and the inefficiency; if compared to the total estimated variance ($\sigma^2 = 4.087$) it is possible to evaluate how the part of inefficiency ($\sigma_u^2/\sigma^2 = 3.367/4.087 = 82\%$) is greater than the random one ($\sigma_v^2/\sigma^2 = 0.719/4.087 = 18\%$), as proof of a good adaptation of the model to the data and a good differentiation of the units in terms of estimated inefficiency. Standard diagnostics, therefore, does not report any warning in the estimated model nor is there any standard test available on the basic assumptions. In other terms, the spatial dependence of inefficiency is not grasped, and this issue may be all the more serious as environmental and contextual factors are important in explaining efficiency differentials among different territories.

To test the spatial autocorrelation among the estimated efficiency, Geary test (through the function `geary.test` of the `spdep` package) can be used; please note that, in this simulation, testing the estimated efficiencies (estimated by the `eff` function, `Benchmarking` package) or residues leads to very similar results since the simulated random part (v) is not spatially autocorrelated.

```
>Wnb <- nb2listw(nb)
>geary.sfa.eff <- geary.test(as.vector(eff(fit.sfa)),listw
= Wnb)
>geary.sfa.eff
##
## Geary C test under randomization
##
## data: as.vector(eff(fit.sfa)) ## weights: Wnb ##
## Geary C statistic standard deviate = 21.235, p-value <
2.2e-16
## alternative hypothesis: Expectation greater than statistic
## sample estimates:
##   Geary C statistic     Expectation        Variance
##       0.7468869459    1.0000000000    0.0001420774
```

Geary test (equal to 0.743) and Figure 9.1 show, as expected, a strong spatial autocorrelation in efficiency estimates: in particular, the first two figures at the top

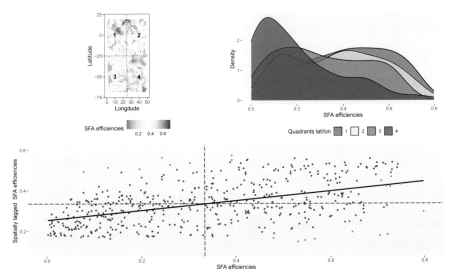

Figure 9.1. SFA efficiencies.

show, respectively, the spatial distribution and the kernel distribution by quadrant of the efficiencies highlighting a strong difference among the quadrants themselves; the graph at the bottom shows the Moran plot.

Once it has been demonstrated that it is the external conditions that led to greater or lesser efficiency and not so much the allocation of the single producer, what method can be used to split the territorial from the individual effect? An effective solution for cross-section data—and a generalization of the SFA—has been proposed by Fusco and Vidoli (2013) with a method called "*Spatial Stochastic Frontier Analysis*" (SSFA); they proposed to "*split the inefficiency term into three parts: the first related to spatial peculiarities of the territory in which each single unit operates, the second related to the specific production features and the third the random error term*". More in depth, Fusco and Vidoli (2013) propose a maximum likelihood solution to this model:[2]

$$y_i = f(x_i; \beta) + v_i - \left(1 - \rho \sum_i w_i\right) \hat{u}_i, \quad i = 1, \ldots, n$$

where $\hat{u}_i$ and v_i are independently distributed of each other and of the regressors. R package ssfa provides the ssfa function that allows this specific model to be estimated; please note, in particular, the need to indicate the spatial weight matrix in the data_w parameter.

```
>library(ssfa)
>fit.ssfa <- ssfa(y_sfa ~ A+L+K, data_w=W, data=df, form =
"production", par_rho=TRUE, intercept = TRUE)
>summary(fit.ssfa)
## Spatial Stochastic frontier analysis model
##
##                  Estimate    Std. Error   z value Pr(>|z|)
## Intercept     1.04612e+01  1.98829e-01 52.61399    < 2e-16***
## A             8.34315e-01  1.57974e-01  5.28134    < 2e-16***
## L             4.76402e-01  1.51928e-01  3.13572  0.001714 **
## K             8.18766e-01  1.90999e-01  4.28675  1.8e-05***
## sigmau2_dmu 9.89902e+00  4.80903e-01 20.58426    < 2e-16***
## sigmav2     2.38643e-06  1.58278e-06  1.50774  0.131620
## ---
## Signif. codes: 0 `***' 0.001 `**' 0.01 `*' 0.05 `.' 0.1 ` ` 1
##
## Pay attention:
 ## 1 - classical SFA sigmau2 = sigmau2_dmu + sigmau2_sar:
10.312152 where sigmau2_sar: 0.413129
## 2 - sigma2 = sigmau2_dmu + sigmau2_sar + sigmav2: 13.767977
##
##   Inefficiency parameter Lambda = sigmau_dmu/sigmav:
4148044.355609
```

```
## Spatial parameter Rho: 0.303932
## Moran I statistic: 0.141358
## Mean efficiency: 0.07876
## LR-test: sigmau2_dmu = 0 (inefficiency has no influence to
the model)
## H0: sigmau2_dmu = 0 (beta_ssfa = beta_ols)
##           Value Log-Lik
## ssfa      -1041.3802
## ols        -914.6648
## Value LR-Test: -253.431 p-value 0
## AIC: 2110.76, (AIC for lm: 1837.33)
```

Even in this case, the β parameters of the simulations have been correctly estimated,[3] but the most interesting issue to consider concerns the opportunity to split the inefficiency variance σ_u^2 into two terms: the individual part (named by authors σ_{dmu}^2) and a neighbourhood related one (named σ_{sar}^2).

In the specific simulation, please note how the individual effect ($\sigma_{dmu}^2 = 1.417$) is more important than the spatial one ($\sigma_{sar}^2 = 0.353$), but this part still represents about 25% of the individual effect. Another key parameter to take into account is the spatial parameter ρ equal to 0.857; in this case, therefore, SSFA diagnostic shows a strong spatial correlation between the inefficiency estimates (please note that the function eff.ssfa estimate the efficiency of each producer without spatial effects). Replicating the Geary test on the SSFA efficiency (equal to 1.033) and analysing Figure 9.2, it can be noted how the SSFA model has succeeded in isolating

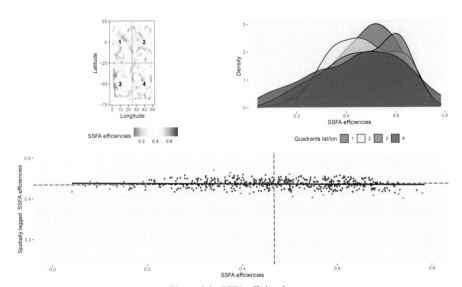

Figure 9.2. SSFA efficiencies.

---

[3] Please note the higher value of the intercept; as stated by Vidoli et al. (2016), there may be a "*shift in the production curve with respect to the SFA as a consequence of the isolation of the spatial effect, transforming the average value of β_0 into a multiplicity of individual effects*".

the territorial effect from the efficiency estimates; of course, the differences between the SFA and SSFA estimates may be of interest in the study of the determinants of local development.

```
>geary.ssfa.eff <- geary.test(as.vector(eff.ssfa(fit.ssfa)),
listw = Wnb)
>geary.ssfa.eff
## Geary C test under randomization
## data: as.vector(eff.ssfa(fit.ssfa))
## weights: Wnb
## Geary C statistic standard deviate = 2.2942, p-value =
0.01089
## alternative hypothesis: Expectation greater than statistic
## sample estimates:
##   Geary C statistic      Expectation         Variance
##        0.9443271280     1.0000000000     0.0005888975
```

9.3 Controlling for unobserved spatial heterogeneity

In this chapter, we explain how to fit and test spatial production models when farm data exhibit spatial heterogeneity. The form of spatial heterogeneity is typically unknown (unobserved), and it can be ascribed either to mean instability or to heteroskedasticity. The omission of such spatial effects leads to biased inference (Anselin 1988; LeSage and Pace 2009; Le Gallo 2014).

Mean instability may imply local clustering of the values of a variable. In the case of agricultural production, mean instability implies that the territory can be divided into clusters of farms, each one using the same production technology. In other words, the production function coefficients differ according to a number of distinct unknown spatial production regimes (groupwise heterogeneity), see Anselin (2010). As explained in Billé et al. (2018), these local clusters in technology can emerge as a result of dynamic interactions among site-specific environmental variables and farmer decision making about technology. For example, farmers usually choose to grow those varieties that are best suited to the environmental conditions in which the farm operates. In turn, the choice of a specific variety is often connected to specific management systems, such as water management or timing of harvesting. The technology prevailing in each local technology cluster is the efficient solution to the specific techno-economic problem faced by farms operating in that portion of the territory. This view is consistent with evolutionary theories (Nelson and Winter 1982; Dosi 1988), according to which firms cannot be assumed to operate using a single common production function.

The zoning of spatial regimes in farm technologies is largely unknown and their identification needs comprehensive spatial modelling of soil, agronomical, and climatic properties, including their changes through time, hence the processing of large quantities of data acquired at a very fine spatial resolution, while researchers can usually rely only on a few control variables. An alternative way to identify the spatial regimes in farm technologies relies on the use of the Earth coordinates of latitude and longitude to proxy the micro-geographic determinants of production that are unknown to the econometrician. In Subsection 9.3.1, we show a possible way to identify unobserved spatial regimes and then explain how to properly estimate more flexible spatial models, see Billé et al. (2017).

Spatial heterogeneity may also arise due to heteroskedasticty, that is when the variances of error terms vary over space causing the instability of the functional form. One solution to this problem has been recently proposed by Chasco et al. (2018). Also, in the case of heteroskedasticity, the territory can then be divided into clusters of farms characterized by significantly different groupwise variances, i.e., unobserved spatial groupwise heteroskedasticity. The identification of such clusters in the error terms is not considered in Billé et al. (2018).

In Subsection 9.3.2, we show how another possible solution to the problem of the identification of contiguous spatial clusters is to use a graph-based approach, which captures the adjacency relations between objects, farms in our case. Differently from the previous case, in this section spatial homogeneous areas are not defined according to a functional relationship, rather on some farm or territorial characteristics likewise to the standard cluster methods.

In both the following subsections, we basically assume the same sample size $n =$ 500, generation of the spatial coordinates, spatial weighting matrices, generation of the regressors $\mathbf{X}$, and of the innovations ϵ of Section 9.3. By setting the same seed we can include in our new database the original $\mathbf{X}$. We also consider the same simulation setup for the identification of the clusters as in the following. Let us first define the dataframe with the coordinates of Section 9.2 and with an indicator variable that associates each point to a cluster.

```
>dataset <- as.data.frame(coords)
>set.seed(3)
>dataset$A <- runif(n,1.5,4)
>dataset$L <- runif(n,1.5,4)
>dataset$K <- runif(n,1.5,4)
>dataset$clu <- ifelse(dataset$long < 20 & dataset$lat < -20, 1, 0)
>dataset$clu <- ifelse(dataset$long > 20 & dataset$lat < -40,
2, dataset$clu)
>dataset$clu <- ifelse(dataset$long < 20 & dataset$lat > -40,
3, dataset$clu)
>dataset$clu <- ifelse(dataset$long > 20 & dataset$long < 40
& dataset$lat > -40 & dataset$lat < 10, 4, dataset$clu)
>dataset$clu <- ifelse(dataset$long > 20 & dataset$lat > 0,
5, dataset$clu)
>dataset$clu <- ifelse(dataset$long > 40 & dataset$lat > -40,
5, dataset$clu)
```

Figure 9.3 shows the generated clusters/regimes in space. Note that these spatial regimes can also be not geographically well-defined, i.e., points that belong to the same cluster can be also sparsely-distributed in space. We need to assume a different spatial production function for each of them.

Suppose we have two different true DGPs: (i) an OLS model with regimes, (ii) a SARAR/SAC model with regimes. Then, the following codes generate different production functions for different clusters

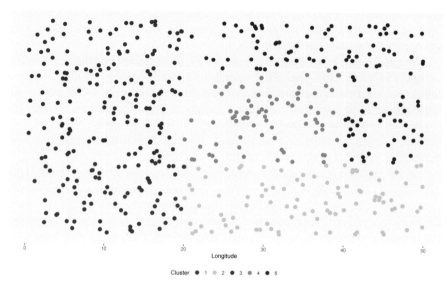

Figure 9.3. Example of spatial regimes.

```
>beta1 <- c(13,0.5,0.3,0.2)
>beta2 <- c(11,0.8,0.1,0.1)
>beta3 <- c(9,0.3,0.2,0.5)
>beta4 <- c(7,0.4,0.3,0.3)
>beta5 <- c(5,0.2,0.6,0.2)
>dataset$y_ols<-ifelse(dataset$clu==1,13+0.5*dataset$A +
0.3*dataset$L+0.2*dataset$K+eps, 0)
>dataset$y_ols <- ifelse(dataset$clu==2,11+0.8*dataset$A
+0.1*dataset$L+0.1*dataset$K+eps, dataset$y_ols)
>dataset$y_ols <- ifelse(dataset$clu==3,9+0.3*dataset$A +
0.2*dataset$L+0.5*dataset$K+eps, dataset$y_ols)
>dataset$y_ols <- ifelse(dataset$clu==4,7+0.4*dataset$A +
0.3*dataset$L+0.3*dataset$K+eps, dataset$y_ols)
>dataset$y_ols <- ifelse(dataset$clu==5,5+0.2*dataset$A +
0.6*dataset$L+0.2*dataset$K+eps, dataset$y_ols)
>dataset$y_sac <- ifelse(dataset$clu==1,as.matrix(A_rho_
inv%*%(X%*%beta1+B_lambda_inv%*%eps)), 0)
>dataset$y_sac <- ifelse(dataset$clu==2,as.matrix(A_rho_
inv%*%(X%*%beta2+B_lambda_inv%*%eps)), dataset$y_sac)
>dataset$y_sac <- ifelse(dataset$clu==3,as.matrix(A_rho_
inv%*%(X%*%beta3+B_lambda_inv%*%eps)), dataset$y_sac)
>dataset$y_sac <- ifelse(dataset$clu==4,as.matrix(A_rho_
inv%*%(X%*%beta4+B_lambda_inv%*%eps)), dataset$y_sac)
>dataset$y_sac <- ifelse(dataset$clu==5,as.matrix(A_rho_
inv%*%(X%*%beta5+B_lambda_inv%*%eps)), dataset$y_sac)
head(dataset)
```

```
##       long      lat      A       L      K clu  y_ols    y_sac
## 1  8.402076 -43.33272 1.920104 2.240758 1.540459   1 12.80053 37.03793
## 2 40.375820 -49.23735 3.518791 2.076740 1.891127   2 12.94834 33.57304
## 3 19.247118  12.29245 2.462356 3.785901 1.758021   3 11.45821 29.45562
## 4 16.386716 -40.08144 2.319336 2.331071 2.664323   1 15.58018 39.85193
## 5 30.105034 -19.92414 3.005252 2.890996 2.832992   4 10.10911 24.67326
## 6 30.219703 -17.64876 3.010985 2.954201 2.368872   4 10.01893 24.28573
```

where the column "`clu`" refers to the specific cluster and the columns "`y_ols`" and "`y_sac`" to the generated dependent variables in the linear and spatial case, respectively. In the following subsection, we formally define a SARAR/SAC model with regimes. Alternative spatial models with regimes can be straightforwardly specified.

9.3.1 Spatial regimes

In this subsection, we explain how to estimate spatial autoregressive models with pre-identified production regimes; the way in which we can identify unobserved spatial regimes can be made with different approaches, like mixture models, see Greene (2005) and Emvalomatis (2012), or iteratively locally weighted regressions, see Billé et al. (2017) and Billé et al. (2018). The procedure in Billé et al. (2017) and Billé et al. (2018) is based on a continuous smoothing updating algorithm of the weights used to repeated local estimations, by making use of the Wald test statistics to simultaneously compare vector of beta coefficients. The algorithm is available upon request.

Let us assume we know the form of the clusters in space. Then, a SARAR/SAC model with regimes can be defined in the following way

$$\dot{y} = \rho W_1 \dot{y} + \dot{X}\dot{\beta} + \dot{u} \; ; \dot{u} = \lambda W_2 \dot{u} + \varepsilon \; ; \varepsilon \sim \mathcal{N}(0, \sigma_\varepsilon^2 I)$$

where $\dot{y} = \{\dot{y}_j\}$ is a partitioned n-dimensional column vector of dependent variables with generic vector element $\dot{y}_j$ for $j = 1,\ldots, c$ and c is the total number of regimes, $\dot{X} = \{\dot{X}_j\}$ is an $n \times (k \times c)$ block-diagonal matrix with generic matrix element $\dot{X}_j$ for $j = 1,\ldots, c, \dot{\beta} = \{\dot{\beta}_j\}$ is $(k \times c)$ partitioned column vector of parameters with generic vector element $\dot{\beta}_j$ for $j = 1,\ldots, c, \varepsilon = \{\varepsilon_j\}$ is a partitioned n-dimensional column vector with generic element $\varepsilon_j \sim \mathcal{N}(0, \sigma_{\varepsilon j}^2 I_{nj})$ for $j = 1,\ldots, c$, while (ρ, λ) and (W_1, W_2) are defined in Section 9.3. By letting $\rho = \lambda = 0$ the model is defined as a linear model (OLS) with regimes.

By using the clusters generated in Figure 9.3, we now estimate the OLS model with and without regimes and the SARAR/SAC model with and without regimes in the following way:

```
>CLU <- as.factor(dataset$clu)
>fit.ols<- lm(dataset$y_ols ~ dataset$A+dataset$L+dataset$K)
>fit.ols.r <- lm(dataset$y_ols ~ (CLU:(dataset$A + dataset$L
+ dataset$K + CLU)) + 0)
>fit.sac <- sacsarlm(dataset$y_sac ~ dataset$A + dataset$L +
dataset$K,data=dataset,listw=listw,type="sac",method="eigen")
>fit.sac.r <- sacsarlm(dataset$y_sac ~ (CLU:(dataset$A +
dataset$L + dataset$K+CLU))+0, data=dataset, listw=listw,
type="sac", method="eigen")
```

where `fit.ols` and `fit.sac.r` are the linear and spatial models with regimes, respectively. By using the function "summary", we can see the difference in the estimates between the spatial model with regimes (correct specification) and the spatial model without regimes as follows:

```
summary(fit.sac)
## Call:
## sacsarlm(formula = dataset$y_sac ~ dataset$A + dataset$L
+ dataset$K,
## data = dataset, listw = listw, type = "sac", method = "eigen")
## Residuals:
##        Min        1Q     Median        3Q       Max
## -8.081323  -1.064716  0.097549  1.097462  7.355392
## Type: sac
## Coefficients: (asymptotic standard errors)
##                Estimate   Std. Error  z value  Pr(>|z|)
## (Intercept)   18.46596    14.81133   1.2467    0.2125
## dataset$A     -0.82323     0.58894  -1.3978    0.1622
## dataset$L     -2.53457     2.13584  -1.1867    0.2354
## dataset$K     -0.18637     0.14083  -1.3234    0.1857
##
## Rho: 0.68998
## Asymptotic standard error: 0.28374
##  z-value: 2.4317, p-value: 0.015026
## Lambda: 0.83349
## Asymptotic standard error: 0.18597
##  z-value: 4.4819, p-value: 7.3987e-06
## LR test value: 463.07, p-value: < 2.22e-16
## Log likelihood: -1116.847 for sac model
## ML residual variance (sigma squared): 4.8025, (sigma: 2.1915)
## Number of observations: 500
## Number of parameters estimated: 7
## AIC: 2247.7, (AIC for lm: 2706.8)
summary(fit.sac.r)
## Call:sacsarlm(formula = dataset$y_sac ~ (CLU:(dataset$A
+ dataset$L +
## dataset$K + CLU)) + 0, data = dataset, listw = listw, type =
"sac", ## method = "eigen")
## Residuals:
##        Min        1Q     Median        3Q       Max
## -3.331635  -0.714640  0.022138  0.723094  2.782272
## Type: sac
## Coefficients: (asymptotic standard errors)
##                Estimate   Std. Error  z value   Pr(>|z|)
## CLU1          36.5398258  3.6448095  10.0252   < 2.2e-16
## CLU2          32.8433967  3.6672349   8.9559   < 2.2e-16
## CLU3          29.2626066  2.8784764  10.1660   < 2.2e-16
```

```
## CLU4              28.3887092  3.7131831   7.6454 2.087e-14
## CLU5              18.6113983  3.5142146   5.2960 1.183e-07
## CLU1:dataset$A -0.1747229  0.5781469 -0.3022    0.76249
## CLU2:dataset$A -0.0168871  0.3791063 -0.0445    0.96447
## CLU3:dataset$A -0.1388111  0.4762281 -0.2915    0.77068
## CLU4:dataset$A -0.0886247  0.6568061 -0.1349    0.89267
## CLU5:dataset$A  0.4818576  0.4812195  1.0013    0.31667
## CLU1:dataset$L  0.9166597  0.9939641  0.9222    0.35641
## CLU2:dataset$L  0.5052904  0.9412775  0.5368    0.59140
## CLU3:dataset$L -0.0339779  0.4444444 -0.0765    0.93906
## CLU4:dataset$L -1.3998376  0.6727553 -2.0808    0.03746
## CLU5:dataset$L -0.2251711  0.5023752 -0.4482    0.65400
## CLU1:dataset$K  0.0816411  0.1743796  0.4682    0.63966
## CLU2:dataset$K -0.0144752  0.1528647 -0.0947    0.92456
## CLU3:dataset$K  0.0105306  0.1235978  0.0852    0.93210
## CLU4:dataset$K  0.0034255  0.1809109  0.0189    0.98489
## CLU5:dataset$K -0.3080458  0.1398329 -2.2030    0.02760
##
## Rho: 0.017777
## Asymptotic standard error: 0.054341
##  z-value: 0.32714, p-value: 0.74356
## Lambda: 0.7
## Asymptotic standard error: 0.079216
##  z-value: 8.9629, p-value: < 2.22e-16
## LR test value: 21.358, p-value: 2.3021e-05
## Log likelihood: -730.6234 for sac model
## ML residual variance (sigma squared): 1.0628, (sigma: 1.0309)
## Number of observations: 500
## Number of parameters estimated: 23## AIC: 1507.2, (AIC for
lm: 1524.6)
```

As it can be observed, the significance of the estimated beta coefficients may change with respect to the type of cluster we consider. In this particular case, according to the results of the spatial model without regimes, only the variable Capital K is significant in all the clusters considered, but the significance level depends on the different clusters. Interestingly, one or more spatial autocorrelation coefficients might be no more statistically significant, as ρ in this case. Details on this aspect have been shown in Billé et al. (2017).

To test if the partition of the spatial data (clusters/regimes) is statistically significant, one can use the Chow test, see Chow (1960), and the Spatial Chow test, see Anselin (1990), for the presence of (spatial) structural breaks. Indeed, identifying clusters in space, by comparing the significance of the beta coefficients, is simply a way to econometrically find structural breaks in a spatial process. The statistics of the Chow test (C) and the Spatial Chow test (C_s) are defined, respectively, as follows

$$C = \frac{(\mathbf{e}'_r \mathbf{e}_r - \mathbf{e}'_u \mathbf{e}_u)/k}{\mathbf{e}'_u \mathbf{e}_u /(n-2k)} \sim F_{k,n-2k}$$

$$C_s = \frac{(\mathbf{e}'_r \mathbf{\Psi}^{-1} \mathbf{e}_r - \mathbf{e}'_u \mathbf{\Psi}^{-1} \mathbf{e}_u)}{\sigma^2} \sim \chi_k$$

where $\mathbf{e}_r$ is the vector of residuals from the restricted model (model without regimes), $\mathbf{e}_u$ is the vector of residuals from the unrestricted model (model with regimes), $\mathbf{\Psi}$ is the variance-covariance matrix of the spatial model and σ^2 is the error variance for either the restricted model, the unrestricted model, or both.

The functions of the Chow test and the Spatial Chow test in R, see Anselin (2005), are written as follows.

```
>chow.test <- function(rest,unrest)
  {
    er <- residuals(rest)
    eu <- residuals(unrest)
    er2 <- sum(er^2)
    eu2 <- sum(eu^2)
    k <- rest$rank
    n2k <- rest$df.residual - k
    c <- ((er2 - eu2)/k) / (eu2 / n2k)
    pc <- pf(c,k,n2k,lower.tail=FALSE)
    list(c,pc,k,n2k)
  }
>spatialchow.test <- function(rest,unrest)
  {
    lrest <- rest$LL
    lunrest <- unrest$LL
    k <- rest$parameters - 2
    spchow <- - 2.0 * (lrest - lunrest)
    pchow <- pchisq(spchow,k,lower.tail=F)
    list(spchow,pchow,k)
  }
```

Both of them compare the restricted model (without regimes) with the unrestricted model (with regimes). The spatial version takes the spatial dependence into account.

```
>Ct.ols <- chow.test(fit.ols,fit.ols.r)[[2]]
>Ct.ols
## [1] 3.837563e-126
>SCt.sac <- spatialchow.test(fit.sac,fit.sac.r)[[2]]
>SCt.sac
## [1] 1.05556e-164
```

As we can observe from the *p*-values, both the tests reject the null hypothesis of absence of spatial regimes.

9.3.2 *Spatially constrained clustering*

In this section, the *Skater* procedure (Spatial K'luster Analysis by TreeEdgeRemoval, Assuncao et al. 2006) for the identification of homogeneous contiguous areas according to a spatial neighborhood is described. In this Section, differently from the previous one, a regressive relationship between the output and the inputs is not assumed *a priori* and spatial homogeneous areas are, therefore, not defined according to a functional relationship, but to some characteristics likewise to the standard cluster methods.

In other terms, the described method can be useful in determining contiguous spatial clusters in which some contextual variables, exogenous to a given production process, describe a similar level of demand and/or supply of the territory.

From an application point of view, starting from the DGP simulated in the previous section, we must, therefore, construct some contextual exogenous variables, correlated with $E(\mathbf{Y}|\mathbf{X})$, that represent some factors external to the farm (such as weather or different production areas).

Given that, the variables $Z1$ and $Z2$ have been calculated according to the following steps: (1) $Z1$ and $Z2$ are generated, (2) the contextual variables are centered and scaled, (3) the correlation among variables is removed according to the Cholesky matrix transformation,[4] (4) a variance-covariance matrix is set and then (5) it is used in order to transform the original $Z1$ and $Z2$; finally (6) the operation of centring and scale is removed.

```
### Step 1
>ZZ <- cbind(dataset$y_sac /dataset$L, Z1=rnorm(dim(dataset)
[1]), Z2=rnorm(dim(dataset)[1]))
### Step 2
>mns <- apply(ZZ, 2, mean)
>sds <- apply(ZZ, 2, sd)
>ZZ2 <- sweep(ZZ, 2, mns, FUN="-")
>ZZ2 <- sweep(ZZ2, 2, sds, FUN="/")
### Step 3
>v.obs <- cor(ZZ2)
>ZZ3 <- ZZ2 %*% solve(chol(v.obs))
### Step 4
>r <- cbind( c(1, 0.7, 0.3),
             c(0.7, 1, 0.03),
             c(0.3, 0.03, 1))
### Step 5
>ZZ4 <- ZZ3 %*% chol(r)
### Step 6
>ZZ4 <- sweep(ZZ4, 2, sds, FUN="*")
>ZZ4 <- sweep(ZZ4, 2, mns, FUN="+")
>dataset$Z1 = ZZ4[,2]
>dataset$Z2 = ZZ4[,3]
```

---

[4] Cholesky matrix transformation allows uncorrelated variables to be transformed into correlated ones according to a set variance-covariance matrix, but it is also useful for the reverse operation.

Several analytical regionalisation methods (also known as "*spatially constrained clustering*") have been proposed in the literature (Murtagh 1985; Duque et al. 2007). *Skater* procedure, proposed in the spdep package, offers many advantages, including simplicity of application, the hierarchical nature of the method and the possibility of binding the minimum number of units within each cluster.

So, let us use the simulated exogenous variables *Z*:

```
>datasetx <- dataset[,c("Z1","Z2")]
```

and load the coordinates using the spdep library; from these coordinates, the neighbours list has been calculated thanks to tri2nb function.

```
>library(spdep)
>coords = coordinates(cbind(dataset$lon,dataset$lat))
>neighbours = tri2nb(coords, row.names = NULL)
```

We, therefore, have all the objects that allow us to practically estimate the contiguous homogeneous areas according to the Skater algorithm; first of all, a cost scheme has to be set given the neighbourhood using the nbcosts function

```
>lcosts <- nbcosts(neighbours, datasetx)
```

After computing the cost of each edge—as the distance between nodes—the neighbours list with spatial weights has to be set as

```
>nb.w <- nb2listw(neighbours, lcosts, style="B")
```

The weighted neighbours list (plotted in Figure 9.4, left plot) is often—in the practical applications—too complex to evaluate; the "*minimum spanning tree*" algorithm—as stated before—allows the structure of the initial graph to be simplified, with the aim of achieving a minimum path among units/nodes.

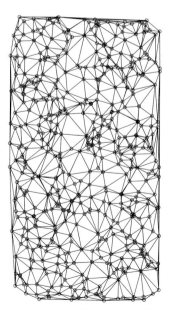
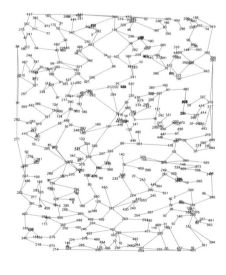

Figure 9.4. Full weighted neighbours and minimum spanning tree.

```
>mst.bh <- mstree(nb.w,5)
```

Figure 9.4 (right plot) highlights how all units are always connected and how these links are reduced compared to the original neighbourhood ones.

```
>par(mar=c(0,0,0,0))
>par(mfrow=c(1,2))
>plot(nb.w, coords)
>plot(mst.bh,coords,col=2,cex.lab=.5, cex.circles=0.035,
fg="blue", bty="n")
```

It is now possible to apply the *Skater* clustering procedure; being a hierarchical procedure derived from the *k-means* algorithm—in addition to the demand/supply variables and the neighbourhood structure—it requires the indication of the preferred number of homogeneous clusters in the `ncuts` argument. Other important options are available in the `skater` function: the most useful one—in our opinion for application purposes—is `crit`, which allows the minimum number of units that must be present in each cluster in order to do not identify clusters too small or be affected by outlier data in the demand/supply variables to be indicated. For the sake of clarity, three thresholds/cuts have been chosen.

```
>ska1 <- skater(mst.bh[,1:2], datasetx, ncuts=4, crit=30)
>ska2 <- skater(mst.bh[,1:2], datasetx, ncuts=5, crit=30)
>ska3 <- skater(mst.bh[,1:2], datasetx, ncuts=6, crit=30)
>dataset$cluster_SKATER1 = ska1$groups
>dataset$cluster_SKATER2 = ska2$groups
>dataset$cluster_SKATER3 = ska3$groups
```

Figure 9.5 shows the units attribution varying parameter *k*; a reasonable stability of the solutions can be noted.

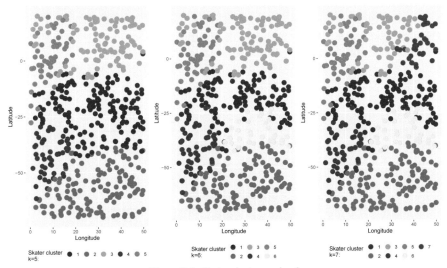

Figure 9.5. Skater cluster varying *k*.

The obtained spatial clusters can, therefore, be used to describe homogeneous demand areas or—for example—be prodromal to subsequent regressive or causal analyses. Similarly to the previous Section, we have chosen to use these spatial clusters in a regressive model to better explain the spatial differences in the relationship between y and X among units.

```
>SKAT <- as.factor(dataset$cluster_SKATER1)
>ly = log(dataset$y_sac)
>lA = log(dataset$A)
>lL = log(dataset$L)
>lK = log(dataset$K)
>fit.ols <- lm(ly ~ lA + lL + lK)
>fit.ols.sk <- lm(ly ~ (SKAT:(lA + lL + lK + SKAT)) + 0)
```

Tables 9.1 and 9.2 and their relative fitting measures (Adjusted R2 respectively equal to 0.6494513 and 0.9993578) show how—for construction—the augmented model can better describe the relationships between inputs and output.

What we are interested in is not only a better adaptation of the model to the data—as it is tautological to expect from our simulation—but also a lower spatial correlation in the estimation residuals. The Moran's I test confirms that the spatial autocorrelation in residuals of the augmented model reduces by half with respect to the linear production model (function `lm.morantest`).

```
>lm.morantest(fit.ols, nb2listw(neighbours, style="W"))
## Global Moran I for regression residuals
## data:
## model: lm(formula = ly ~ lA + lL + lK)
## weights: nb2listw(neighbours, style = "W")
##
## Moran I statistic standard deviate = 24.608, p-value < 2.2e-16
## alternative hypothesis: greater
## sample estimates:
##   Observed Moran I      Expectation        Variance
##      0.6320120827    -0.0060215765    0.0006722517
>lm.morantest(fit.ols.sk, nb2listw(neighbours, style="W"))
##
## Global Moran I for regression residuals
## data:
## model: lm(formula = ly ~ (SKAT:(lA + lL + lK + SKAT)) + 0)
## weights: nb2listw(neighbours, style = "W")
```

Table 9.1. OLS.

	Estimate	Std. Error	t value	Pr(>|t|)
(Intercept)	4.217	0.037	112.457	0.000
lA	−0.378	0.021	−17.592	0.000
lL	−0.527	0.021	−25.184	0.000
lK	0.004	0.022	0.193	0.847

Table 9.2. OLS with cluster.

	Estimate	Std. Error	t value	Pr(>|t|)
SKAT1	3.722	0.087	42.625	0.000
SKAT2	3.715	0.052	71.245	0.000
SKAT3	3.636	0.186	19.507	0.000
SKAT4	4.292	0.113	37.920	0.000
SKAT5	3.291	0.206	15.989	0.000
SKAT1:lA	−0.544	0.029	−19.015	0.000
SKAT2:lA	−0.105	0.034	−3.071	0.002
SKAT3:lA	−0.468	0.035	−13.275	0.000
SKAT4:lA	0.013	0.101	0.129	0.898
SKAT5:lA	0.046	0.105	0.435	0.664
SKAT1:lL	0.025	0.070	0.360	0.719
SKAT2:lL	−0.068	0.060	−1.127	0.260
SKAT3:lL	−0.025	0.135	−0.188	0.851
SKAT4:lL	−0.929	0.120	−7.731	0.000
SKAT5:lL	0.048	0.176	0.270	0.787
SKAT1:lK	0.020	0.026	0.785	0.433
SKAT2:lK	0.011	0.027	0.401	0.688
SKAT3:lK	0.043	0.030	1.417	0.157
SKAT4:lK	−0.032	0.046	−0.699	0.485
SKAT5:lK	0.020	0.048	0.415	0.678

```
##
## Moran I statistic standard deviate = 11.918, p-value < 2.2e-16
## alternative hypothesis: greater
## sample estimates:
##  Observed Moran I      Expectation        Variance
##      0.2744713564    -0.0228614607    0.0006223683
```

References

Abdulai, A. and W.E. Huffman. 2005. The diffusion of new agricultural technologies: The case of crossbred-cow technology in Tanzania. *American Journal of Agricultural Economics* 87: 645–59.

Ackerberg, D.A., K. Caves and G. Frazer. 2015. Identification properties of recent production function estimators. *Econometrica* 83: 2411–51.

Aida, T. 2018. Neighbourhood effects in pesticide use: Evidence from the rural Philippines. *Journal of Agricultural Economics* 69: 163–81.

Aigner, D., C.A.K. Lovell and P. Schmidt. 1977. Formulation and estimation of stochastic frontier production function models. *Journal of Econometrics* 6: 21–37.

Anselin, L. 1988. *Spatial Econometrics: Methods and Models*. Vol. 4. Springer Science & Business Media.

Anselin, L. 1990. Spatial dependence and spatial structural instability in applied regression analysis. *Journal of Regional Science* 30: 185–207.

Anselin, L., R. Bongiovanni and J. Lowenberg-DeBoer. 2004. A spatial econometric approach to the economics of site-specific nitrogen management in corn production. *American Journal of Agricultural Economics* 86: 675–87.

Anselin, L. 2005. *Spatial Regression Analysis in R: A Workbook*. Urbana 51.

Anselin, L. 2010. Thirty years of spatial econometrics. *Papers in Regional Science* 89: 3–25.

Arbia, G. 2014. *A Primer for Spatial Econometrics: With Applications in R*. Springer.

Areal, F.J., K. Balcombe and R. Tiffin. 2012. Integrating spatial dependence into stochastic frontier analysis. *Australian Journal of Agricultural and Resource Economics* 56: 521–41.

Assuncao, R.M., M.C. Neves, G. Camara and C. Da Costa Freitas. 2006. Efficient regionalization techniques for socio-economic geographical units using minimum spanning trees. *International Journal of Geographical Information Science* 20: 797–811.

Billé, A.G., R. Benedetti and P. Postiglione. 2017. A two-step approach to account for unobserved spatial heterogeneity. *Spatial Economic Analysis* 12: 452–71.

Billé, A.G., C. Salvioni and R. Benedetti. 2018. Modelling spatial regimes in farms technologies. *Journal of Productivity Analysis* 49: 173–85.

Bivand, R. and G. Piras. 2015. Comparing implementations of estimation methods for spatial econometrics. *Journal of Statistical Software* 63: 1–36.

Case, A. 1992. Neighborhood influence and technological change. *Regional Science and Urban Economics* 22: 491–508.

Cavailhès, J. and P. Wavresky. 2003. Urban influences on periurban farmland prices. *European Review of Agricultural Economics* 30: 333–57.

Chambers, G. 1998. *Applied Production Analysis: A Dual Approach*. Cambridge University Press, UK.

Charnes, A., W. Cooper and W. Rhodes. 1978. Measuring the efficiency of decision making units. *European Journal of Operational Research* 2: 429–44.

Chasco, C., J. Le Gallo and F.A. López. 2018. A scan test for spatial groupwise heteroscedasticity in cross-sectional models with an application on houses prices in Madrid. *Regional Science and Urban Economics* 68: 226–38.

Chow, G.C. 1960. Tests of equality between sets of coefficients in two linear regressions. *Econometrica: Journal of the Econometric Society* 591–605.

Corral, P. and N. Radchenko. 2017. What's so spatial about diversification in Nigeria? *World Development* 95: 231–53.

Debreu, G. 1951. The coefficient of resource utilization. *Econometrica* 19: 273–92.

Dosi, G. 1988. Sources, procedures, and microeconomic effects of innovation. *Journal of Economic Literature* 26: 1120–71.

Druska, V. and W.C. Horrace. 2004. Generalized moments estimation for spatial panel data: Indonesian rice farming. *American Journal of Agricultural Economics* 86: 185–98.

Duque, J.C., R. Ramos and J. Suriñach. 2007. Supervised regionalization methods: A survey. *International Regional Science Review* 30: 195–220.

Elhorst, J.P. 2014. *Spatial Econometrics: From Cross-Sectional Data to Spatial Panels*. Springer.

Emvalomatis, G. 2012. Adjustment and unobserved heterogeneity in dynamic stochastic frontier models. *Journal of Productivity Analysis* 37: 7–16.

Fang, D. and T.J. Richards. 2018. New maize variety adoption in Mozambique: A spatial approach. *Canadian Journal of Agricultural Economics* 66: 469–488.

Farrell, M.J. 1957. The measurement of productive efficiency. *Journal of the Royal Statistic Society* 120: 253–81.

Feichtinger, P. and K. Salhofer. 2016. The Fischler reform of the common agricultural policy and agricultural land prices. *Land Economics* 92: 411–32.

Ferrara, G. and F. Vidoli. 2017. Semiparametric stochastic frontier models: A generalized additive model approach. *European Journal of Operational Research* 258: 761–77.

Florax, R.J., R.L. Voortman and J. Brouwer. 2002. Spatial dimensions of precision agriculture: A spatial econometric analysis of millet yield on Sahelian coversands. *Agricultural Economics* 27: 425–43.

Fried, H.O. and C.A.K. Lovell. 2008. *The Measurement of Productive Efficiency and Productivity Growth*. Oxford University Press: USA.

Fruh-Muller, A., M. Bach, L. Breuer, S. Hotes, T. Koellner, C. Krippes and V. Wolters. 2019. The use of agri-environmental measures to address environmental pressures in Germany: Spatial mismatches and options for improvement. *Land Use Policy* 84: 347–62.

Fusco, E. and F. Vidoli. 2013. Spatial stochastic frontier models: Controlling spatial global and local heterogeneity. *International Review of Applied Economics* 27: 679–94.

Greene, W. 2005. Reconsidering heterogeneity in panel data estimators of the stochastic frontier model. *Journal of Econometrics* 126: 269–303.

Greene, W.H. 2008. The econometric approach to efficiency analysis. *The Measurement of Productive Efficiency and Productivity Growth* 1: 92–250.

Griffin, R.C., J.M. Montgomery and M.E. Rister. 1987. Selecting functional form in production function analysis. *Western Journal of Agricultural Economics* 12: 216–27.

Grogan, K.A. and R.E. Goodhue. 2012. Spatial externalities of pest control decisions in the California citrus industry. *Journal of Agricultural and Resource Economics* 37: 156–79.

Holloway, G. and M.L.A. Lapar. 2007. How big is your neighbourhood? Spatial implications of market participation among Filipino smallholders. *Journal of Agricultural Economics* 58: 37–60.

Kelejian, H. and G. Piras. 2017. *Spatial Econometrics*. Academic Press.

Kostov, P. 2009. Spatial dependence in agricultural land prices: Does it exist? *Agricultural Economics* 40: 347–53.

Krishna, V.V. and M. Qaim. 2012. Bt cotton and sustainability of pesticide reductions in India. *Agricultural Systems* 107: 47–55.

Krishnan, P. and M. Patnam. 2013. Neighbors and extension agents in Ethiopia: Who matters more for technology adoption? *American Journal of Agricultural Economics* 96: 308–27.

Kumbhakar, S.C., C.F. Parmeter and V. Zelenyuk. 2017. Stochastic frontier analysis: Foundations and advances. *Working Papers 2017–10*. University of Miami, Department of Economics.

Kuosmanen, T. and M. Kortelainen. 2012. Stochastic non-smooth envelopment of data: Semi-parametric frontier estimation subject to shape constraints. *Journal of Productivity Analysis* 38: 11–28.

Lapple, D. and H. Kelley. 2015. Spatial dependence in the adoption of organic drystock farming in Ireland. *European Review of Agricultural Economics* 42: 315–37.

Lapple, D., G. Holloway, D.J. Lacombe and C. O'Donoghue. 2017. Sustainable technology adoption: A spatial analysis of the Irish dairy sector. *European Review of Agricultural Economics* 44: 810–35.

Latruffe, L., B.E. Bravo-Ureta, A. Carpentier, Y. Desjeux and V.H. Moreira. 2017. Subsidies and technical efficiency in agriculture: Evidence from European dairy farms. *American Journal of Agricultural Economics* 99: 783–99.

Le Gallo, J. 2014. Cross-section spatial regression models. pp. 1511–1533. *In*: Fischer, M.M. and P. Nijkamp (eds.). *Handbook of Regional Science*. Berlin: Springer.

LeSage, J.P. and R.K. Pace. 2009. *Introduction to Spatial Econometrics*. Chapman & Hall/CRC: Boca Raton, FL.

Levinsohn, J. and A. Petrin. 2003. Estimating production functions using inputs to control for unobservables. *The Review of Economic Studies* 70: 317–41.

Livanis, G., C.B. Moss, V.E. Breneman and R.F. Nehring. 2006. Urban sprawl and farmland prices. *American Journal of Agricultural Economics* 88: 915–29.

Maddison, D. 2009. A spatio-temporal model of farmland values. *Journal of Agricultural Economics* 60: 171–89.

Marconi, V., M. Raggi and D. Viaggi. 2015. Assessing the impact of Rdp agri-environment measures on the use of nitrogen-based mineral fertilizers through spatial econometrics: The case study of Emilia-Romagna. *Ecological Indicators* 59: 27–40.

Marschak, J. and W. Andrews. 1944. Random simultaneous equations and the theory of production. *Econometrica* 12: 143–205.

Meeusen, W. and J. van den Broeck. 1977. Efficiency estimation from Cobb-Douglas production functions with composed error. *International Economic Review* 18: 435–44.

Moran, P.A.P. 1950. Notes on continuous stochastic phenomena. *Biometrika* 37: 17–23.

Murtagh, F. 1985. A survey of algorithms for contiguity-constrained clustering and related problems. *The Computer Journal* 28: 82–88.

Nelson, R. and S.G. Winter. 1982. *An Evolutionary Theory of Economic Change*. Harvard Business School Press: Cambridge.

Nickerson, C.J. and L. Lynch. 2001. The effect of farmland preservation programs on farmland prices. *American Journal of Agricultural Economics* 83: 341–51.

Olley, G. and A. Pakes. 1996. The dynamics of productivity in the telecommunications equipment industry. *Econometrica* 64: 1263–97.

Park, B.U. and L. Simar. 1994. Efficient semiparametric estimation in a stochastic frontier model. *Journal of the American Statistical Association* 89: 929–36.

Patton, M. and S. McErlean. 2003. Spatial effects within the agricultural land market in Northern Ireland. *Journal of Agricultural Economics* 54: 35–54.

Pede, V.O., F.J. Areal, A. Singbo, J. McKinley and K. Kajisa. 2018. Spatial dependency and technical efficiency: An application of a Bayesian stochastic frontier model to irrigated and rainfed rice farmers in Bohol, Philippines. *Agricultural Economics* 49: 301–12.

Saint-Cyr, L.D., H. Storm, T. Heckelei and L. Piet. 2018. Heterogeneous impacts of neighbouring farm size on the decision to exit: Evidence from Brittany. *European Review of Agricultural Economics* 46: 237–66.

Schmidtner, E., C. Lippert, B. Engler, A.M. Hearing, J. Aurbacher and S. Dabbert. 2012. Spatial distribution of organic farming in Germany: Does neighbourhood matter? *European Review of Agricultural Economics* 39: 661–83.

Shee, A. and S.E. Stefanou. 2014. Endogeneity corrected stochastic production frontier and technical efficiency. *American Journal of Agricultural Economics* 97: 939–52.

Skevas, T., I. Skevas and S.M. Swinton. 2018. Does spatial dependence affect the intention to make land available for bioenergy crops? *Journal of Agricultural Economics* 69: 393–412.

Storm, H., K. Mittenzwei and T. Heckelei. 2014. Direct payments, spatial competition, and farm survival in Norway. *American Journal of Agricultural Economics* 97: 1192–1205.

Tsionas, E.G. and P.G. Michaelides. 2016. A spatial stochastic frontier model with spillovers: Evidence for Italian regions. *Scottish Journal of Political Economy* 63: 243–57.

Vidoli, F., C. Cardillo, E. Fusco and J. Canello. 2016. Spatial nonstationarity in the stochastic frontier model: An application to the Italian wine industry. *Regional Science and Urban Economics* 61: 153–64.

Wang, H. 2018. The spatial structure of farmland values: A semiparametric approach. *Agricultural and Resource Economics Review* 47: 568–91.

Wilson, P.W. 2008. FEAR 1.0: A software package for frontier efficiency analysis with R. *Socio-Economic Planning Sciences* 42: 247–54.

Wollni, M. and C. Andersson. 2014. Spatial patterns of organic agriculture adoption: Evidence from Honduras. *Ecological Economics* 97: 120–28.

Yang, X., M. Odening and M. Ritter. 2019. The spatial and temporal diffusion of agricultural land prices. *Land Economics* 95: 108–23.

CHAPTER 10
Areal Interpolation Methods
The Bayesian Interpolation Method

Domenica Panzera

10.1 Introduction

The availability of reliable agricultural and rural statistics serves different purposes and involves relevant needs of both the private and the public sectors. Agricultural and rural data can be useful to identify trends and future prospects for agricultural commodities markets, as well as to assess the impact of agricultural activities on economic development and the environment. Furthermore, these data are especially necessary to inform policy decisions and to monitor the efficacy of commitments and policies at national, regional, and sub-regional levels.

The main approaches to the generation of agricultural statistics include complete enumeration and sample survey. Cost effectiveness and time saving motivate the widespread use of sample survey. Sample survey data certainly can be used to derive direct estimators of characteristics of interest in large areas or domains. Moreover, the growing demand for reliable agricultural statistics at a local level contributed to the increasing importance of small area estimation in survey sampling (Ghosh and Rao 1994). Small area estimation (SAE) relates to the production of reliable estimates of characteristics of interest for small geographical areas or domains. The usual direct sample estimators for a small area, based on the only sample data in the area, are likely to yield large standard errors, because of the small sample size. When direct estimation based on area-specific sample data is not possible, one should rely on indirect, or model-based, SAE methods. These methods, making use of auxiliary information on related areas, allow more accurate estimates for a given area or for several areas to be obtained (Ghosh and Rao 1994). See Chapter 11 for further details about SAE.

When reliable and quality aggregate data are available for a large area that includes the small area of interest, areal interpolation techniques can be applied.

"G. d'Annunzio" University of Chieti-Pescara, Department of Economic Studies.
Email: domenica.panzera@unich.it

Areal interpolation refers to the transformation of spatial data from one zonal system to another (Goodchild and Lam 1980; Sadahiro 1999). Areal interpolation can be considered as the conversion of spatial data from a set of *source zones* into a set of *target zones* (Ford 1976). A number of different areal interpolation methods have been proposed in the literature, and they can be classified according to the variable they apply to, the mass preserving property satisfaction, the use of auxiliary information, and the use of simplifying assumptions (Goodchild et al. 1993; Sadahiro 1999).

For accurate areal interpolations, the special features of spatial data should also be considered. Spatial autocorrelation reflects a situation where values observed at nearby locations are similar, whereas those more widely separated are less so (Cressie 1993). Spatial autocorrelation has been considered in some contributions on areal interpolation. Bayesian spatial statistics-based methods (Murakami and Tsutsumi 2012) consider spatial autocorrelation, making assumptions on the spatial data generating process. A conditional autoregressive process is considered by Benedetti and Palma (1994) and Mugglin et al. (1999), and a hierarchical Bayesian framework has been developed by Wikle and Berliner (2005). Further procedures for areal interpolation, that account for spatial autocorrelation, have been proposed by Kyriakidis (2004), and Murakami and Tsutsumi (2011), among others.

This chapter provides an overview of the most common areal interpolation methods. The Bayesian interpolation method introduced by Benedetti and Palma (1994) will be illustrated as a special case of areal interpolation method that accounts for spatial autocorrelation and applied to the areal conversion of agricultural data. The chapter is organised as follows: Section 10.2 is focused on illustrating the main areal interpolation methods. Section 10.3 concerns the introduction of spatial autocorrelation in the areal interpolation, and Section 10.4 describes the Bayesian Interpolation Method (BIM) as a solution to the areal interpolation problem that exploits this general property of spatial data. The methodology is illustrated trough an empirical application focused on agricultural data related to Italian regions. Suggestions to perform the empirical analysis by the R software are given in Section 10.5, and Section 10.6 concludes the chapter.

10.2 An overview of areal interpolation methods

Areal interpolation consists of converting spatial data from a given zonal system to another. Specifically, it refers to the process of estimating one or more variables for a set of target zones, based upon the known values in the set of source zones. Target units can be either finer-scale or misaligned with respect to the source zones.

Formally, let us consider n_s source zones s_i, $i = 1,2,..,n_s$, and n_t target zones t_j, $j = 1,2,..,n_t$, with n_s generally not equal to n_t. Assume that source and target zones cover the same geographical domain, and that the values of the variable Y, known for the source zones, are unobserved for the target zones. Solving the areal interpolation problem consists of obtaining $\mathbf{y}_t$, a n_t-dimensional vector with elements the target zones estimates $\hat{y}_{tj}$, $j = 1,2,..,n_t$, from the n_s-dimensional vector $\mathbf{y}_s$, with elements the observations on Y for the n_s source zones, y_{si}, $i = 1,2,..,n_s$ (Goodchild and Lam 1980). Many areal interpolation methods use spatial units whose boundaries are created by an intersection between source and target zones. These units are termed intersection units and could be denoted by st_k, with $k = 1,2,..., n_k$. A representation of source, target and intersection zones is reported in Figure 10.1.

The variable under consideration, Y, may depend on the size of the spatial units. Counts, totals or measures, such as perimeters or areas, are summary statistics for the unit, and are only true when they represent the area as a whole. In other words, the value of these variables extends with the extension of the area, and they are defined as *spatially extensive* variables (Goodchild and Lam 1980; Lam 1983). In contrast, a *spatially intensive* variable is one which describes any part of the unit, and thus is expected to have the same value in each part of the area as in the whole area. These variables do not depend on the size of the units. Population density, ratios or averages are examples of spatially intensive variables (Goodchild and Lam 1980; Lam 1983).

Areal interpolation methods can be classified into two broad categories corresponding to non-volume-preserving methods and volume-preserving methods (Lam 1983) and apply to the estimation of either spatially extensive variables or spatially intensive variables.

The non-volume-preserving methods, also known as point-based methods, imply overlaying a grid on the map of source units, and assigning a control point, usually the centroid, to represent each source zone. A representative value for the region is then assigned to each point. For spatially extensive data, this value would be the region's value y_i divided by the region's area; for spatially intensive data, this value would be y_i itself. Point interpolation schemes are then applied to interpolate the values at each node in the grid. The interpolated values lying within each target zone are then summed, in the case of spatially extensive data, or averaged, for spatially intensive data, in order to obtain the final target zone estimate (Goodchild and Lam 1980; Lam 1983). Different point interpolation methods can be applied to assign values to grid points. For a review of such methods see, among others, Ripley (1981) and Cressie (1993). The point-based areal interpolation methods are affected by the limitations of the point interpolation methods upon which they are based. A further drawback of these approaches is that they do not preserve the total value within each source zone.

In contrast, when volume-preserving methods are applied, the estimates of the variable in the target units are such that the total value of the variable within each source zone is preserved. This feature makes the volume-preserving approach more reliable than the non-volume-preserving method. Furthermore, unlike the non-

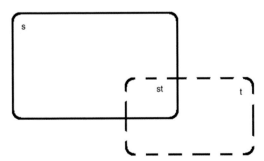

Figure 10.1. Source zone *s* (full line), target zone *t* (dashed line) and their intersection st.

volume-preserving approach, the volume-preserving areal interpolation method uses the zone itself as the unit of operation instead of the arbitrarily assigned control points. Because of this feature the volume-preserving approach is also termed area-based areal interpolation. Two different methods based upon this approach can be distinguished, such as the overlay (or areal weighting) method, and the pycnophylactic method.

In the overlay approach (Goodchild and Lam 1980), source and target zones are superimposed, and an $n_t \times n_s$ matrix $\mathbf{A}$, with elements a_{ts} identifying the areas of intersection between the n_t target units and the n_s source zones, is constructed. The matrix $\mathbf{A}$ is then standardised to obtain the weight matrix $\mathbf{W}$. The matrix is standardised by columns for spatially extensive data, and by rows, when spatially intensive variables are considered.

The values of the variable of interest for the target zones, y_{tj}, can be then estimated as follows

$$\hat{y}_{tj} = \sum_i y_{si} w_{ji} \qquad (10.1)$$

with w_{ji} denoting the (j, i)-th element of the matrix $\mathbf{W}$, for $j = 1, 2, .., n_t$ and $i = 1, 2, .., n_s$.

For a recent application of areal weighting to deal with regional data comparability, see Netrdová et al. (2020).

The main drawback of the areal weighting method is the underlying assumption of homogeneity within each source zone. This assumption is relaxed in the pycnophylactic approach proposed by Tobler (1979). This method assumes the existence of a smooth density function which takes into account the effect of adjacent source zones (Lam 1983). According to the pycnophylactic method, the average value z of the source zone is first assigned to each grid cell superimposed on the source zone. Then, each cell is averaged with its neighbours, and the predicted z values in each source zone are compared with the actual z values, and adjusted to satisfy the following pycnophylactic condition

$$\iint_{xy} Z(x, y) dx \, dy = y_{si} \qquad (10.2)$$

where $Z(x, y)$ is the density function, (x, y) defines a location for which no direct observation is available, and y_{si} is the value of the variable in the source zone. This iterative procedure continues until the predicted values z are not significantly different from the actual values z, or the values for each cell do not greatly change between subsequent iterations.

The aforementioned areal interpolation methods do not use auxiliary information available for the target zones. The availability of observations on the variable X for target zones could improve the accuracy of the estimate of the unobserved values of Y. According to Langford et al. (1991), areal interpolation methods that use auxiliary information can be defined as *intelligent* interpolation methods, as opposed to *simple* methods, that do not consider ancillary data. Some contributions focused on intelligent areal interpolation that rely on establishing a regression relationship between the variable of interest and one or more auxiliary variables have been proposed (Flowerdew and Green 1989, 1991; Flowerdew et al. 1991; Goodchild et al. 1993). Methods for the estimation of the assumed relationships have been developed, accounting for the probability distribution of the variable of interest.

A linear regression-based areal interpolation method that uses the EM (expectation-maximization) algorithm (Dempster et al. 1977) has been developed by Flowerdew and Green (1992). The EM algorithm is a statistical approach that has been primarily developed to deal with missing data. Flowerdew and Green (1992) consider data for the intersections between target and source zones as missing values, to be estimated by the EM algorithm. Given $\bar{y}_{si}$, the means of a continuous variable observed for n_s source zones, the means $\bar{y}_{tj}$ of the same variable for n_t target zones are unobserved. To estimate these unknown values, first, the value of the variable of interest for the n_k areas of intersection between source and target zones, y_{stk}, is considered.

At the initial step, the value of y_{stk} is set as follows:

$$\hat{y}_{stk} = \sum_p x_{kp}\beta_p + [\bar{y}_{s_i} - \sum_k \frac{n_{st_k}}{n_{s_i}} \sum_p x_{kp}\beta_p] \tag{10.3}$$

where $\bar{y}_{si}$ is the observed objective variable of the source unit s_i that contains the intersection unit st_k, n_{st_k}, and n_{s_i} are the weights assigned to source and intersection units, respectively, p is the index for the explanatory variables, x_{kp} is the p-th explanatory variable observed for the intersection unit k, and β_p is the associated parameter.

The following maximization (M) and expectation (E) steps are then iterated until convergence. In the M step the parameter β_p is estimated by applying weighted least squares, according to the following assumption:

$$\hat{y}_{st_k} = \sum_p x_{kp}\beta_p + \varepsilon_k \qquad \varepsilon_k \sim N\left(0, \frac{\sigma^2}{n_{st}}\right) \tag{10.4}$$

In the E-step, by substituting the estimated β_p into equation (10.3), $\hat{y}_{st_k}$ is interpolated.

The final step consists of obtaining the target values $\bar{y}_{tj}$ as the weighted means of the $\hat{y}_{st_k}$ from the E-step, that is

$$\bar{y}_{tj} = \sum_k n_{st_k} \hat{y}_{st_k}/n_t \tag{10.5}$$

Despite the number of areal interpolation techniques that have been proposed, little attention has been paid to quantify the error introduced in the resulting data. In fact, the accuracy of areal interpolation methods has been mainly treated at an empirical level (Goodchild et al. 1993; Flowerdew and Green 1994). Fisher and Langford (1995) used a Monte Carlo simulation to compare different areal interpolation methods. For all the compared methods, the Authors found that accuracy improves as the number of target zones decline. For regression-based approaches, they found that the complexity of the regression model used could improve accuracy. Sadahiro (1999) compared alternative areal interpolation methods using stochastic modelling and concluded that intelligent methods are generally more accurate than simple methods. A recent review of the main approaches to areal interpolation with or without ancillary information can be found in Comber and Zeng (2019).

10.3　Areal interpolation methods accounting for spatial autocorrelation

Among the different areal interpolation methods, only few consider spatial autocorrelation. Spatial autocorrelation is a general property of spatial data which implies that data collected at near locations tend to be more similar than more widely separated observations (Cressie 1993). Following Murakami and Tsutsumi (2012), the areal interpolation methods accounting for spatial autocorrelation can be classified into three main categories, such as the *standard spatial-statistic-based* methods, the *Bayesian spatial-statistic-based* methods, and the *spatial-econometric-based* methods.

The first category includes area-to-point kriging (Kyriakidis 2004), and its extensions (Yoo and Kyriakidis 2006, 2009; Goovaerts 2006, 2010; Gotway and Young 2007; Liu et al. 2008). Conventional kriging is a geostatistical method for spatial prediction that uses information on spatial autocorrelation of different attributes (Matheron 1971). In its original formulation, kriging focuses on the spatial prediction of an areal value, using available areal data for the same or another variable. Kyriakidis (2004) proposed the use of kriging in the context of areal interpolation, by introducing the area-to-point kriging, where both source and target data pertain to the same attribute that is defined over areal units and points, respectively. In area-to-point kriging, both the unobserved point support values, and the known observations at source areal units are viewed as realizations of a random function, that is a collection of spatially correlated random variables. A functional relationship is established between point and areal data, with the latter expressed as a weighted linear combination of point support values. The functional relationship between point and areal data ensures that the area-to-area covariance and the cross-covariance between the areal data and the point values, that are used in the kriging approach, are completely specified in terms of the point covariance model.

The Bayesian spatial-statistics-based methods consider spatial autocorrelation, assuming some spatial stochastic process as the data generating process. Benedetti and Palma (1994) and Mugglin et al. (1999) used a Conditional Autoregressive (CAR) prior (Besag 1974) to incorporate similarities in neighbouring regions, and quantified the posterior distribution, that allows inference to be made on the variable of interest at the desired spatial scale. Winkle and Berliner (2005) developed a Bayesian hierarchical framework by which information at different spatial scales can be combined. This approach is based on conditioning a continuous spatial process on an areal average of the process defined at some resolution for which inference is desired and/or prior information is available. Then, data observed at larger, the same, or smaller resolutions are modelled conditionally on the process at the resolution of interest.

Spatial econometrics has been applied for areal interpolation by Murakami and Tsutsumi (2011). The proposed approach extends the linear regression areal interpolation methods to include spatial autocorrelation. The authors proposed combining the basic linear regression model with a spatial process model (Banerjee et al. 2004) in order to take into account spatial dependence.

The standard spatial-statistic-based methods have as their main advantage that their predictors minimize the expected square error, whereas the main merit of

Bayesian spatial-statistic-based methods is that they allow predictive distributions to be identified. Finally, a benefit of spatial-econometrics-based methods concerns their adaptability to socio-economic data (Murakami and Tsutsumi 2012).

10.4 The Bayesian Interpolation Method (BIM)

A Bayesian solution to the areal interpolation problem which uses spatial autocorrelation is given by the Bayesian Interpolation Method (BIM) introduced by Benedetti and Palma (1994).

The BIM requires some assumptions on the spatial data generating process. Commonly, spatially referenced data are considered to be a realization from a spatial stochastic process or random field, that is a collection of random variables indexed by their locations. When dealing with the areal interpolation problem, data related to both source and target zones can be interpreted as realizations of spatial stochastic processes. The spatial stochastic process generating the data related to the target zones (i.e., the areal units corresponding to the finer spatial scale) is referred to as the original process. The spatial stochastic process generating the data for the source zones (i.e., the areal units corresponding to the aggregated spatial level) is referred to as the aggregated process. Assuming that data are available only at the aggregated spatial level, the objective becomes to restore the realizations of the original process given the realization of the aggregated one.

The assumption on which the BIM is based concerns the joint probability distribution of the original process, which is assumed to be a Gaussian distribution. The spatial dependence effect is taken into account by modelling the Gaussian random field by the Conditional Autoregressive (CAR) specification (Besag 1974). This assumption does not entail any loss of generality since any Gaussian process on a finite set of sites can be modelled according to this specification (Ripley 1981). CAR specification introduces the spatial dependence effect in the covariance structure of the process as a function of a scalar parameter of spatial autocorrelation and of a spatial weight matrix, which summarizes the proximity between any pairs of spatial units. Following a Bayesian approach, the prior information on the distribution of the original process is combined with the data available at the aggregated spatial level to derive the posterior probability distribution of the original process. Benedetti and Palma (1994) derive the parameters of this posterior probability distribution, that are the BIM estimates. Any inference on the original process can be based upon the specified posterior distribution.

To formalize the described methodology, consider n areal units which form a partition Ω over a geographical domain. Denote by $\mathbf{y} = (y_1, y_2, \ldots, y_n)'$ the data related to a variable of interest Y observed on the n areal units. The vector $\mathbf{y}$ can be interpreted as a realization of the original process expressed by the random vector $\mathbf{Y} = (Y_1, Y_2, \ldots, Y_n)'$. By grouping the n units into larger areas, we obtain a set of $m < n$ areal units which define a new partition Ω^* over the same geographical domain. The data observed for this new partition can be denoted by $\mathbf{y}^* = (y_1^*, y_2^*, \ldots, y_m^*)'$, and the underlying spatial stochastic process, that is expressed by the random vector $\mathbf{Y}^* = (Y_1^*, Y_2^*, \ldots, Y_m^*)'$, represents the aggregated process. Assume that data are only available for the partition Ω^*, while we are interested in the spatial scale corresponding to

Ω. The issue becomes to restore the realizations of the original process given the realization $\mathbf{y}^*$ of the aggregated one.

According to Benedetti and Palma (1994), the joint distribution of the random vector $\mathbf{Y}$ is assumed to be a multivariate normal distribution, and $\mathbf{Y}$ is expressed in an additive form as:

$$\mathbf{Y} = \mathbf{S} + \boldsymbol{\varepsilon} \tag{10.6}$$

where $\mathbf{S} = (S_1, S_2, \ldots, S_n)'$ refers to the variable of interest at the n sites, and $\boldsymbol{\varepsilon}$ is a random vector of error terms. As an additional assumption, the random vectors $\mathbf{S}$ and $\boldsymbol{\varepsilon}$ are modeled through the Conditional Autoregressive (CAR) specification (Besag 1974).

The CAR specification introduces the spatial dependence effect in the covariance structure of the process as a function of the scalar parameter ρ and of the spatial weight matrix $\mathbf{W}$. Assuming a CAR specification, the random vector $\mathbf{S}$ has a multivariate normal distribution, with the following parameters:

$$\mathbf{S} \sim MVN(\boldsymbol{\mu}, (\mathbf{I} - \rho\mathbf{W})^{-1} \boldsymbol{\Sigma}_S) \tag{10.7}$$

where $\boldsymbol{\mu} = (\mu_1, \mu_2, \ldots, \mu_n)'$ is the mean vector, and $\boldsymbol{\Sigma}_S$ is the variance-covariance matrix expressed as:

$$\boldsymbol{\Sigma}_S = (\mathbf{I} - \rho\mathbf{W})^{-1} \mathbf{M} \tag{10.8}$$

with $\mathbf{I}$ denoting a n-dimensional identity matrix and $\mathbf{M} = diag(\sigma_1^2, \ldots, \sigma_n^2)$. The matrix $\mathbf{W}$ is a $n \times n$ matrix commonly specified by normalizing a contiguity matrix $\mathbf{C}$ with elements $c_{ij} = 1$ if $j \in N(i)$, with $N(i)$ denoting the set of units that are neighbours of unit i, $c_{ij} = 0$ otherwise, and $c_{ii} = 0$. The contiguity matrix can be defined according to the geographical arrangement of the observations or alternatively using their economic or social distances (Conley and Topa 2002).

Let $c_{i+} = \sum_{j=1}^{n} c_{ij}$ denote the cardinality of $N(i)$, for all i. Constructing the n-dimensional diagonal matrix $\mathbf{D} = diag(c_{1+}, \ldots, c_{n+})$, the matrix $\mathbf{W}$ can be expressed as follows:

$$\mathbf{W} = \mathbf{D}^{-1} \mathbf{C} \tag{10.9}$$

When the weighting scheme in (10.9) is used, the conditional variances have to be inversely proportional to c_{i+}, for all i (Clayton and Berardinelli 1992; Wall 2004). Then, by setting $\sigma_i^2 = \sigma_S^2/c_{i+}$ where σ_S^2 is a scalar parameter measuring the overall variability of $\mathbf{S}$, the covariance matrix in (10.8) becomes:

$$\boldsymbol{\Sigma}_S = \sigma_S^2 (\mathbf{D} - \rho\mathbf{C})^{-1} \tag{10.10}$$

where the matrices $\mathbf{D}$ and $\mathbf{C}$ are defined as above.

Similar considerations hold for the random vector $\boldsymbol{\varepsilon}$ which is modeled through a zero-centered CAR, so that its joint distribution is given by:

$$\boldsymbol{\varepsilon} \sim MVN(\mathbf{0}, \boldsymbol{\Sigma}_\varepsilon) \tag{10.11}$$

where $\mathbf{0}$ denotes a $n \times 1$ vector of zeros and $\boldsymbol{\Sigma}_\varepsilon = \sigma_\varepsilon^2(\mathbf{D} - \rho_\varepsilon\mathbf{C})^{-1}$, with σ_ε^2 and ρ_ε denoting scalar parameters. From the formulated assumptions, it follows that the joint distribution of the original process can be expressed as:

$$\mathbf{Y} \sim MVN(\mathbf{S}, \boldsymbol{\Sigma}_\varepsilon) \tag{10.12}$$

where $\mathbf{S}$ and $\mathbf{\Sigma}_\varepsilon$ are specified as above. The distribution in (10.12) is assumed for the original process and a correspondence between the original process and the aggregated one can be established by introducing a linear transformation operator $\mathbf{G}$. The operator $\mathbf{G}$ is constructed as a $m \times n$ matrix whose elements can be specified according to any averaging or sum operations, so that:

$$\mathbf{Y}^* = \mathbf{GY} = \mathbf{G}(\mathbf{S} + \varepsilon) = \mathbf{GS} + \mathbf{G}\varepsilon \tag{10.13}$$

Since the observed aggregated data is derived from the unobserved disaggregated data through the operator $\mathbf{G}$, Benedetti and Palma (1994) give a solution to the areal data conversion problem based on identifying the posterior probability distribution of $\mathbf{S}|\mathbf{Y}^*$. According to the Bayes' rule, this posterior probability distribution can be derived as follows:

$$P(\mathbf{S}|\mathbf{Y}^*) \propto P(\mathbf{S})P(\mathbf{Y}^*|\mathbf{S}) \tag{10.14}$$

where $P(\mathbf{S})$ is the prior probability distribution of the random vector $\mathbf{S}$, and $P(\mathbf{Y}^*|\mathbf{S})$ is its likelihood function, given the observed data.

Based upon the assumption in (10.7), $\mathbf{S}$ has a multivariate normal distribution. Furthermore, the conditional distribution of $\mathbf{Y}^*|\mathbf{S}$ can be derived by the distribution of $\mathbf{Y}$ (see equation 10.12) as follows (Anderson 1958):

$$\mathbf{Y}^*|\mathbf{S} \sim MVN(\mathbf{GS}, \mathbf{G}\mathbf{\Sigma}_\varepsilon \mathbf{G}') \tag{10.15}$$

From these results, it follows that the posterior distribution of $\mathbf{S}|\mathbf{Y}^*$ derived as in (10.14) is again multivariate normal. Under the additional hypothesis of known covariance matrix of the original process, we obtain:

$$\mathbf{S}|\mathbf{Y}^* \sim MVN(\hat{\mathbf{S}}, \mathbf{V}_{\hat{S}}) \tag{10.16}$$

where $\hat{\mathbf{S}}$, and $\mathbf{V}_{\hat{S}}$ are the BIM estimates defined as follows:

$$\mathbf{V}_{\hat{S}} = [\mathbf{G}'(\mathbf{G}\frac{(\mathbf{D}-\rho_\varepsilon\mathbf{C})}{\sigma_\varepsilon^2}\mathbf{G}')^{-1}\mathbf{G} + \frac{(\mathbf{D}-\rho\mathbf{C})}{\sigma_S^2}]^{-1} \tag{10.17}$$

$$\hat{\mathbf{S}} = \mathbf{V}_{\hat{S}}[\frac{(\mathbf{D}-\rho\mathbf{C})}{\sigma_S^2}\mathbf{\mu} + \mathbf{G}'(\mathbf{G}\frac{(\mathbf{D}-\rho_\varepsilon\mathbf{C})}{\sigma_\varepsilon^2}\mathbf{G}')^{-1}\mathbf{Y}^*] \tag{10.18}$$

The solutions in (10.17) and (10.18) can be conditioned to the linear constrain $\mathbf{G}\hat{\mathbf{S}} = \mathbf{Y}^*$ obtaining the constrained BIM estimates, specified as:

$$\tilde{\mathbf{S}} = \hat{\mathbf{S}} + \mathbf{V}_{\hat{S}}\mathbf{G}'[\mathbf{GV}_{\hat{S}}\mathbf{G}']^{-1}(\mathbf{Y}^* - \mathbf{G}\hat{\mathbf{S}}) \tag{10.19}$$

$$\mathbf{V}_{\tilde{S}} = \mathbf{V}_{\hat{S}} - \mathbf{V}_{\hat{S}}\mathbf{G}'[\mathbf{GV}_{\hat{S}}\mathbf{G}']^{-1}\mathbf{GV}_{\hat{S}}. \tag{10.20}$$

The imposed constraint allows the pycnophylactic property discussed above to be preserved (Tobler 1979).

BIM estimates completely summarize the available information on the original process, which can be then inferred. Point estimates for $\mathbf{S}$ can be obtained by using $\hat{\mathbf{S}}$ or the constrained solution given in (10.20). Confidence intervals and hypothesis tests can be performed as usual using multivariate normal distributions.

Some contributions in the recent literature proposed the application of the described method. The BIM has been applied by Panzera and Viñuela (2018) to estimate GDP data for the Local Labour Markets (LLMs) in Spain. In Panzera and Postiglione (2014) the method has been applied to address the lack of data on investment for Italian provinces. The authors derived the BIM estimates assuming a Simultaneous Autoregressive (SAR) specification (Whittle 1954) for the data generating process. Furthermore, in Panzera et al. (2016) the BIM has been combined with a multiple imputation procedure in order to develop an approach for handling missing data.

10.5 An example of application of the BIM

We consider an application of the BIM in the disaggregation of agricultural data. Specifically, we consider, as variable to disaggregate, the production per hectare (in euros), of firms specialized in olive growing. This variable, available from the 2010 Agricultural Census, is considered with reference to the 20 Italian regions and to the 110 Italian provinces, corresponding to the NUTS 2 and the NUTS 3 levels of the official European classification, respectively. The BIM is used to disaggregate the production (in euros) per hectare of firms specialized in olive growing, available for Italian regions, at the provincial level. The observations at the NUTS 3 level, available from the 2010 Agricultural Census, are assumed to be the target values. After performing the BIM, the actual values at the NUTS 3 level are used to assess the performance of the method.

In our empirical application, the 20 Italian regions represent the source zones, and the 110 Italian provinces are the target zones. Source and target units can be depicted on a map by using the R software. The maps depicted in Figure 10.2 are obtained by plotting the shape files that, for Italian regions and provinces, are available on the ISTAT website at the following link: https://www.istat.it/it/archivio/222527. In this empirical analysis, the shape files with administrative boundaries referred to 2010 are considered.

The package rgdal offers the readOGR() function that allows to read the shape files using the following syntax:

```
> library(rgdal)
> REG_shape<-readOGR("Reg01012010_WGS84.shp")
> PROV_shape<-readOGR("Prov01012010_WGS84.shp")
```

The basic plot() function can be thus used to plot the shape files as follows:

```
> par(mfrow=c(1,2))
> plot(REG_shape)
> plot(PROV_shape)
# The par(mfrow) function allows a multi-panelled plot to be
created. The first argument in mfrow specifies the number of rows,
the second argument indicates the number of columns of the plot.
```

To build the variable used in this empirical application (i.e., the production, in euros, per hectare of firms specialized in olive growing), we consider the production in euros of firms specialized in olive growing divided by the agricultural area used by these firms for production, in hectares. These data are available on the ISTAT web site, for both NUTS 2 and NUTS 3 regions, at the following link: http://dati-censimentoagricoltura.istat.it/index.aspx?queryid=10164.

Figure 10.2. Italian regions (NUTS 2) and provinces (NUTS 3).

The available data for Italian regions and provinces can be stored in separate files, that should also contain the identification codes of regions and provinces. These identification codes could correspond to the codes that for both regions and provinces are included in the shape files.

Specifically, data on the variables of interest can be stored in the following files "Prod_reg.csv", "Hectares_reg.csv", "Prod_prov.csv" and "Hectare_prov.csv", including data on production and agricultural area, for both regions and provinces. These data can be uploaded as follows:

```
> A_p<-read.csv(file=" Prod_reg.csv", sep=";",dec=",",header=T)
>  A_h<-  read.csv(file=" Hectares_reg.csv ",  sep=";",dec=
",",header=T)
> D_p<-read.csv(file=" Prod_prov.csv", sep=";",dec=",",header=T)
>  D_h<-read.csv(file=" Hectares_prov.csv ",  sep=";",dec=
",",header=T)
```

The variable of interest can be calculated for both regions and provinces as:

```
> Ay<-A_p/A_h
> Dy<-D_p/D_h
# The vectors Ay and Dy contain aggregated (regional) data and
disaggregated (provincial) data on the production, in euros,
per hectares of firms specialized in olive growing.
```

The matrices for aggregated and disaggregated data can, thus, be constructed as follows:

```
> data_A<-cbind(REG_shape@data$COD_REG, Ay)
> data_D<-cbind(PROV_shape@data$COD_REG, PROV_shape@data
$COD_PROV, Dy)
# The cbind function allows for merging different data frames
with the same number of rows. The vectors of observations
on the variable of interest are combined with data that are
extracted from the shape files using the slot operator. The
dollar is an operator that allows to extract specific columns
from the data extracted from the shape files.
```

The variables included in the dataset A are related to 20 areal units and correspond to the codes of regions and to the production in euros of firms specialized in olive growing for 2010 (Ay). The variables included in the dataset D are related to 110 areal units and correspond, for each province, to the code of the region to which it belongs, the code of province, and the production per hectare (in euro) of firms specialized in olive growing for 2010 (Dy).

The correspondence between aggregated and disaggregated data is defined by specifying an aggregation matrix **G**, that is a 20 × 110 matrix with elements defined according to administrative criteria. Specifically, the element g_{ji} of the matrix **G** can be defined as:

$$g_{ji} = b_{ji}\frac{h_i}{h_j} \qquad i = 1,2,\ldots,n \qquad j = 1,2,\ldots,m \qquad (10.21)$$

where the element b_{ji} is such that $b_{ji} = 1$ if the province i belongs to region j, and $b_{ji} = 0$ otherwise, and h_i and h_j denote the total agricultural area (in hectares) utilized by the firm specialized in olive growing in the province i and in the region j, respectively.

The aggregation matrix can be constructed based on the codes of regions that are included in the data matrices. Using the R software, we can construct the matrix **G** as follows:

```
> G<-matrix(data=0,nrow=nrow(data_A),ncol=nrow(data_D))
> for(j in 1:nrow(G))
+    for(i in 1:ncol(G))
+      if(data_A[j,1]==data_D[i,1])
+    {G[j,i]<-D_h[i]/A_h[j]}
# This syntax returns a matrix with elements specified as in
equation (10.21).
```

The observations at NUTS 3 level are assumed to be unknown and are estimated by applying the BIM. The mean vector and the variance of disaggregated data, that are known by assumption, are computed as follows:

```
> mu <-rep (mean (Dy), length (Dy))
> varD <-var (Dy).
```

The variance of aggregated data is computed as

```
> varA <-var (Ay).
```

Since the data generating process at the NUTS 3 level is modelled through a CAR specification, the covariance matrix $\Sigma_S = \sigma_S^2 (\mathbf{D} - \rho\mathbf{C})^{-1}$ has to be computed (see Section 10.4). The scalar parameter σ_S^2 is computed as the variance of disaggregated data. The contiguity matrix **C** can be calculated using the shape file of Provinces. The library spdep offers the functions knearneigh () and knn2nb () that can be applied to obtain a list of neighbours for each areal unit, specified according to the k nearest neighbours' criterion. A number $k = 5$ of neighbours is considered.

```
> library(spdep)
> col.knn <- knearneigh (coordinates (PROV_shape), k = 5)
> nb<-knn2nb (col.knn, sym=TRUE)
```

```
#The function knn2nb() converts the  knn  object returned
by knearneigh() into a neighbours list of class nb with a list of
integer vectors containing the neighbour province number ids.
```

The function `nb2mat ()` is then used to build the proximity matrix. By setting `style=` "B" in this function, a binary proximity matrix is obtained.

```
> C<-nb2mat(nb, style="B").
```

The diagonal matrix **D** with elements the number of neighbours for each areal unit is obtained as follows:

```
> D<-diag(1,nrow(C),ncol(C))*colSums(C).
```

After considering different values of the spatial autocorrelation parameter ρ the BIM estimates, in their unconstrained and constrained versions, are computed as follows:

```
> rho<-seq(0.1,0.99,0.05)
# The vector rho includes different values of the spatial
autocorrelation parameter varying within the interval [0.10;
0.99], with an increment step of 0.05.
> V<-array(NA,c(length(Dy),length(Dy),length(rho)))
> S<-array(NA,c(1,length(Dy),length(rho)))
# The arrays V and S are constructed to include the unconstrained
BIM estimates computed for different values of the spatial
autocorrelation parameter
> SS<-array(NA,c(1,length(Dy),length(rho)))
> VV<-array(NA,c(length(Dy),length(Dy),length(rho)))
# The arrays VV and SS are constructed to include the
constrained BIM estimates that satisfy the pycnophylactic
property and are computed for different values of the spatial
autocorrelation parameter.
> for(i in 1:length(rho))
+ {
+ V [,,i] <- solve(t(G)%*%solve(G%*%((D-rho[i]*C)/varA)
%*%t(G))%*%G+((D-rho[i]*C)/varD))
+  S  [,,i]  <-  V[,,i]%*%(((D-rho[i]*C)/varD)%*%mu+t(G)
%*%solve(G%*%((D-rho[i]*C)/varA)%*%t(G))%*%Ay)
+ SS[,,i]<- S[,,i]+V[,,i]%*%t(G)%*%(solve(G%*%V[,,i]%*%t(
G)))%*%(Ay-G%*%S[,,i])
+ VV[,,i]<- V[,,i]-V[,,i]%*%t(G)%*%(solve(G%*%V[,,i]%*%t(
G)))%*%G%*%V[,,i]
+ }
```

Point estimates for the production per hectare of firm specialized in olive growing at NUTS 3 level are obtained by considering the array SS, which includes the estimated values that correspond to different value of ρ. All the estimates derived for the provincial data are such that by applying the aggregation matrix, the data at the regional level are again obtained (i.e., the pycnophylactic property is satisfied).

The accuracy of the estimates can be assessed by comparing the estimated values and the actual values of the variable of interests. The comparison between the estimated values $\hat{y}_i$ and the actual values y_i has been performed by using the mean absolute percentage error (*MAPE*) and the coefficient of determination, R^2, which are, respectively, specified as follows:

$$MAPE = \frac{1}{n}\sum_{i=1}^{n}\left|\frac{y_i - \hat{y}_i}{y_i}\right|$$

(10.22)

$$R^2 = \frac{VAR(y_i)}{VAR(\hat{y}_i)}$$

(10.23)

By using the R software, the indicators in (11.22) and (11.23) can be calculated as follows:

```
> MAPE<-rep(NA,length(rho))
> for(i in 1:length(rho))
+ {
+   MAPE[i]<-(1/length(Dy))*(sum(abs((Dy-SS[,,i])/Dy)))
+ }
# The vector MAPE includes the mean percentage error computed
for  different  values  of  the  spatial  autocorrelation
parameter.
> R2<-rep(NA,length(rho))
> for(i in 1:length(rho))
+ {
+   R2[i]<-var(SS[,,i])/VarD
+ }
# The vector R2 includes the coefficient of determination
computed for different values of the spatial autocorrelation
parameter.
```

The values of the mean absolute percentage error (*MAPE*) and of the coefficient of determination calculated for the estimates obtained for different values of ρ are depicted in Figure 10.3 and Figure 10.4, respectively.

These figures are obtained by using the following syntaxes:

```
> plot(rho,MAPE,cex=0.4,pch=1,type="p",lty=2,xlab="rho",
ylab="MAPE",col="black")
>    plot(rho,R2,cex=0.4,pch=1,type="p",lty=1,xlab="rho",
ylab="R^2",col="black")
```

As shown in Figure 10.3, the values of *MAPE* decrease as the value of the spatial autocorrelation increases. This result indicates that the more the spatial dependence is exploited, the more the accuracy of estimates improves. This result is confirmed by the computation of R^2, that, as shown in Figure 10.4, reaches its maximum value when $\rho = 0.99$.

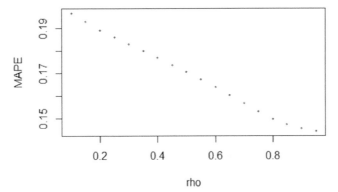

Figure 10.3. Mean absolute percentage error (*MAPE*) for different values of the spatial autocorrelation parameter (ρ).

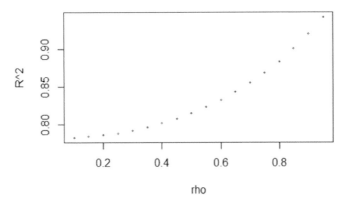

Figure 10.4. Coefficient of determination (R^2) for different values of the spatial autocorrelation parameter (ρ).

10.6 Conclusion

This chapter offered a short description of the main areal interpolation methods, with a close look at the methodologies accounting for the spatial dependence among spatial units. Spatial dependence is a peculiar feature of georeferenced data by which observations on a variable collected at near sites tend to be similar. This property could be usefully incorporated in methodologies used to estimate data that are not available at the desired spatial scale.

Spatial dependence is considered in areal data disaggregation by the Bayesian Interpolation method (BIM) introduced by Benedetti and Palma (1994). The main assumption on which this methodology is based is the Gaussian distribution of the data generating process, that is modelled through a CAR specification (Besag 1974). The CAR specification introduces the spatial dependence effect in the covariance structure of the process, as a function of a proximity matrix and of a spatial autocorrelation parameter. Based on this assumption and on the Bayes' rule, Benedetti and Palma (1994) derived the posterior probability distribution of the data generating process at the disaggregated level given the aggregated data. This posterior probability distribution is a multivariate normal distribution with the BIM estimates as parameters.

The BIM is used to disaggregate data on the production per hectare of firms specialized in olive growing in Italy, available at regional level, at the provincial level. Suggestions to perform the methodology using the R software are presented. The method is applied using different values of the spatial autocorrelation parameter. The obtained results revealed that the more the spatial dependence is considered, the more the accuracy of estimates improves.

References

Anderson, T.W. 1958. *An Introduction to Multivariate Statistical Analysis*. New York: John Wiley.

Banerjee, S., B.P. Carlin and A.E. Gelfand. 2004. *Hierarchical Modelling and Analysis for Spatial Data*. Boca Raton: Chapman & Hall.

Benedetti, R. and D. Palma. 1994. Markov random field-based image subsampling method. *Journal of Applied Statistics* 21: 495–509.

Besag, J. 1974. Spatial interaction and the statistical analysis of lattice systems. *Journal of the Royal Statistical Society Series B* 36: 192–225.

Clayton, D. and L. Berardinelli. 1992. Bayesian methods for mapping disease risk. pp. 205–220. *In*: Elliott, P., J. Cuzick, D. English and D. Stern (eds.). *Geographical and Environmental Epidemiology: Methods for Small-area Studies*. Oxford: Oxford University Press.

Comber, A. and W. Zeng. 2019. Spatial interpolation using areal features: A review of methods and opportunities using new forms of data with coded illustrations. *Geography Compass* 13: 10. https://doi.org/10.1111/gec3.12465.

Conley, T.G. and G. Topa. 2002. Socio-economic distance and spatial patterns in unemployment. *Journal of Applied Econometrics* 17: 303–327.

Cressie, N. 1993. *Statistics for Spatial Data*. New York: John Wiley.

Dempster, A.P., N.M. Laird and D.B Rubin. 1977. Maximum likelihood from incomplete data via the EM algorithm. *Journal of the Royal Statistical Society: Series B* 39: 1–38.

Fisher, P.F. and M. Langford. 1995. Modelling the errors in areal interpolation between zonal systems by Monte Carlo simulation. *Environment and Planning A* 27: 211–224.

Flowerdew, R. and M. Green. 1989. Statistical methods for inference between incompatible zonal systems. pp. 239–243. *In*: Goodchild, M. and S. Gopal (eds.). *Accuracy of Spatial Databases*. London: Taylor & Francis.

Flowerdew, R. and M. Green. 1991. Data integration: statistical methods for transferring data between zonal systems. pp. 38–54. *In*: Masser, I. and M. Bakemore (eds.). *Handling Geographical Information*. London: Longman.

Flowerdew, R., M. Green and E. Kehris. 1991. Using areal interpolation methods in geographical information systems. *Papers in Regional Science* 70: 303–315.

Flowerdew, R. and M. Green. 1994. Areal interpolation and types of data. pp. 121–145. *In*: Fotheringham, A.S. and P.A. Rogerson (eds.). *Spatial Analysis and GIS*. London: Taylor and Francis.

Ford, L. 1976. Contour reaggregation: another way to integrate data. *Papers, Thirteenth Annual URISA Conference* 11: 528–575.

Gelfand, A.E., L. Zhu and B.P. Carlin. 2001. On the change of support problem for spatio-temporal data. *Biostatistics* 2: 31–45.

Ghosh, M. and J.N.K. Rao. 1994. Small area estimation: an appraisal. *Statistical Science* 9: 55–93.

Goodchild, M.F. and N.S. Lam. 1980. Areal interpolation: a variant of the traditional spatial problem. *Geo-Processing* 1: 297–312.

Goodchild, M.F., L. Anselin and U. Deichmann. 1993. A framework for the areal interpolation of socioeconomic data. *Environment and Planning A* 25: 383–397.

Goovaerts, P. 2006. Geostatistical analysis of disease data: accounting for spatial support and population density in the isopleth mapping of cancer mortality risk using area-to-point Poisson kriging. *International Journal of Health Geographics* 5: 52. https://doi.org/10.1186/1476-072X-5-52.

Goovaerts, P. 2010. Combining areal and point data in geostatistical interpolation: applications to soil science and medical geography. *Mathematical Geosciences* 42: 535–554.

Gotway, C. and L. Young. 2007. A geostatistical approach to linking geographically aggregated data from different sources. *Journal of Computational and Graphical Statistics* 16: 115–135.

Kyriakidis, P.C. 2004. A geostatistical framework for area-to-point spatial interpolation. *Geographical Analysis* 36: 259–289.

Langford, M., D. Maguire and D. Unwin. 1991. The areal interpolation problem: estimating population using remote sensing in a GIS framework. pp. 55–77. *In*: Masser, I. and M. Blakemore (eds.). *Handling Geographic Information: Methodology and Potential Application.* London: Longman.

Liu, X.H., P.C. Kyriakidis and M.F. Goodchild. 2008. Population-density estimation using regression and area-to-point residual kriging. *International Journal of Geographical Information Science* 22: 431–447.

Matheron, G. 1971. The theory of regionalized variables and its applications. *Cahiers du Centre de Morphologie Mathématique* 5. Fontainebleau, France.

Mugglin, A.S., B.P. Carlin, L. Zhu and E. Conlon. 1999. Bayesian areal interpolation, estimation and smoothing: an inferential approach for geographic information systems: population interpolation over incompatible zones. *Journal of Agricultural, Biological and Environmental Statistics* 3: 117–130.

Murakami, D. and M. Tsutsumi. 2011. A new areal interpolation method based on spatial statistics. *Procedia Social and Behavioral Sciences* 21: 230–239.

Murakami, D. and M. Tsutsumi. 2012. Practical spatial statistics for areal interpolation. *Environment and Planning B: Planning and Design* 39: 1016–1033.

Netrdová, P., V. Nosek and P. Hurbánek. 2020. Using areal interpolation to deal with differing regional structures in international research. *International Journal of Geo-Information* 9: 126. https://doi.org/10.3390/ijgi9020126.

Panzera, D. and P. Postiglione. 2014. Economic growth in Italian NUTS3 provinces. *The Annals of Regional Science* 53: 273–293.

Panzera, D., R. Benedetti and P. Postiglione. 2016. A Bayesian approach to parameter estimation in the presence of spatial missing data. *Spatial Economic Analysis* 11: 201–218.

Panzera, D. and A. Viñuela. 2018. A Bayesian interpolation method to estimate per capita GDP at the sub-regional level: Local labor markets in Spain. pp. 9–31. *In*: Thill, J.-C. (ed.). *Spatial Analysis and Location Modeling in Urban and Regional Systems.* Berlin: Springer, Advances in Geographic Information Science.

Ripley, B.D. 1981. *Spatial Statistics.* New York: John Wiley.

Sadahiro, Y. 1999. Accuracy of areal interpolation: a comparison of alternative methods. *Journal of Geographical Systems* 1: 323–346.

Wall, M.M. 2004. A close look at the spatial structure implied by the CAR and SAR models. *Journal of Statistical Planning and Inference* 121: 311–324.

Wikle, C.K. and L.M. Berliner. 2005. Combining information across spatial scales. *Technometrics* 47: 80–91.

Yoo, E.H. and P.C. Kyriakidis. 2006. Area-to-point kriging with inequality-type data. *Journal of Geographical Systems* 8: 357–390.

Yoo, E.H. and P.C. Kyriakidis. 2009. Area-to-point kriging in spatial hedonic pricing models. *Journal of Geographical Systems* 11: 381–406.

CHAPTER 11
Small Area Estimation of Agricultural Data

Gaia Bertarelli, Francesco Schirripa Spagnolo,* Nicola Salvati
and *Monica Pratesi*

11.1 Introduction

Improving social inclusion and life conditions is the object of many international project ("InGRID-2: supporting expertise in inclusive growth", "MAKSWELL: Making Sustainable development and WELL-being frameworks works for policy analysis"). In the evidence-based policy making to combat poverty, agricultural data assume a key role. In fact, mainly in developing countries, it is out of doubt that monitoring agricultural indicators at the local level allows for an appropriate policy intervention. In general, local/small area level is the geographical level from which data are requested in order to plan sub-regional policies and/or evaluate the results of policy. At the local/small area level, direct estimates of agricultural and rural statistics are not accurate because sample surveys are usually designed so that direct estimators lead to reliable estimates only for larger domains (states, regions). By direct estimators, here we mean estimators that use only the sample data from the domain of interest. However, direct estimation is typically inefficient for smaller domains where sample sizes are small and cannot be used when there are no sample units in the domain.

Small area estimation (SAE) comprises the methods for obtaining more precise estimators in local/small areas, making use of the common features of the areas. In particular, we focus on SAE methods based on area-level models, which are used to smooth out the variability in the unstable area-level direct estimates. Area-level modelling is typically used when unit-level data are unavailable, or, as is often the case, where model covariates (e.g., census variables) are only available in aggregate form. Fay and Herriot (1979) model is an area-level SAE model (hereafter the FH

University of Pisa, Department of Economics and Management.
Emails: gaia.bertarelli@ec.unipi.it; nicola.salvati@unipi.it; monica.pratesi@unipi.it
* Corresponding author: francesco.schirripa@ec.unipi.it

model) that relates small area direct survey estimates to area-level covariates. The FH model is widely used because of its flexibility in combining different sources of information with different error structures.

In this chapter, we present a review of the most important and used area level models, which mostly take advantage of the spatial information, to provide estimates of agricultural and rural statistics at a local/small area level. The famous first law of geography 'everything is related to everything else, but near things are more related than distant things' (Tobler 1970) is valid also for small local areas: close areas are more likely to have similar values of the target parameter than areas which are far from each other. This evidence suggests that an adequate use of geographic information and geographical modelling can help in producing more accurate estimates for small area parameters. Several authors (Cressie 1991; Singh et al. 2005; Petrucci and Salvati 2006; Pratesi and Salvati 2008; Longford 2010; D'Alo et al. 2012; Porter et al. 2014; Chandra et al. 2015; Pratesi 2015) have already demonstrated in the literature that borrowing strength over space offers some gains in terms of efficiency in small area estimation. In many situations, additional spatial information, such as coordinates and distances between small areas, is available and can be incorporated in the model. The location of a small area leads to two different types of spatial effects: spatial dependence and spatial heterogeneity. The former is the coincidence of value similarity with locational similarity. The dependence in spatial data is often referred to as spatial correlation. The latter is simply structural instability in the form of non-constant error variances (heteroscedasticity) or model coefficients (variable coefficients).

Cressie (1991), Singh et al. (2005), and Pratesi and Salvati (2008) extend the mixed model to allow for spatially correlated random effects using conditional autoregressive (CAR) and simultaneous autoregressive (SAR) specifications for the random effects (Anselin 1988). These models allow for spatial correlation in the area effects, while keeping the fixed effects parameters spatially invariant. Under the area-level version of this spatial mixed model, Singh et al. (2005), Petrucci and Salvati (2006) and Pratesi and Salvati (2008) define the spatial empirical best linear unbiased predictor (SEBLUP) for a small area mean and also derive an approximately unbiased estimator of the MSE of the SEBLUP. The spatial mixed model will be presented in Section 11.2.

An approach to include the spatial heterogeneity assumes that the regression coefficients vary spatially across the small area of interest. This situation is referred to as *spatial non-stationarity* (Brunsdon et al. 1996; Fotheringham et al. 2002). Recently, Chandra et al. (2015) proposed a spatially non-stationary Fay-Herriot model (Fay and Herriot 1979) for this situation and it will be revised in Section 11.3.

A third approach to incorporate the spatial structure in the data is based on a non-parametric spatial P-spline model for small area estimation (Opsomer et al. 2008; Giusti et al. 2012). The spatial relationship is captured by spatially varying auxiliary information, whereas the regression coefficients are spatially invariant. These models based on P-splines can handle situations where the data are supposed to be affected by spatial proximity effects. In these cases, P-spline bivariate smoothing can introduce spatial effects in the model. This approach will be presented in Section 11.4.

In each section, we will show the R command for running the models for small area estimation using a well-known data set available in the sae package (Molina and Marhuenda 2015) in R: the grape data set. This data set is based on the Italian Agricultural Census of year 2000 for the Italian region of Tuscany. The goal of this example is to estimate the mean agrarian surface area used for production of grape at the municipality level in Tuscany. The data contains information for 274 municipalities considered as small areas. For each municipality, the following information is available: the direct estimator of the mean agrarian surface area used for production of grape in hectares, its sampling variance, the latitude and the longitude of the centroids of each municipality, the agrarian surface area used for production in hectares and the average number of working days in the reference year (the last two variables will be used as covariates in the models). The data can be regarded as lattice data. The centroids of the small areas are taken as spatial reference points. For more details, see Molina et al. (2009). A comparison of the point estimates and their MSE estimates will be showed in Section 11.5.

11.2 Spatial Fay-Herriot model for small area estimation

The basic and most popular model to produce small area estimates using area level data is the FH model introduced by Fay and Herriot (1979) to obtain small area estimates of median income in U.S. small places. Let ϑ be the $d \times 1$ vector of the parameters of inferential interest (small area totals or small area means with $j = 1,\ldots, d$). Moreover, we assume that the design unbiased direct estimator $\hat{\vartheta}$ is available and the follow model holds:

$$\hat{\vartheta} = \vartheta + \mathbf{e}, \tag{11.1}$$

where $\mathbf{e}$ is a vector of independent sampling errors with mean vector $\mathbf{0}$ and known diagonal variance matrix $\mathbf{R} = diag(\psi_j)$, with ψ_j representing the sampling variances of the direct estimators of the area parameters of interest. Usually, ψ_j is unknown and is estimated by a *generalized variance function* applied not only to the specific area sample, but to the whole sample as well; for details see Wolter (1985, Ch. 5) and Wang and Fuller (2003).

Petrucci and Salvati (2006) proposed the spatial autocorrelation in SAE under the FH model by specifying a linear mixed model with spatially correlated random area effects for the vector of the parameters of interest. Area level random effects model implies that a matrix of $d \times p$ area-specific auxiliary variables, $\mathbf{X}$, and the parameters of interest, ϑ are linked by a linear relationship:

$$\vartheta = \mathbf{X}\boldsymbol{\beta} + \mathbf{Zv} \tag{11.2}$$

where $\boldsymbol{\beta}$ is the $p \times 1$ vector of regression parameters, $\mathbf{Z}$ is a $d \times d$ matrix of known positive constants and $\mathbf{v}$ is a $d \times 1$ vector of spatially correlated random area effects defined by the following autoregressive process (see Anselin 1988; Cressie 1993):

$$\mathbf{v} = \rho\mathbf{Wv} + \mathbf{u} = (\mathbf{I} - \rho\mathbf{W})^{-1}\mathbf{u}. \tag{11.3}$$

Here, $\mathbf{W}$ is a $d \times d$ spatial interaction matrix which indicates whether the areas are neighbours or not (one way to define $\mathbf{W}$ is to set $w_{ij} = 1$ if small area i and j are neighbour or 0 otherwise, however, there are other ways to define $\mathbf{W}$); ρ is the spatial autoregressive coefficient which defines the strength of the spatial relationship among the random effects associated with the neighbouring areas; $\mathbf{u}$ is a $d \times 1$ vector of independent random area effects with zero mean and variance-covariance matrix $\Sigma_u = \sigma_u^2 \mathbf{I}_d$ and $\mathbf{I}_d$ is an $d \times d$ identity matrix (as in FH models).

Combining Eq. 1 and Eq. 2, we obtain the model with spatially correlated errors:

$$\hat{\vartheta} = \mathbf{X}\boldsymbol{\beta} + \mathbf{Z}(\mathbf{I} - \rho\mathbf{W})^{-1}\mathbf{u} + \mathbf{e}.$$

The $d \times d$ Simultaneously Autoregressive (SAR) covariance matrix of the error term $\mathbf{v}$ is:

$$G(\delta) = \sigma_u^2[(\mathbf{I} - \rho\mathbf{W}^T)(\mathbf{I} - \rho\mathbf{W})]^{-1},$$

where $\delta = (\sigma_u^2, \rho)$. The Spatial Empirical Best Linear Unbiased Predictor (SEBLUP) estimator of ϑ_j, i.e., the parameter of interest in the small area j, is:

$$\hat{\vartheta}_j^{SFH}(\hat{\sigma}_u^2, \hat{\rho}) = \mathbf{x}_j\hat{\boldsymbol{\beta}} + \mathbf{b}_j^T\,\hat{\mathbf{G}}\mathbf{Z}^T(\mathbf{R} + \mathbf{Z}\hat{\mathbf{G}}\mathbf{Z}^T)^{-1}\,(\hat{\vartheta} - \mathbf{X}\hat{\boldsymbol{\beta}}), \qquad (11.4)$$

where $\hat{\boldsymbol{\beta}}$ the weighted least squares estimator of $\boldsymbol{\beta}$ equal to $(\mathbf{X}^T\,\hat{\mathbf{V}}^{-1}\mathbf{X})^{-1}\,\mathbf{X}^T\,\hat{\mathbf{V}}^{-1}\,\hat{\vartheta}$ with $\hat{\mathbf{V}} = \mathbf{R} + \mathbf{Z}\hat{\mathbf{G}}\mathbf{Z}^T$ and $\mathbf{R} = \mathbf{diag}(\psi_j)$; $\mathbf{b}_j^T$ is an $1 \times d$ vector with value 1 in the j-th position; $\hat{\sigma}_u^2$ and $\hat{\rho}$ are asymptotically consistent estimators of σ_u^2 and ρ obtained by Maximum Likelihood (ML) or Restricted Maximum Likelihood (REML) methods based on the normality assumption of the random effects (for more details see Singh et al. 2005; Pratesi and Salvati 2008).

Concerning the MSE, an analytical estimator has been proposed by Singh et al. (2005): their proposal is a second order approximation of the MSE of the SEBLUP. Another analytical formula of the MSE can be found in Petrucci and Salvati (2006). Analytical approximations may require strong model assumptions and many small areas to approximate well the true values, therefore, Molina et al. (2009) proposed parametric and non-parametric bootstrap procedures for estimation of the MSE under the SFH model. The authors provided naïve and bias-corrected bootstrap estimators.

11.2.1 *Example of the estimation of the EBLUP-SFH*

The goal of this example is to estimate the mean agrarian surface area used for production of grape at the municipality level in Tuscany by using the R software. In order to estimate the mean agrarian surface area used for production of grape (θ_j) the agrarian surface area used for production in hectares (x_{1j}) and the average number of working days in the reference year (x_{2j}) are used as covariates. First of all, the research has to load data stored in the file grapes_all.txt:

```
> data.grapes <- read.table("grapes_all.txt", header = T,
sep="")
```

The dataset contains the following variables: the municipality code (ID); the direct estimates of the mean agrarian surface area used for production of grape (grapehect); the agrarian surface area used for production in hectares (area); the average number of working days in the reference year (workdays) and the variance of the direct estimates (var).

In this example, the spatial relation between the contiguous areas is described by a Spatial Autoregressive process (SAR) process. In order to run the SFH model, the proximity matrix ($\mathbf{W} = (w_{ij})$) has to be specified. The proximity matrix may be calculated using the shape file available on the ISTAT website[1] and taking the centroids of the small areas (the municipalities) as spatial reference points.

In this study, the shapefile with administrative boundaries referring to the year 2001 has been used. After loading and merging the shapefile with grapes dataset, the library spdep (Bivand and Piras 2015) and the functions poly2nb() and nb2mat() are applied to build the proximity matrix. In particular, a binary proximity matrix (w_{ij} equal to 1 if municipality i shares an edge with municipality j, and 0 otherwise) with row standardised has been computed.

```
> library(maptools)
> italy3<-readShapePoly("Com2001_ED50.shp")
#select Tuscany (COD_REG=="9")
> tuscany<-italy3[italy3$COD_REG=="9",]
# Tuscany municipalities are 287 (administrative boundaries
2001)
# Merging the shapefile with our dataset using the municipality
codes
# to obtain the shapefile for the 274 municipalities
>  tuscany1<-merge(tuscany, data.grapes,by.x= "PRO_COM",
by.y="ID", all.x = F)
# calculate the proximity matrix using the library "spdep"
> library("spdep")
>     neigh<-poly2nb(tuscany1,row.names=tuscany1$PRO_COM,
queen=TRUE)
# nb2mat() generates a weights matrix. Setting style="W" a
row standardised proximity matrix is obtained.
>sp_w<-nb2mat(neigh,glist=NULL,style="W", zero.policy=TRUE)
```

In order to fit the spatial FH model, the researcher has to use the function eblupSFH(). This function gives small area estimates based on a spatial FH model, where area effects follow a SAR process. In the function eblupSFH(), the following arguments have to be specified: formula: the formula of the model; vardir: the sampling variances of direct estimators for each domain; proxmat: the proximity matrix; method: the type of fitting method; MAXITER: maximum number of iterations; PRECISION: the convergence tolerance limit; data: the (optional) data frame containing the variables.

```
> SFH.eblup<-eblupSFH (grapehect ~ area + workdays - 1,
vardir=var, proxmat=SP_W, data=data.grapes, MAXITER = 500)
```

The function returns a list with the following objects: eblup, a vector with the values of the estimators for the domains; fit, a list containing the following objects: method: type of fitting method applied; convergence: a logical value equal to TRUE if Fisher-scoring algorithm converges in less than MAXITER iterations; iterations: number of iterations performed by the Fisher-scoring algorithm; estcoef: a data frame with the estimated model coefficients; refvar: estimated

---

[1] http://www.istat.it/storage/cartografia/confini_amministrativi/archivio-confini/non_generalizzati/
Limiti_2001_ED50.zip.

random effects variance; `spatialcorr`: estimated spatial correlation parameter; `goodness`: vector containing three goodness-of-fit measures.

The analytical MSE estimates (Singh et al. 2005; Pratesi and Salvati 2008) may be obtained using the function `mseSFH()`. Alternatively, parametric and non-parametric bootstrap MSE estimates (Molina et al. 2009) may be obtained from the SFH model by the functions `pbmseSFH()` and `npbmseSFH()`, respectively.

The arguments of `mseSFH()` are the same of the function `eblupSFH()`. The function returns: a list with the results of the SFH estimation process (`est`); a vector with the analytical MSE (`mse`).

```
> SFH.mse<-mseSFH(grapehect ~ area + workdays - 1, vardir=var,
proxmat=SP_W,
data=data.grapes, MAXITER = 500)
```

The two functions using bootstrap methods (`pbmseSFH()` and `npbmseSFH()`), in addition to previous arguments, require the number of bootstrap replicates (`B`) to be specified. By default, `B` is set to 100.

```
> SFH.pbmse<-pbmseSFH(grapehect ~ area + workdays - 1, vardir
=var, proxmat=SP_W, data=data.grapes, MAXITER = 500, B=100)
> SFH.npbmse<-npbmseSFH(grapehect ~ area + workdays - 1, vardir
=var, proxmat=SP_W, data=data.grapes, MAXITER = 500, B=100)
```

These two functions return a list with the results of the SFH estimation process (`est`), the naïve nonparametric bootstrap MSE estimates (`mse`) and the bias-corrected nonparametric bootstrap MSE estimates (`msebc`) (for more detail on the bias correction see Molina et al. 2009).

11.3 Spatially non-stationary Fay-Herriot Model for small area estimation

Let the spatial location of area j correspond to the coordinates of an arbitrarily defined spatial location in the area, e.g., its centroid, denoted by loc_j. Let $f(loc_i, loc_j)$ be an appropriate measure of the distance between the spatial locations of areas i and j, and define the spatial contiguity of these two locations to be $\omega_{ij} = (1 + f(loc_i, loc_j))^{-1}$. Let $\Omega = [\omega_{ij}]$ denote the positive definite $d \times d$ matrix of spatial contiguities defined by loc_j. The spatially non-stationary extension of the proposed FH model for area j proposed by Chandra et al. (2015) can be written as:

$$\hat{\vartheta}_j - \vartheta_j = e_j \text{ and } \vartheta_j - \mathbf{x}_j^T \boldsymbol{\beta}(loc_j) = u_j,$$

where $\boldsymbol{\beta}(loc_j) = \boldsymbol{\beta} + \boldsymbol{\gamma}(loc_j)$ where $\boldsymbol{\gamma}(loc) = (\gamma_k(loc); k = 1,\ldots, p)$ is a spatially varying multivariate random process of dimension p. Put $\boldsymbol{\Gamma} = (\boldsymbol{\gamma}^T(loc_1),\ldots, \boldsymbol{\gamma}^T(loc_d))^T$ and $\mathbf{loc} = \{loc_1,\ldots, loc_d\}$, i.e., the set of locations for the d areas. Then

$$E(\boldsymbol{\Gamma}|\mathbf{X}, \mathbf{loc}) = \mathbf{0}_{\{pd \times 1\}}$$

and

$$Var(\boldsymbol{\Gamma}|\mathbf{X}, \mathbf{loc}) = \boldsymbol{\Sigma} = \boldsymbol{\Omega} \otimes \mathbf{C},$$

where $\boldsymbol{\Omega} = [\omega_{ij}]$, as previously defined, and $\mathbf{C} = [c_{kl}]$ is a $p \times p$ covariance matrix that characterizes the correlations between the components of $\boldsymbol{\gamma}$ at an arbitrary location *loc* and $\otimes$ denotes the Kronecker product. It is now possible to re-write as an area-level mixed model the nonstationary FH proposed by Chandra et al. (2015) as:

$$\hat{\vartheta} = \mathbf{X}\boldsymbol{\beta} + \boldsymbol{\Psi}\boldsymbol{\Gamma} + \mathbf{u} + \mathbf{e}, \tag{11.5}$$

where $\boldsymbol{\Psi} = diag(\mathbf{x}_j^T, j = 1, \ldots, d)$. Then, it follows:

$$E(\hat{\vartheta}|\mathbf{X}, \mathbf{loc}) = \mathbf{X}\boldsymbol{\beta}$$

$$Var(\hat{\vartheta}|\mathbf{X}, \mathbf{loc}) = \mathbf{V} = \boldsymbol{\Psi}\boldsymbol{\Sigma}\boldsymbol{\Psi}^{\mathsf{T}} - \sigma_u^T \mathbf{I}_m + diag\{\psi_j; j = 1, \ldots, d\}$$

$$Cov(\vartheta, \hat{\vartheta}|\mathbf{X}, \mathbf{loc}) = \boldsymbol{\Psi}_j \boldsymbol{\Sigma}\boldsymbol{\Psi}^{\mathsf{T}} + \sigma_u^2 \delta_j^T,$$

where $\boldsymbol{\Psi}_j$ is the *j*-th row of $\boldsymbol{\Psi}$. The component Ω and $\mathbf{C}$ of the covariance structure, coming from $Var(\boldsymbol{\Gamma}|\mathbf{X}, \mathbf{loc})$, can be seen as the geographic scale and intensity of the spatial correlation in the population of interest. In practice, not all the components of γ will be random, corresponding to the regression parameters that are not spatially varying. As a consequence, it is possible that the covariance matrix $\mathbf{C}$ is not full rank, and in many applications $\mathbf{C}$ will have rank p_1 smaller than p, corresponding to the p_1 components of γ that are spatially varying. Without loss of generality, Chandra et al. (2015) assume that the first p_1 components of γ are non-zero (that is spatially varying). Then, $\mathbf{C} = diag(\mathbf{C}_{p_1}, \mathbf{0}_{p-p_1})$, where $\mathbf{C}_{p_1}$ is the location-specific covariance of the p_1 spatially varying parameters in the model, and $\mathbf{0}_{p-p_1}$ is a zero matrix of order $p - p_1$.

The minimum mean squared error (MMSE) predictor of ϑ_j under the proposed model is the expected value of ϑ_j given $\hat{\vartheta}, \mathbf{X}, \mathbf{loc}$. Under a Gaussian error assumption, and assuming the inverse of $\mathbf{V}$ exists, this is

$$\tilde{\vartheta}_j = \mathbf{x}_j^T\boldsymbol{\beta} + Cov(\vartheta_j, \hat{\vartheta}|\mathbf{X}, \mathbf{loc})\mathbf{V}^{-1}(\hat{\vartheta} - \mathbf{X}\boldsymbol{\beta}). \tag{11.6}$$

For small area *j*, the EBLUP of ϑ_j, that is the nonstationary EBLUP (NSEBLUP) is:

$$\hat{\vartheta}_j^{NSFH}(\hat{\sigma}_u^2, \hat{\mathbf{C}}) = \mathbf{x}_j^T\hat{\boldsymbol{\beta}} + \boldsymbol{\Psi}_j \hat{\boldsymbol{\gamma}} + \hat{u}_j, \tag{11.7}$$

where $\hat{\boldsymbol{\beta}} = [\hat{\mathbf{X}}^{\mathsf{T}} \hat{\mathbf{V}}^{-1}\mathbf{X}]^{-1} [\mathbf{X}^{\mathsf{T}} \hat{\mathbf{V}}^{-1}\hat{\vartheta}]$, $\hat{\boldsymbol{\gamma}} = \hat{\boldsymbol{\Sigma}}\boldsymbol{\Psi}^{\mathsf{T}} \hat{\mathbf{V}}^{-1} (\hat{\vartheta} - \mathbf{X}\hat{\boldsymbol{\beta}})$ and $\hat{u}_j$ is the *j*-th components of the vector $\hat{\mathbf{u}} = \hat{\sigma}_u^2 \hat{\mathbf{V}}^{-1}(\hat{\vartheta} - \mathbf{X}\hat{\boldsymbol{\beta}})$. The unknown parameters σ_u^2 and $\mathbf{C}$ can be estimated from the data by ML or REML methods. Regarding the MSE of the nonstationary EBLUP, Chandra et al. (2015) proposed an analytical approach. Moreover, they also proposed an alternative procedure for the estimation of the MSE based on a parametric bootstrap developing the naïve parametric MSE estimator. For more details about the procedure to estimate the MSE see Chandra et al. (2015). Using the same bootstrap procedure, Chandra et al. (2015) proposed a procedure to test the spatial non-stationarity.

11.3.1 Example of the estimation of the NSEBLUP

The estimation of the NSEBLUP can be obtained using the R functions `nseblup()`. This function allows the EBLUP to be estimated under the nonstationary FH model and the MSE.

The function `nseblup()` requires the following arguments: `direct`: the (numeric) response vector for the sampled units; `Xpop`: the matrix of the auxiliary variables; `variance`: the sampling variances of direct estimators for each domain; `lon.s`: vector contains the longitude of the (centroids of the) sampled small areas; `lat.s`: vector contains the latitude of the (centroids of the) sampled small areas; `m.sample`: the number of sampled small areas; `seed`: argument to set.seed; `boot`: logical, indicating whether the bootstrap procedure for the estimation of the naïve and the bias-corrected MSE estimator has to be performed (default is TRUE); `n.boot`: number of the replicates of the two bootstrap procedure for the MSE (default value is 100).

The researcher has to load the function `nseblup()` stored in the file 'nseblup.R' (see the Appendix 1 for the code):

```
> source("nseblup.R")
```

Then, data stored in the file `grapes_all.txt` are loaded and the researcher has to define the matrix of auxiliary covariates: surface area used for production in hectares (x_{1j}) and the average number of working days in the reference year (x_{2j}).

```
> data.grapes<-read.table("grapes_all.txt", header = T, sep="")
> data.grapes<-data.grapes[order(data.grapes$ID),]
> Xvar<-cbind(1,data.grapes$area,data.grapes$workdays)
```

As spatial reference points, the dataset contains the latitude (utm_lat) and the longitude (utm_lon) of the centroids of the small areas (standardised UTM).

The researcher can now run the `nseblup()` function:

```
> NSEBLUP_results <- nseblup(direct=data.grapes$grapehect,
                    Xpop=Xvar, variance=data.grapes$var,
                    lon.S=data.grapes$utm_lon,
                    lat.S=data.grapes$utm_lat,
                    m.sample=274,
                    seed=1973, boot = T, n.boot=100)
```

The function returns a list with the following objects: the SFH estimation process (`est`), the analytical MSE estimates (`mse`) and the naïve parametric bootstrap MSE estimates (`mse_boot`).

11.4 Semiparametric Fay-Herriot Model for small area estimation

Giusti et al. (2012) proposed a semiparametric specification of the FH model obtained by Psplines, which allows non-linearities in the relationship between the response variable and the auxiliary variables.

A semiparametric additive model with one covariate x_1 can be written as $\tilde{g}(x_1)$, where the function $\tilde{g}(\cdot)$ is unknown, but assumed to be sufficiently well approximated by the function

$$g(x_1; \boldsymbol{\beta}, \boldsymbol{\eta}) = \beta_0 + \beta_1 x_1 + \dots + \beta_p x_1^p + \sum_{k=1}^{K} \eta_k (x_1 - \kappa_k)_+^p, \tag{11.8}$$

where $\boldsymbol{\beta} = (\beta_0, \beta_1, \dots, \beta_p)^T$ is the $(p+1) \times 1$ vector of the coefficients of the polynomial function; $\boldsymbol{\eta} = (\eta_1, \eta_2, \dots \eta_K)^T$ is the coefficient vector of the truncated polynomial spline basis (P-spline) and p is the degree of the spline $(t)_+^p = t^p$ if $t > 0$, and 0 otherwise. The latter part of the model allows for handling departures from a p-polynomial t in the structure of the relationship. In this portion, κ_k for $k = 1,\ldots, K$ is a set of fixed knots and, if K is the class of functions in Eq. 11.8, is very large and can approximate most smooth functions. Details on bases and knots choice can be found in Ruppert et al. (2003).

We can handle geographically referenced variables that need to be converted to maps by using bivariate smoothing $g(x_1, x_2) = g(x_1, x_2; \boldsymbol{\beta}, \boldsymbol{\eta})$ and specify a semiparametric bivariate additive model:

$$g(x_1, x_2; \boldsymbol{\beta}, \boldsymbol{\eta}) = \beta_0 + \beta_1 x_1 + \beta_2 x_2 + \mathbf{z}_j^P \boldsymbol{\eta}, \tag{11.9}$$

where $\mathbf{z}_j^P$ is the j-th row of the following $n \times K$ matrix

$$\mathbf{Z}^P = [C(\tilde{\mathbf{x}}_j - \kappa_k)]_{1 \leq j \leq n; 1 \leq k \leq K} \, [C(\kappa_k - \kappa_k')]_{1 \leq k \leq K'}^{-1/2},$$

where $C(\mathbf{t}) = \|t\|^2 \, \log\|t\|$, $\tilde{\mathbf{x}}_j = (x_{1j}, x_{2j})$ and κ_k, $k = 1,\ldots, K$ is a set of knots. The use of $\mathbf{Z}^P$ allows the estimation procedure to be simplified (more details on the matrix $\mathbf{Z}^P$ may be found in Ruppert et al. 2003).

In general, the P-spline model can be viewed as a random effects model (Ruppert et al. 2003; Opsomer et al. 2008), therefore it can be combined with the standard Fay-Herriot model in order to obtain the following semiparametric FH model:

$$\hat{\boldsymbol{\vartheta}} = \mathbf{X}\boldsymbol{\beta} + \mathbf{Z}^P \boldsymbol{\eta} + \mathbf{Z}\mathbf{u} + \mathbf{e}, \tag{11.10}$$

where $\boldsymbol{\beta}$ is a vector of regression coefficients; $\boldsymbol{\eta}$ can be treated as a vector of independent and identically distributed random variables with mean $\mathbf{0}$ and $K \times K$ variance matrix $\boldsymbol{\Sigma}_\eta = \sigma_\eta^2 \, \mathbf{I}_K$; the variance-covariance matrix of the model is $\boldsymbol{\Sigma}(\sigma_y^2, \sigma_u^2) = \mathbf{Z}^P \, \boldsymbol{\Sigma}_\eta \, (\mathbf{Z}^P)^T + \mathbf{Z}\boldsymbol{\Sigma}_\mathbf{u} \, \mathbf{Z}^T + \mathbf{R}$.

Under the model in Eq. (14), it is possible to obtain the non-parametric (NP) EBLUP:

$$\hat{\boldsymbol{\vartheta}}^{NPFH} = \mathbf{X}\hat{\boldsymbol{\beta}} + \hat{\boldsymbol{\Lambda}}(\hat{\sigma}_y^2, \hat{\sigma}_u^2)(\hat{\boldsymbol{\vartheta}} - \mathbf{X}\hat{\boldsymbol{\beta}}),$$

where $\hat{\boldsymbol{\Lambda}}(\hat{\sigma}_y^2, \hat{\sigma}_u^2) = \mathbf{Z}^P \, \hat{\boldsymbol{\Sigma}}_\eta \, (\mathbf{Z}^P)^T + \mathbf{Z}\hat{\boldsymbol{\Sigma}}_\mathbf{u} \, \mathbf{Z}^T + \mathbf{R}$ and $\hat{\boldsymbol{\beta}} = (\mathbf{X}^T \, \hat{\boldsymbol{\Sigma}}^{-1}\mathbf{X})^{-1} \, \mathbf{X}^T \, \hat{\boldsymbol{\Sigma}}^{-1} \, \hat{\boldsymbol{\vartheta}}$. Regarding the MSE of the NPEBLUP, Giusti et al. (2012) proposed an analytical approximation. Moreover, they also proposed an alternative procedure for the estimation of the MSE based on a nonparametric bootstrap developing the naïve nonparametric MSE estimator and the bias-corrected combined analytical bootstrap MSE estimator. For more details about the procedure to estimate the MSE, see the paper by Giusti et al. (2012).

11.4.1 *Example of the estimation of the EBLUP-NPFH*

The goal of the example is still to estimate the mean agrarian surface area used for production of grape at the municipality level in Tuscany. The data of the previous examples will be used.

The estimation of the NPFH with spatial effects can be obtained using the R functions `eblup_npfh()`. This function allows the EBLUP to be estimated under the semiparametric FH model and the MSE.

The function `eblup_npfh()` requires the following arguments: `y`: the (numeric) response vector for the sampled units; `Xpop`: the matrix of the auxiliary variables; `vardir`: the sampling variances of direct estimators for each domain; `lon`: vector contains the longitude of the (centroids of the) small areas; `lat`: vector contains the latitude of the (centroids of the) small areas; `m.sample`: the number of small areas; n.knots: the number of knots; seed: argument to set.seed; boot: logical, indicating whether the bootstrap procedure for the estimation of the naïve and the bias-corrected MSE estimator has to be performed (default is TRUE); n.boot: number of the replicates of the two bootstrap procedure for the MSE (default value is 100).

First of all, the researcher has to load the function eblup_npfh() stored in the file 'npfh.sae.R' (see the Appendix 2 for the code):

```
> source("npfh.sae.R")
```

Then the data stored in the file grapes_all.txt are loaded and the matrix of auxiliary covariates is defined: as in the previous examples, the agrarian surface area used for production in hectares (x_{1j}) and the average number of working days in the reference year (x_{2j}) are used as auxiliary information.

```
> data.grapes<-read.table("grapes_all.txt", header = T, sep="")
> data.grapes<-data.grapes[order(data.grapes$ID),]
> Xvar<-cbind(1,data.grapes$area,data.grapes$workdays)
```

As spatial reference points, the dataset contains the latitude (utm_lat) and the longitude (utm_lon) of the centroids of the small areas (standardised UTM). The researcher can now run the eblup_npfh() function:

```
> NPFH_results <- eblup_npfh(y=data.grapes$grapehect,
               Xpop=Xvar, vardir=data.grapes$var,
               lon=data.grapes$utm_lon,
               lat=data.grapes$utm_lat,m.sample=274,
               n.knots=80, seed=88, boot = T, n.boot=100)
```

The function returns a list with the following objects: the NPFH estimation process (EBLUP), the analytical MSE estimates (MSE), the naïve parametric bootstrap MSE estimates (MSE.boot) and the bias-corrected combined analytical bootstrap MSE estimator (MSE.BOOT.BC)

11.5 Estimating mean agrarian surface area used for production of grape at the municipality level in Tuscany

In this section, in order to assess the results achieved by introducing spatial information in the small area estimation, we comment on the estimates of the small area estimation procedure outlined in the previous sections. Figure 11.1 maps the estimated levels of mean agrarian surface area used for production of grape for municipalities in the Tuscany region by using SAE methodologies, namely direct estimation, SFH estimation, NSFH estimation and NPFH estimation. The emerging patterns of mean agrarian surface area used for production of grape for municipalities in the Tuscany region produced by the methodologies that use spatial information are consistent with the patterns that are reported by the direct estimates. Clusters of higher levels of mean agrarian surface area used for production of grape are located in municipalities in the north-east of Tuscany (Chianti region) and in the central south of Tuscany. The value of the estimated spatial autocorrelation coefficient $\hat{\rho}$ of SFH is 0.601, which suggests a high level of spatial relationship.

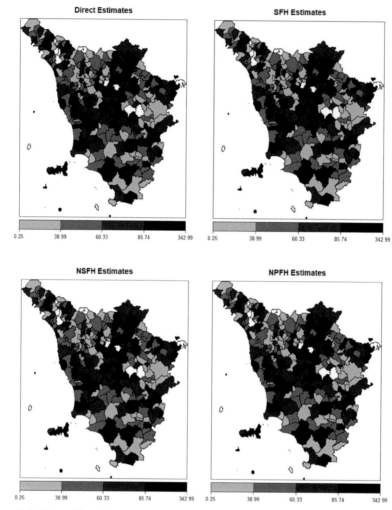

Figure 11.1. Maps of the estimated levels of mean agrarian surface area used for production of grape for municipalities in the Tuscany region by using SAE methodologies.

Although maps offer an effective tool for summarizing small area estimates, they do not answer an important question. How good are the estimates produced by using the models that use spatial information? To answer this question, we note that model-based estimates should be (i) 'close' to the direct estimates and (ii) more precise than direct estimates.

Following Brown et al. (2001), we assess condition (i) by computing the correlation between direct and model-based estimates and a goodness-of-fit diagnostic. From the values of correlation, we note that the small area estimates appear to be generally consistent with the direct estimates, with the correlation between the two sets of estimates being 0.77 for SFH, 0.78 for NSFH and 0.77 for NPFH.

The goodness-of-fit diagnostic is based on the idea that, if model-based estimates are close to the small area value of interest, then unbiased direct estimates can be considered as random variables whose expected values are equal to the values of the corresponding model-based estimates. The goodness-of-fit diagnostic is computed as the value of the following Wald statistic for each model-based estimator:

$$T = \sum_{j=1}^{d} \frac{(\hat{\vartheta}_j - \hat{\vartheta}_j^{model})^2}{\hat{V}(\hat{\vartheta}_j) - \widehat{MSE}(\hat{\vartheta}_j^{model})}$$

The realized value of T can then be compared against the 0.95-quantile of a $\chi2$-distribution with 273 degrees of freedom, i.e., 312.53. Table 11.1 reports the T and the *p-values* for different small area estimators and analytic and bootstrap methods of MSE estimation. No set of model-based estimates is statistically different from the direct estimates.

To assess condition (ii) above, i.e., the potential gains in precision from using model-based estimates instead of the direct estimates, we examine the distribution of the ratios of the estimated CVs of the direct and the model-based estimates for the grapes data. Table 11.2 reports the number of municipalities with values of CV less than 16.6%, between 16.6% and 33.3% and over 33.3% for direct and model-based estimators using different MSE estimation methods. These values of CV are suggested by Statistics Canada to provide quality level guidelines for publishing tables: estimates with a CV less than 16.6% are considered reliable for general use. Estimates with CV between 16.6% and 33.3% should be accompanied by a warning to users. Estimates with coefficients of variation larger than 33.3% are deemed to be unreliable. Table 11.2 shows that the model-based estimators that use spatial information provide reliable estimates, especially SFH and NPFH.

Spatial small area models require the access to survey estimates. They need to know the direct estimates at the area level. They do not require the access to micro-data. Also, for this reason, the methods are frequently applied and routines to implement them can be downloaded for free from several websites. The methodologies presented allow one to make use of all the informative components of the survey data, including the geographical ones. However, spatial contiguity matrices, centroids of the areas and their coordinates and distances between them are easily obtained when maps of the studied areas are stored and managed by a Geographic Information System (GIS). This is a relevant opportunity for environmental and agro-environmental studies where geographic information plays a fundamental role for a better understanding of the spatial pattern of the phenomena under analysis.

Table 11.1. *T* and the *p-values* for FH estimation, SFH estimation, NSFH estimation and NPFH estimation and analytic and bootstrap MSE estimation.

	SFH		NPFH			NSFH	
	Analytical MSE	Non-parametric Bootstrap Bias Corrected MSE	Analytical MSE	Naïve Bootsrap MSE	Bias Corrected MSE	Analytical MSE	Bootstrap MSE
	113.351	113.211	104.994	105.415	104.943	93.327	93.423
p-value	≈ 1	≈ 1	≈ 1	≈ 1	≈ 1	≈ 1	≈ 1

Table 11.2. Number of municipalities with values of CV less than 16.6%, between 16.6% and 33.3% and over 33.3% for direct estimator and model-based predictors and different MSE estimation methods.

Estimators	CV of the estimates		
	< 16.6%	from 16.6% to 33.3%	> 33.3%
Direct	117	53	104
SFH Analytical MSE	224	48	2
SFH Non-parametric Bootstrap Bias Corrected MSE	222	49	3
NPFH Analytical MSE	211	58	5
NPFH Naïve Bootsrap MSE	214	56	4
NPFH Bias Corrected MSE	208	61	5
NSFH Analytical MSE	186	78	10
NSFH Bootstrap MSE	187	78	9

Given the presence of spatial dependence, better estimates can be obtained by using the spatial information both in the fixed part of the models (NSFH) and in the random part (SFH and NPFH), even by specifying models with spatially correlated random area effects. The explicit modelling of spatial effects in the SFH is strongly suggested when (i) we have no geographic covariates that are able to take into account the spatial interaction in the target variable, (ii) we have some geographic covariates, but the spatial interaction is so important (autoregressive spatial coefficient $> |0.5|$) that the small area random effects are presumably still correlated. However, the inclusion of varying covariates that capture the spatial effects may be useful even when the strength of spatial link is weak (Chandra et al. 2015).

Acknowledgment

The work of Salvati, Schirripa Spagnolo and Pratesi has been carried out with the support of the project PRA2018-9 (From survey-based to register-based statistics: a paradigm shift using latent variable models). The work of Bertarelli has been carried out with the support of the project InGRID-2, European project G.A. no. 730998.

Appendix 1 'nseblup.R'

```
>library(MASS)
>library(nlme)
>library(spgwr)
>library(pps)
>    nseblup<-function(direct,Xpop,variance,lon.S,lat.S,
m.sample, seed, boot=T, n.boot=100)
+ {
+   set.seed(seed)
+   # collections of functions
+   ##################################################################
+
+ #Functions
+ #--------------------------------
+ #log-likelihood function
+ logl=function(delta){
+   area=m.sample
+   psi=matrix(c(vardir),area,1)
+   Y=matrix(c(direct),area,1)
+   X=Xpop
+   p<-ncol(X)
+   Z.area=diag(1,area)
+   lambda<-delta[1]
+   sigma.u<-delta[2]
+   I<-diag(1,area)
+   #V is the variance covariance matrix
+   C<-lambda*diag(1,p)
+   Cov<-(X%*%C%*%t(X))*W+sigma.u*Z.area%*%t(Z.area)
+   V<-Cov+I*psi[,1]
+   Vi<-solve(V)
+   Xt=t(X)
+   XVi<-Xt%*%Vi
+   Q<-solve(XVi%*%X)
+   P<-Vi-(Vi%*%X%*%Q%*%XVi)
+   b.s<-Q%*%XVi%*%Y
+   ee=eigen(V)
+    -(area/2)*log(2*pi)-0.5*sum(log(ee$value))-(0.5)*log
(det(t(X)%*%Vi%*%X))-(0.5)*t(Y)%*%P%*%Y
+ }
+
+ grr=function(delta){ #this function contains the first
derivatives of the log-likelihood, s
+   lambda<-delta[1]
+   sigma.u<-delta[2]
+   area=m.sample
```

```
+    psi=matrix(vardir,area,1)
+    Y=matrix(c(direct),area,1)
+    X=Xpop
+    p<-ncol(X)
+    Z.area=diag(1,area)
+    I<-diag(1,area)
+    #V is the variance covariance matrix
+    C<-lambda*diag(1,p)
+    Cov<-(X%*%C%*%t(X))*W+sigma.u*Z.area%*%t(Z.area)
+    V<-Cov+I*psi[,1]
+
+    Vi<-solve(V)
+    Xt=t(X)
+    XVi<-Xt%*%Vi
+    Q<-solve(XVi%*%X)
+    P<-Vi-(Vi%*%X%*%Q%*%XVi)
+
+    derLambda<-(X%*%t(X))*W
+    derSigmau<-Z.area%*%t(Z.area)
+    s<-matrix(0,2,1)
+
+    PG<-P%*%derLambda
+    PU<-P%*%derSigmau
+    Pdir<-P%*%Y
+    s[1,1]<-((-0.5)*sum(diag(PG)))+((0.5)*(t(Y)%*%PG%*%Pdir))
+    s[2,1]<-((-0.5)*sum(diag(PU)))+((0.5)*(t(Y)%*%PU%*%Pdir))
+    c(s[1,1],s[2,1])
+ }
+
+ #####boot
+
+ logl.boot=function(delta){
+    area=m.sample
+    psi=matrix(c(vardir),area,1)
+    Y=matrix(c(direct.boot),area,1)
+    X=Xpop
+    p<-ncol(X)
+    Z.area=diag(1,area)
+    lambda<-delta[1]
+    sigma.u<-delta[2]
+    I<-diag(1,area)
+    #V is the variance covariance matrix
+    C<-lambda*diag(1,p)
+    Cov<-(X%*%C%*%t(X))*W+sigma.u*Z.area%*%t(Z.area)
+    V<-Cov+I*psi[,1]
+    Vi<-solve(V)
```

```
+   Xt=t(X)
+   XVi<-Xt%*%Vi
+   Q<-solve(XVi%*%X)
+   P<-Vi-(Vi%*%X%*%Q%*%XVi)
+   b.s<-Q%*%XVi%*%Y
+   ee=eigen(V)
+ -(area/2)*log(2*pi)-0.5*sum(log(ee$value))-(0.5)*log(det
(t(X)%*%Vi%*%X))-(0.5)*t(Y)%*%P%*%Y
+ }
+
+grr.boot=function(delta){ #this function contains the first
derivatives of the log-likelihood, s
+   lambda<-delta[1]
+   sigma.u<-delta[2]
+   area=m.sample
+   psi=matrix(vardir,area,1)
+   Y=matrix(c(direct.boot),area,1)
+   X=Xpop
+   p<-ncol(X)
+   Z.area=diag(1,area)
+   I<-diag(1,area)
+   #V is the variance covariance matrix
+   C<-lambda*diag(1,p)
+   Cov<-(X%*%C%*%t(X))*W+sigma.u*Z.area%*%t(Z.area)
+   V<-Cov+I*psi[,1]
+
+   Vi<-solve(V)
+   Xt=t(X)
+   XVi<-Xt%*%Vi
+   Q<-solve(XVi%*%X)
+   P<-Vi-(Vi%*%X%*%Q%*%XVi)
+
+   derLambda<-(X%*%t(X))*W
+   derSigmau<-Z.area%*%t(Z.area)
+   s<-matrix(0,2,1)
+
+   PG<-P%*%derLambda
+   PU<-P%*%derSigmau
+   Pdir<-P%*%Y
+   s[1,1]<-((-0.5)*sum(diag(PG)))+((0.5)*(t(Y)%*%PG%*%Pdir))
+   s[2,1]<-((-0.5)*sum(diag(PU)))+((0.5)*(t(Y)%*%PU%*%Pdir))
+   c(s[1,1],s[2,1])
+ }
+
+ set.seed(seed)
+
```

```
+ Xpopt=t(Xpop)
+ set.seed(seed)
+
+ vardir<-matrix(variance,m.sample,1)
+
+
+ distance<-matrix(0,m.sample,m.sample)
+     distance<-as.matrix(dist(cbind(as.vector(lon.S),as.
vector(lat.S))))
+
+ I=diag(1,m.sample)
+ p=c(dim(Xpop))[2]
+
+ #-----------------------------------------------------
+ #bandwidth selection
+ #This function finds a bandwidth for a given geographically
weighted regression by optimzing a selected function.
+ #For cross-validation, this scores the root mean square
prediction error for the geographically weighted regressions,
+ #choosing the bandwidth minimizing this quantity
+ #-----------------------------------------------------
+
+ band<- gwr.sel(direct ~ 1+Xpop[,2]+Xpop[,3], coords=cbind
(lon.S, lat.S))
+
+ #-----------------------------------------------------
+ #gaussian spatial weights
+ #The gwr.gauss function returns a vector of weights using
the Gaussian scheme: w(g) = e^{{-(d/h)}^2}
+ #where d are the distances between the observations and h is
the bandwidth.
+ #distance^2 means vector of squared distances between
observations and fit point
+ #-----------------------------------------------------
+
+ W<-gwr.Gauss(distance^2,band)
+ #test for checking the stationarity
+
+ mod<-gwr(direct ~ 1+Xpop[,2]+Xpop[,3],bandwidth=band,
coords=cbind(lon.S, lat.S),hatmatrix=TRUE)
+ BFC99.gwr.test(mod)
+
+ Z<-cbind(diag(Xpop[,1]),diag(Xpop[,2]),diag(Xpop[,3]))
+
+ Xpop=as.matrix(Xpop)
+
```

```
+ ottimo=constrOptim(c(0.1,0.2),logl,grr,method="Nelder-
Mead",ui=rbind(c(1,0),c(0,1)),ci=c(0.001,0),control=lis
t(fnscale=-1))
+
+ lambda.stim.S=ottimo$par[1]
+ sigma2.u.stim.S=ottimo$par[2]
+
+ #Small area point estimates
+ ###================================================
+
+ # Computation of the coefficients estimator (Bstim)
+ D=diag(1,m.sample)
+ C.est<-lambda.stim.S*diag(1,p)
+ Sigma.l<-kronecker(C.est,W)
+ Cov.est<-Z%*%Sigma.l%*%t(Z)+sigma2.u.stim.S*D%*%t(D)
+
+ V<-Cov.est+I*c(vardir)
+
+ Vi<-solve(V)
+ Q<-solve(t(Xpop)%*%Vi%*%Xpop)
+ Beta.hat<-Q%*%t(Xpop)%*%Vi%*%direct
+ P<-Vi-Vi%*%Xpop%*%solve(t(Xpop)%*%Vi%*%Xpop)%*%t(Xpop)%*%Vi
+
+ # Computation of the Nonparametric EBLUP
+ res<-direct-c(Xpop%*%Beta.hat)
+ Sigma.u=sigma2.u.stim.S*I
+ spatial.hat=Sigma.l%*%t(Z)%*%Vi%*%res
+ u.hat=Sigma.u%*%t(D)%*%Vi%*%res
+ GWR.EBLUP.Mean<-Xpop%*%Beta.hat+Z%*%spatial.hat+D%*%u.hat
+
+ Synth<-Xpop%*%Beta.hat+Z%*%spatial.hat
+
+ Sigma.w<-matrix(0,(4*m.sample),(4*m.sample))
+ Sigma.w[1:(3*m.sample),1:(3*m.sample)]<-Sigma.l
+ Sigma.w[(3*m.sample+1):(4*m.sample),(3*m.sample+1):(4*m.
sample)]<-Sigma.u
+
+ w.i<-cbind(Z,D)
+ c.i<-Xpop-w.i%*%Sigma.w%*%t(w.i)%*%Vi%*%Xpop
+
+ # AA
+ #
+
+ AA<-matrix(0,m.sample,1)
+ for (i in 1:m.sample){
+ AA[i,1]<-c.i[i,]%*%solve(t(Xpop)%*%Vi%*%Xpop)%*%cbind(
c.i[i,])
```

```
+ }
+ AA
+
+ # BB
+ #
+ BB<-matrix(0,m.sample,1)
+ for(i in 1:m.sample){
+ BB[i,1]<-w.i[i,]%*%Sigma.w%*%(diag(4*m.sample)-t(w.i)%
*%Vi%*%w.i%*%Sigma.w)%*%cbind(w.i[i,])
+ print(i)
+ }
+
+ BB
+
+ # CC (scalar, one for each smallarea)
+ #
+ CC<-matrix(0,m.sample,1)
+ Ds.1<-matrix(0,(4*m.sample),(4*m.sample))
+ Ds.1[1:(3*m.sample),1:(3*m.sample)]<-kronecker(diag(1,p),W)
+    Ds.1[(3*m.sample+1):(4*m.sample),(3*m.sample+1):(4*m.
sample)]<-0
+ Ds.2<-diag(c(rep(0,3*m.sample),rep(1,m.sample)))
+ B.1<-Z%*%(kronecker(diag(1,p),W))%*%t(Z)
+ B.2<-D%*%t(D)
+
+ B<-list(B.1,B.2)
+ Dv.1<--Vi%*%B.1%*%Vi
+ Dv.2<--Vi%*%B.2%*%Vi
+ #
+ II<-matrix(0,2,2)
+ P<-Vi-Vi%*%Xpop%*%solve(t(Xpop)%*%Vi%*%Xpop)%*%t(Xpop)
%*%Vi
+ for(i in 1:2){
+   for(j in 1:2){
+     II[i,j]<--0.5*sum(diag(P%*%B[[i]]%*%P%*%B[[j]]))
+   }
+ }
+ II<--II
+ #
+ ESS<-matrix(0,2,m.sample)
+ for (i in 1:m.sample){
+   ESS[1,]<-w.i[i,]%*%(Ds.1%*%t(w.i)%*%Vi+Sigma.w%*%t(w
.i)%*%Dv.1)
+   ESS[2,]<-w.i[i,]%*%(Ds.2%*%t(w.i)%*%Vi+Sigma.w%*%t(w
.i)%*%Dv.2)
+   CC[i,1]<-2*t(direct-Xpop%*%Beta.hat)%*%t(ESS)%*%solve
(II)%*%ESS%*%(direct-Xpop%*%Beta.hat)
```

```
+  print(i)
+ }
+ CC/2
+ GWR.EBLUP.MSE.PR<-c(AA+BB+CC)
+
+
+ ###goodness of fit
+
+  gof<-sum((direct-GWR.EBLUP.Mean)^2/(GWR.EBLUP.MSE.PR+
vardir))
+
+ a<-(sqrt(GWR.EBLUP.MSE.PR)/GWR.EBLUP.Mean)
+ b<-sqrt(vardir)/direct
+ b/a
+
+
+ ####Bootstrap estimate
+ # Parametric Bootstrap
+
+ if (boot==T) {
+  Bstim.boot=Beta.hat
+  sigma2l.boot=sqrt(lambda.stim.S)
+  sigma2u.boot=sqrt(sigma2.u.stim.S)
+  thetaNSEBLUP.boot=matrix(0,m.sample,n.boot)
+  thetaSynth.boot=matrix(0,m.sample,n.boot)
+  theta.boot=matrix(0,m.sample,n.boot)
+  W.tot<-W
+  Z.tot<-Z
+
+  for (boot in 1:n.boot)
+  {
+
+    matrix.C.boot<-(sigma2l.boot^2)*diag(1,p)
+    var.lambda.boot<-Z.
      tot%*%kronecker(matrix.C.boot,W.tot)%*%t(Z.tot)
+    mu=matrix(0,m.sample,1)
+ spatial.boot<-mvrnorm(n =1,mu, Sigma=var.lambda.boot,
tol = 1e-6, empirical = FALSE)
+
+
+
+  u.boot=rnorm(m.sample,0,sigma2u.boot)
+  e.boot=rnorm(m.sample,0,sqrt(vardir[,1]))
+
+  #STEP 4
+ theta.boot[1:m.sample,boot]=Xpop%*%Bstim.boot+spatial.
boot[1:(m.sample)]+D%*%u.boot
```

```
+       direct.boot=matrix(c(theta.boot[1:m.sample,boot]+e.
boot),m.sample,1)
+
+
+    #STEP 5 estimation procedure
+
+    # Fit of the model to the data
+
+    ottimo=constrOptim(c((sigma2l.boot^2),(sigma2u.boot^2)),
logl.boot,grr.boot,method="Nelder-Mead",ui=rbind(c(1,0),c(
0,1)),ci=c(0.001,0),control=list(fnscale=-1))
+
+    lambda.stim.S=ottimo$par[1]
+    sigma2.u.stim.S=ottimo$par[2]
+
+    # Computation of the coefficients estimator (Bstim)
+    D=diag(1,m.sample)
+    C.est<-lambda.stim.S*diag(1,p)
+    Sigma.l<-kronecker(C.est,W)
+    Cov.est<-Z%*%Sigma.l%*%t(Z)+sigma2.u.stim.S*D%*%t(D)
+    V<-Cov.est+I*c(vardir)
+
+    Vi<-solve(V)
+    Q<-solve(t(Xpop)%*%Vi%*%Xpop)
+    Beta.hat<-Q%*%t(Xpop)%*%Vi%*%direct.boot
+    P<-Vi-Vi%*%Xpop%*%solve(t(Xpop)%*%Vi%*%Xpop)%*%t(Xpop)%*%Vi
+
+    # Computation of the Nonparametric EBLUP
+    res<-direct.boot-c(Xpop%*%Beta.hat)
+    Sigma.u=sigma2.u.stim.S*I
+    spatial.hat=Sigma.l%*%t(Z)%*%Vi%*%res
+    u.hat=Sigma.u%*%t(D)%*%Vi%*%res
+    thetaNSEBLUP.boot[1:m.sample,boot]<-Xpop%*%Beta.
hat+Z%*%spatial.hat+D%*%u.hat
+    thetaSynth.boot[1:m.sample,boot]<-Xpop%*%Beta.
hat+Z%*%spatial.hat
+
+
+    } # End of bootstrap cycle
+
+
+    GWR.EBLUP.MSE.boot<-NULL
+    SynthGWR.EBLUP.MSE.boot<-NULL
+
+    for (i in 1:m.sample)
+    {GWR.EBLUP.MSE.boot[i]=mean((theta.boot[i,]-
thetaNSEBLUP.boot[i,])^2)
```

```
+   SynthGWR.EBLUP.MSE.boot[i]=mean((theta.boot[i,]-
thetaSynth.boot[i,])^2)}
+
+
+ ###goodness of fit
+
+   gof1<-sum((direct-GWR.EBLUP.Mean)^2/(GWR.EBLUP.MSE.
boot +vardir))
+   gof2<-sum((direct-Synth)^2/(SynthGWR.EBLUP.MSE.boot
+vardir))
+   gof2bis<-sum((GWR.EBLUP.Mean-Synth)^2/(SynthGWR.
EBLUP.MSE.boot +GWR.EBLUP.MSE.boot))
+   cv_boot=(sqrt(GWR.EBLUP.MSE.boot)/GWR.EBLUP.Mean)*100
+   list(est=GWR.EBLUP.Mean,mse_boot=GWR.EBLUP.MSE.boot)}
+
+ else{
+   cv_pr=(sqrt(GWR.EBLUP.MSE.PR)/GWR.EBLUP.Mean)*100
+   list(est=GWR.EBLUP.Mean,mse=GWR.EBLUP.MSE.PR)
+   }
+ }
```

Appendix 2 'npfh.sae.R'

```
>require(SemiPar)
>require(MASS)
>require(nlme)
> # function
>
>   eblup_npfh<-function(y,Xpop,vardir,lon,lat,  m.sample,
n.knots, boot=T, n.boot=100, seed)
+ {
+   set.seed(seed)
+   # collections of functions
+   ################################################################
+   my.default.knots.2D<-function (x1, x2, num.knots)
+   {
+       require("cluster")
+       if (missing(num.knots))
+       num.knots <- max(10, min(50, round(length(x1)/4)))
+       X <- cbind(x1, x2)
+       dup.inds <- (1:nrow(X))[dup.matrix(X) == T]
+       if (length(dup.inds) > 0)
+       X <- X[-dup.inds, ]
+       knots <- clara(X, num.knots)$medoids
+       return(knots)
+ }
```

```
+  #
+  #
+  #
+  Z.matrix<-function(lon,lat,knots){
+    K<-nrow(knots)
+    dist.knot<-matrix(0,K,K)
+    dist.knot[lower.tri(dist.knot)]<-dist(knots)
+    dist.knot<-dist.knot+t(dist.knot)
+    Omega<-tps.cov(dist.knot)
+    dist.lon<-outer(lon,knots[,1],"-")
+    dist.lat<-outer(lat,knots[,2],"-")
+    dist.x<-sqrt(dist.lon^2+dist.lat^2)
+    svd.Omega<-svd(Omega)
+    sqrt.Omega<-t(svd.Omega$v  %*%  (t(svd.Omega$u)  *
       sqrt(svd.Omega$d)))
+    Z<- t(solve(sqrt.Omega,t(tps.cov(dist.x))))
+    return(Z)
+  }
+
+  logl=function(delta){
+    area=m.sample
+    psi=matrix(c(vardir),area,1)
+    Y=matrix(c(direct),area,1)
+    X=Xpop
+    Z.spline=Z
+    Z.area=diag(1,area)
+    sigma.g<-delta[1]
+    sigma.u<-delta[2]
+    I<-diag(1,area)
+    #V is the variance covariance matrix
+    V<-sigma.g*Z.spline%*%t(Z.spline)+sigma.u*Z.
       area%*%t(Z.area)+I*psi[,1]
+    Vi<-solve(V)
+    Xt=t(X)
+    XVi<-Xt%*%Vi
+    Q<-solve(XVi%*%X)
+    P<-Vi-(Vi%*%Xpop%*%Q%*%XVi)
+    b.s<-Q%*%XVi%*%Y
+    ee=eigen(V)
+    -(area/2)*log(2*pi)-0.5*sum(log(ee$value))-
       (0.5)*log(det(t(X)%*%Vi%*%X))-(0.5)*t(Y)%*%P%*%Y
+  }
+
+ grr=function(delta){ #this function contains the first
derivatives of the log-likelihood, s
+    sigma.g<-delta[1]
```

```
+       sigma.u<-delta[2]
+       area=m
+       psi=matrix(vardir,area,1)
+       Y=matrix(c(direct),area,1)
+       X=Xpop
+       Z.spline=Z
+       Z.area=diag(1,area)
+       I<-diag(1,area)
+       #V is the variance covariance matrix
+       V<-sigma.g*Z.spline%*%t(Z.spline)+sigma.u*Z.
area%*%t(Z.area)+I*psi[,1]
+       Vi<-solve(V)
+       Xt=t(X)
+       XVi<-Xt%*%Vi
+       Q<-solve(XVi%*%X)
+       P<-Vi-(Vi%*%Xpop%*%Q%*%XVi)
+
+       derSigmag<-Z.spline%*%t(Z.spline)
+       derSigmau<-Z.area%*%t(Z.area)
+       s<-matrix(0,2,1)
+
+       PG<-P%*%derSigmag
+       PU<-P%*%derSigmau
+       Pdir<-P%*%Y
+       s[1,1]<-((-0.5)*sum(diag(PG)))+((0.5)*(t(Y)%*%PG%*%Pdir))
+       s[2,1]<-((-0.5)*sum(diag(PU)))+((0.5)*(t(Y)%*%PU%*%Pdir))
+       c(s[1,1],s[2,1])
+    }
+
+   # set longitudine and latitude
+   lon_s=lon
+   lat_s=lat
+
+   # knots and Z matrix
+   #n.knots=80
+
+   knots<-my.default.knots.2D(lon_s,lat_s,n.knots)
+   Z<-Z.matrix(lon_s,lat_s,knots)
+
+   # sample data
+   #m.sample=274
+   vardir<-matrix((c(vardir)),m.sample,1)
+   direct=matrix((c(y)),m.sample,1)
+
+   # auxiliary variables
+   # Xpop<-cbind(1,data.grapes$area,data.grapes$workdays)
```

```
+    Xpopt<-t(Xpop)
+    I=diag(1,m.sample)
+    p=c(dim(Xpop))[2]
+    #AreaCode<- data.grapes$ID
+
+    ottimo=constrOptim(c(1,2),logl,grr,
+                       method="Nelder-Mead",
+                       ui=rbind(c(1,0),c(0,1)),ci=c(0,0),
+                       control=list(fnscale=-1))
+
+    estsigma2g=ottimo$par[1]
+    estsigma2u=ottimo$par[2]
+
+    # Computation of the coefficients estimator (Bstim)
+    D=diag(1,m.sample)
+    V<-estsigma2g*Z%*%t(Z)+estsigma2u*D%*%t(D)+I*c(vardir)
+    Vi<-solve(V)
+    Q<-solve(Xpopt%*%Vi%*%Xpop)
+    Beta.hat<-Q%*%Xpopt%*%Vi%*%direct
+    P<-Vi-Vi%*%Xpop%*%solve(t(Xpop)%*%Vi%*%Xpop)%*%t(Xpop)
%*%Vi
+
+    # Computation of the Nonparametric EBLUP
+    K<-nrow(knots)
+    res<-direct-c(Xpop%*%Beta.hat)
+    Sigma.g=estsigma2g*diag(1,K)
+    Sigma.u=estsigma2u*I
+    g.hat=Sigma.g%*%t(Z)%*%Vi%*%res
+    u.hat=Sigma.u%*%t(D)%*%Vi%*%res
+    #Small area point estimates
+    ###===============================================
+    NP.EBLUP.Mean<-Xpop%*%Beta.hat+Z%*%g.hat+D%*%u.hat
+
+    #---------------
+    # ANALYTICAL MSE
+    #---------------
+
+    Sigma.w<-diag(c(rep(estsigma2g,K),rep(estsigma2u,m.
sample)))
+    W<-cbind(Z,D)
+    w.t<-cbind(Z,D)
+    c.t<-Xpop-w.t%*%Sigma.w%*%t(W)%*%Vi%*%Xpop
+
+    g1<-matrix(0,m.sample,1)
+    #
+    for(i in 1:m.sample){
```

```
+   g1[i,1]<-w.t[i,]%*%Sigma.w%*%(diag(K+m.sample)-t(W)%
*%Vi%*%W%*%Sigma.w)%*%cbind(w.t[i,])
+   }
+   M1<-g1
+
+   g2<-matrix(0,m.sample,1)
+   for (i in 1:m.sample){
+      g2[i,1]<-c.t[i,]%*%solve(t(Xpop)%*%Vi%*%Xpop)%*%cb
      ind(c.t[i,])
+   }
+
+   Ds.1<-diag(c(rep(1,K),rep(0,m.sample)))
+   Ds.2<-diag(c(rep(0,K),rep(1,m.sample)))
+   B.1<-Z%*%t(Z)
+   B.2<-D%*%t(D)
+   B<-list(B.1,B.2)
+   Dv.1<--Vi%*%B.1%*%Vi
+   Dv.2<--Vi%*%B.2%*%Vi
+   #
+   II<-matrix(0,2,2)
+   P<-Vi-Vi%*%Xpop%*%solve(t(Xpop)%*%Vi%*%Xpop)%*%t(Xpop)%*%Vi
+   for(i in 1:2){
+      for(j in 1:2){
+      II[i,j]<--0.5*sum(diag(P%*%B[[i]]%*%P%*%B[[j]]))
+      }
+   }
+   II<--II
+   ESS<-matrix(0,2,m.sample)
+   g3<-matrix(0,m.sample,1)
+   for (i in 1:m.sample){
+   ESS[1,]<-w.t[i,]%*%(Ds.1%*%t(W)%*%Vi+Sigma.w%*%t(W)%*%Dv.1)
+   ESS[2,]<-w.t[i,]%*%(Ds.2%*%t(W)%*%Vi+Sigma.w%*%t(W)%*%Dv.2)
+   g3[i,1]<-t(direct[,1]-Xpop%*%Beta.hat)%*%t(ESS)%*%so
lve(II)%*%ESS%*%(direct[,1]-Xpop%*%Beta.hat)
+   }
+
+   M3<-g3
+
+   NP.EBLUP.MSE.PR<-g1+g2+2*g3
+   #---------------
+
+   # PARAMETRIC BOOTSTRAP MSE
+   #---------------
+
+   if (boot==T) {
+
```

```
+   logl.boot=function(delta){
+     area=m
+     psi=matrix(c(vardir),area,1)
+     Y=matrix(c(direct.boot),area,1)
+     X=Xpop
+     Z.spline=Z
+     Z.area=diag(1,area)
+     sigma.g<-delta[1]
+     sigma.u<-delta[2]
+     I<-diag(1,area)
+     #V is the variance covariance matrix
+     V<-sigma.g*Z.spline%*%t(Z.spline)+sigma.u*Z.
      area%*%t(Z.area)+I*psi[,1]
+     Vi<-solve(V)
+     Xt=t(X)
+     XVi<-Xt%*%Vi
+     Q<-solve(XVi%*%X)
+     P<-Vi-(Vi%*%Xpop%*%Q%*%XVi)
+     b.s<-Q%*%XVi%*%Y
+     ee=eigen(V)
+     -(area/2)*log(2*pi)-0.5*sum(log(ee$value))-
      (0.5)*log(det(t(X)%*%Vi%*%X))-(0.5)*t(Y)%*%P%*%Y
+   }
+
+   grr.boot=function(delta){ #this function contains the
first derivatives of the log-likelihood, s
+     sigma.g<-delta[1]
+     sigma.u<-delta[2]
+     area=m
+     psi=matrix(vardir,area,1)
+     Y=matrix(c(direct.boot),area,1)
+     X=Xpop
+     Z.spline=Z
+     Z.area=diag(1,area)
+     I<-diag(1,area)
+     #V is the variance covariance matrix
+     V<-sigma.g*Z.spline%*%t(Z.spline)+sigma.u*Z.
      area%*%t(Z.area)+I*psi[,1]
+     Vi<-solve(V)
+     Xt=t(X)
+     XVi<-Xt%*%Vi
+     Q<-solve(XVi%*%X)
+     P<-Vi-(Vi%*%Xpop%*%Q%*%XVi)
+
+     derSigmag<-Z.spline%*%t(Z.spline)
+     derSigmau<-Z.area%*%t(Z.area)
```

```
+    s<-matrix(0,2,1)
+
+    PG<-P%*%derSigmag
+    PU<-P%*%derSigmau
+    Pdir<-P%*%Y
+    s[1,1]<-((-0.5)*sum(diag(PG)))+((0.5)*(t(Y)%*%PG%*%Pdir))
+    s[2,1]<-((-0.5)*sum(diag(PU)))+((0.5)*(t(Y)%*%PU%*%Pdir))
+    c(s[1,1],s[2,1])
+    }
+
+
+    sigma2.g.stim.S=ottimo$par[1]
+    sigma2.u.stim.S=ottimo$par[2]
+    m=m.sample
+
+    Bstim.boot=Beta.hat
+    sigma2g.boot=sqrt(sigma2.g.stim.S)
+    sigma2u.boot=sqrt(sigma2.u.stim.S)
+    g3.boot=matrix(0,m,n.boot)
+    g1g2.boot=matrix(0,m,n.boot)
+    thetaNPEBLUP.boot=matrix(0,m,n.boot)
+    theta.boot=matrix(0,m,n.boot)
+    tmp=Sigma.g%*%t(Z)%*%P%*%Z%*%Sigma.g
+    svd.tmp=svd(tmp)
+    sqrt.tmp<-t(svd.tmp$v %*% (t(svd.tmp$u) * sqrt(svd.tmp$d)))
+    g.hat.st=solve(sqrt.tmp)%*%g.hat
+
+    g.std=0
+    sdg=sqrt(sigma2.g.stim.S)
+    for (i in 1:K)
+    {g.std[i]=(sdg*(g.hat.st[i,1]-mean(g.hat.st)))/
sqrt(mean((g.hat.st-mean(g.hat.st))^2))}
+
+    tmp=Sigma.u%*%t(D)%*%P%*%D%*%Sigma.u
+    svd.tmp=svd(tmp)
+    sqrt.tmp<-t(svd.tmp$v %*% (t(svd.tmp$u) * sqrt(svd.tmp$d)))
+    u.hat.st=solve(sqrt.tmp)%*%u.hat
+
+    u.std=0
+    sdu=sqrt(sigma2.u.stim.S)
+    for (i in 1:m)
+    {u.std[i]=(sdu*(u.hat.st[i,1]-mean(u.hat.st)))/
sqrt(mean((u.hat.st-mean(u.hat.st))^2))}
+
+
+    tmp=(vardir[,1]*I)%*%P%*%(vardir[,1]*I)
```

```
+   svd.tmp=svd(tmp)
+   sqrt.tmp<-t(svd.tmp$v%*%(t(svd.tmp$u)*sqrt(svd.tmp$d)))
+
+   estim<-solve(sqrt.tmp)%*%(direct[,1]-Xpop%*%Beta.
hat-D%*%u.hat-Z%*%g.hat)
+
+   e.std=0
+   for (i in 1:m)
+   {e.std[i]=(estim[i,1]-mean(estim))/sqrt(mean((estim-
mean(estim))^2))}
+
+
+   for (boot in 1:n.boot)
+   {
+
+     g.boot=sample(g.std,K,replace=TRUE)
+     u.boot=sample(u.std,m,replace=TRUE)
+     e.samp=sample(e.std,m,replace=TRUE)
+     e.boot=sqrt(vardir[,1])*e.samp
+
+     #STEP 4
+     theta.boot[,boot]=Xpop%*%Bstim.boot+Z%*%g.
boot+D%*%u.boot
+     direct.boot=matrix(c(theta.boot[,boot]+e.boot),m,1)
+
+
+     #STEP 5 estimation procedure
+
+     # Fit of the model to the data
+
+     ottimo=constrOptim(c(1,2),logl.boot,grr.
boot,method="Nelder-Mead",ui=rbind(c(1,0),c(0,1)),ci=c(0,
0),control=list(fnscale=-1))
+
+     sigma2.g.stim.S=ottimo$par[1]
+     sigma2.u.stim.S=ottimo$par[2]
+
+     V<-sigma2.g.stim.S*Z%*%t(Z)+sigma2.u.stim.S*D%*%t(D
)+I*vardir[,1]
+     Vi<-solve(V)
+     Q<-solve(Xpopt%*%Vi%*%Xpop)
+     Beta.hat<-Q%*%Xpopt%*%Vi%*%direct.boot[,1]
+
+     #STEP 6
+     res<-direct.boot[,1]-Xpop%*%Beta.hat
+     Sigma.g=sigma2.g.stim.S*diag(1,K)
```

```
+    Sigma.u=sigma2.u.stim.S*D
+    g.hat=Sigma.g%*%t(Z)%*%Vi%*%res
+    u.hat=Sigma.u%*%t(D)%*%Vi%*%res
+
+    thetaNPEBLUP.boot[,boot]<-Xpop%*%Beta.hat+Z%*%g.
hat+D%*%u.hat
+
+    Vblup<-(sigma2g.boot^2)*Z%*%t(Z)+(sigma2u.
boot^2)*D%*%t(D)+I*vardir[,1]
+    Vblupi<-solve(Vblup)
+    Q<-solve(Xpopt%*%Vblupi%*%Xpop)
+    Beta.blup<-Q%*%Xpopt%*%Vblupi%*%direct.boot[,1]
+
+    res.blup<-direct.boot[,1]-Xpop%*%Beta.blup
+    Sigma.g.blup=(sigma2g.boot^2)*diag(1,K)
+    Sigma.u.blup=(sigma2u.boot^2)*D
+    g.hat.blup=Sigma.g.blup%*%t(Z)%*%Vblupi%*%res.blup
+    u.hat.blup=Sigma.u.blup%*%t(D)%*%Vblupi%*%res.blup
+    thetaNPEBLUP.blup.boot<-Xpop%*%Beta.blup+Z%*%g.hat.
blup+D%*%u.hat.blup
+    g3.boot[,boot]=(thetaNPEBLUP.boot[,boot]-
thetaNPEBLUP.blup.boot)^2
+
+    Sigma.w.boot<-diag(c(rep(sigma2.g.stim.S,K),rep(sig
ma2.u.stim.S,m)))
+    W<-cbind(Z,D)
+    w.t<-cbind(Z,D)
+    c.t<-Xpop-w.t%*%Sigma.w.boot%*%t(W)%*%Vi%*%Xpop
+
+
+    g1.c<-matrix(0,m,1)
+    #
+    for(i in 1:m){
+    g1.c[i,1]<-w.t[i,]%*%Sigma.w.boot%*%(diag(K+m)-t(W)
%*%Vi%*%W%*%Sigma.w.boot)%*%cbind(w.t[i,])
+    }
+
+    g2.c<-matrix(0,m,1)
+    for (i in 1:m){
+     g2.c[i,1]<-c.t[i,]%*%solve(t(Xpop)%*%Vi%*%Xpop)%*%c
bind(c.t[i,])
+    }
+
+    g1g2.boot[,boot]=g1.c+g2.c
+    print(boot)
+    }
```

```
+   # End of bootstrap cycle
+   #---------------
+   # bootstrap mse
+   mse.boot<-{}
+   for (i in 1:m)
+   {mse.boot[i]=mean((theta.boot[i,]-thetaNPEBLUP.
boot[i,])^2)}
+
+   G3.boot<-{}
+   for (i in 1:m)
+   {G3.boot[i]=mean(g3.boot[i,])}
+
+   G1G2.boot<-{}
+   for (i in 1:m)
+   {G1G2.boot[i]=mean(g1g2.boot[i,])}
+
+   mse.hat.boot=2*(g1+g2)-G1G2.boot+G3.boot
+
+
+   list(EBLUP=NP.EBLUP.Mean,MSE=NP.EBLUP.MSE.PR,     MSE.
    boot=mse.boot,
+   MSE.BOOT.BC=mse.hat.boot)
+ }
+   else {
+   list(EBLUP=NP.EBLUP.Mean,MSE=NP.EBLUP.MSE.PR)
+   }
+ }
```

References

Anselin, L. 1988. *Spatial Econometrics: Methods and Models*. Boston: Kluwer Academic Publishers.

Bivand, R. and G. Piras. 2015. Comparing implementations of estimation methods for spatial econometrics. *Journal of Statistical Software* 63: 18.

Bivand, R. and D. Yu. 2017. *spgwr: Geographically Weighted Regression. R Package Version 0.6-32*. https://CRAN.R-project.org/package=spgwr.

Brown, G., R. Chambers, P. Heady and D. Heasman. 2001. Evaluation of small area estimation methods— An application to unemployment estimates from the UK LFS. *In: Proceedings of Statistics Canada Symposium 2001. Achieving Data Quality in a Statistical Agency: A Methodological Perspective*. Statistics Canada, Hull.

Brunsdon, C., A.S. Fotheringham and M.E. Charlton. 1996. Geographically weighted regression: a method for exploring spatial nonstationarity. *Geographical Analysis* 28: 281–298.

Brunsdon, C., A.S. Fotheringham and M.E. Charlton. 1999. Some notes on parametric significance tests for geographically weighted regression. *Journal of Regional Science* 39: 497–524.

Chandra, H., N. Salvati and R. Chambers. 2015. A spatially nonstationary Fay-Herriot model for small area estimation. *Journal of Survey Statistics and Methodology* 3: 109–135.

Cressie, N. 1991. Small-area prediction of undercount using the general linear model. pp. 93–105. *In: Proceedings of Statistics Symposium 90: Measurement and Improvement of Data Quality*. Statistics Canada, Ottawa, Canada.

Cressie, N. 1993. *Statistics for Spatial Data*. John Wiley & Sons, New York.

D'Alo, M., L. Di Consiglio, S. Falorsi, M.G. Ranalli and F. Solari. 2012. Use of spatial information in small area models for unemployment rate estimation at sub-provincial areas in Italy. *Journal of the Indian Society of Agricultural Statistics* 66: 43–53.

Fay III, R.E. and R.A. Herriot. 1979. Estimates of income for small places: An application of James-Stein procedures to census data. *Journal of the American Statistical Association* 74: 269–277.

Fotheringham, A.S., C. Brunsdon and M. Charlton. 2003. *Geographically Weighted Regression: The Analysis of Spatially Varying Relationships.* John Wiley & Sons, West Sussex.

Giusti, C., S. Marchetti, M. Pratesi and N. Salvati. 2012. Semiparametric Fay-Herriot model using penalized splines. *Journal of the Indian Society of Agricultural Statistics* 66: 1–14.

InGRID-2: supporting expertise in inclusive growth. http://www.inclusivegrowth.eu/.

Longford, N.T. 2010. Small area estimation with spatial similarity. *Computational Statistics & Data Analysis* 54: 1151–1166.

MAKSWELL: Making Sustainable development and WELL-being frameworks works for policy analysis https://www.makswell.eu/.

Molina, I., N. Salvati and M. Pratesi. 2009. Bootstrap for estimating the MSE of the spatial EBLUP. *Computational Statistics* 24: 441–458.

Molina, I. and Y. Marhuenda. 2015. sae: An R package for small area estimation. *The R Journal* 7: 81–98.

Opsomer, J.D., G. Claeskens, M.G. Ranalli, G. Kauermann and F.J. Breidt. 2008. Non-parametric small area estimation using penalized spline regression. *Journal of the Royal Statistical Society: Series B* 70: 265–286.

Petrucci, A. and N. Salvati. 2006. Small area estimation for spatial correlation in watershed erosion assessment. *Journal of Agricultural, Biological, and Environmental Statistics* 11: 169–182.

Porter, A.T., S.H. Holan, C.K. Wikle and N. Cressie. 2014. Spatial Fay–Herriot models for small area estimation with functional covariates. *Spatial Statistics* 10: 27–42.

Pratesi, M. and N. Salvati. 2008. Small area estimation: the EBLUP estimator based on spatially correlated random area effects. *Statistical Methods and Applications* 17: 113–141.

Pratesi, M. 2015. *Spatial Disaggregation and Small Area Estimation Methods for Agricultural Surveys: Solutions and Perspectives.* FAO Technical Report in the Global Strategy Publications.

Ruppert, D., M.P. Wand and R. Carrol. 2003. *Semiparametric Regression.* Cambridge University Press.

Singh, B.B., G.K. Shukla and D. Kundu. 2005. Spatio-temporal models in small area estimation. *Survey Methodology* 31: 183–195.

Tobler, W.R. 1970. A computer movie simulating urban growth in the Detroit region. *Economic Geography* 46: 234–240.

Wang, J. and W.A. Fuller. 2003. The mean squared error of small area predictors constructed with estimated area variances. *Journal of the American Statistical Association* 98: 716–723.

Wolter, K. 1985. *Introduction to Variance Estimation.* Springer-Verlag, New York.

CHAPTER 12

Cross-sectional Spatial Regression Models for Measuring Agricultural β-convergence

Alfredo Cartone* and *Paolo Postiglione*

12.1 Introduction

Over the past few decades, several contributions focusing on agricultural convergence have been introduced in the literature (Suhariyanto and Thirtle 2001; Sassi 2010; Akram et al. 2020). In the European Union (EU), the importance of those analyses is backed by the relevance that agricultural policies have had (Crescenzi et al. 2016). In fact, the EU spends a considerable amount of its annual budget on Common Agricultural Policy, which in 2000 measured about 25% of overall structural policies. Those policies promote economic growth, especially in regions whose income per capita is considerably below the European average.

Numerous studies have focused on investigating the hypothesis of catching-up in agricultural productivity among countries (see, for example, Rezitis 2010). However, this chapter aims for a deeper exploration of the factors at the basis of agricultural growth capacity and focusing on the regional level.

Generally, the β-convergence hypothesis expresses a cross-sectional test on the existence of a negative link between the initial level and the growth rate of product per capita (Solow 1956). The β-convergence refers to a dynamic process in which poorer regions catch up the richer ones. This approach assumes exogenous saving rates and a production function based on decreasing productivity of capital and constant returns. From an empirical point of view, β-convergence occurs if the regression yields a negative coefficient for the initial level of GDP per capita income. Besides, the highest speed of convergence is associated with the magnitude of the β-convergence coefficient (Bivand and Brunstad 2003).

"G. d'Annunzio" University of Chieti-Pescara, Department of Economic Studies.
Email: postigli@unich.it
* Corresponding author: alfredo.cartone@unich.it

When discussing convergence, a substantial difference lies between absolute convergence and conditional convergence (Barro and Sala-i-Martin 1995). In fact, if regional economies are expected to converge towards the same steady state, β-convergence is denoted as absolute. Conversely, the focus may be on conditional β-convergence (Mankiw et al. 1992), which takes into account the initial level of the income but also includes a set of control variables expected to determine the steady-state growth of the output. In this case, economies with similar structural characteristics converge to a common steady state.

Further, a relevant aspect emerges from the peculiarity of regional data. The need for including spatial attributes in regional economic convergence has been highlighted by a wide literature (among others, Ertur and Koch 2007). Hence, those authors highlighted drawbacks in the econometric modelling linked to the use of standard OLS (Anselin 1988).

Spatial econometric models have offered a large contribution in the field of convergence and agricultural economics (Anselin et al. 2004). The increasing use of spatial econometrics tools has been theoretically and empirically motivated by the fact that not considering spatial dependence may imply severe consequences on both model estimation and interpretation. On the one hand, ignoring spatial dependence in the dependent variable is equivalent to the omission of relevant explanatory variables in the regression equation. This omission results in biased and inconsistent estimators of the coefficients for the remaining explanatory variables. On the other hand, discarding spatial dependence on the disturbances may result in a severe loss of efficiency (Elhorst 2014).

Advances in spatial econometrics resulted in a wide range of model specifications that may be adopted to introduce spatial effects in regression models. Different modelling strategies typically lead to include spatially lagged terms as a weighted average of neighbouring observations collected at a given location (Anselin 2010). Spatially lagged variables can be added for the case of the dependent variable, the explanatory variables, the error term, as well as for combinations of these (Elhorst 2014).

In this chapter, we provide an application to β-convergence models for the agricultural sector in Europe, with particular attention to the consequences of spatial effects and their treatment in the R environment. For this reason, a series of spatially augmented models are estimated using `spdep` and `spatialreg` packages (Bivand and Piras 2015).

The chapter is structured as follows: in Section 12.2, some of the most relevant spatial regression specifications are presented. In Section 12.3 all the models are applied to test for cross-sectional β-convergence in the agricultural sector. Outputs from the R Software are reported. In Section 12.4, concluding remarks are outlined.

12.2 Spatial models for the analysis of agricultural β-convergence

Modelling growth dynamics often requires the adoption of spatial modelling (Cartone et al. 2021). Particularly, in the convergence model, spatial dependence implies that values observed at one location depend on the observations at nearby locations (LeSage and Fischer 2008). Therefore, standard linear regression models should be augmented to incorporate spatial dependence.

For cross-sectional data, we can start by adopting a standard linear regression as follows:

$$g_i = \beta_0 + \mathbf{x}_i' \boldsymbol{\beta} + \varepsilon_i \qquad\qquad \varepsilon_i \sim N(0, \sigma^2) \qquad\qquad (12.1)$$

where $i = 1, 2, ..., N$ denotes the spatial units, $g_i = \dfrac{1}{T}\ln\left(\dfrac{y_{it}}{y_{it} - T}\right)$ is the average growth rate of Gross Value Added (GVA) of the agricultural sector which is the dependent variable, $\mathbf{x}_i$ is a vector of p explanatory variables, and ε_i is an error term, with zero mean and variance σ^2.

In our case, the explanatory variables vector, $\mathbf{x}_i$, for each unit i, includes the initial level of the natural logarithm of the agricultural output measure in terms of the agricultural GVA per capita (i.e., $\ln y_{it-T}$) and other variables to control for regional differences in the agricultural sector.

According to Lusigi et al. (1998), we include, as covariates, investments in agriculture and the growth of the working population in agriculture to assess labour market in agriculture (Deller et al. 2003). Hence, this model also offers an empirical application which is similar to those estimated for global economic growth (Panzera and Postiglione 2014).

To estimate spatial models, we need to consider the effects of proximity. The definition of neighbours is typically carried out through the specification of an exogenous spatial connectivity weight matrix $\mathbf{W}$. Different connectivity matrices may be selected in the definition of neighbour structure. The spatial weight matrix is a square matrix of dimension $N \times N$, where N is the number of regions, and the entries $w_{ij} = 0$ if locations i and j are not neighbours and $w_{ij} = 1$ if i and j are neighbours, according to a specified proximity criterion. Typically, the diagonal elements of $\mathbf{W}$ are zero. Furthermore, most applications in spatial econometrics use a row-standardised weight matrix, where the individual rows of $\mathbf{W}$ are scaled by the row totals, so that rows of $\mathbf{W}$ sum to 1. The weight matrices are often based on the k nearest neighbours' as well as queen and rook criteria. Inverse distance matrices can also be defined according to some transformation of the geographical distance between regions (Getis 2009). In this chapter, we use the k nearest neighbours' specification of $\mathbf{W}$, defined so that each region has the same number of neighbour equal to k.

To identify the most appropriate spatial model specification, two completely different approaches can be followed (Elhorst 2014). In the specific-to-general approach, we start from the standard linear regression model and test whether the model needs to be extended with the spatial effects. Conversely, one can start with a more general model to successively achieve simpler models (i.e., top-down approach).

The most general spatial specification is known as General Nesting Spatial (GNS) or Manski model (Manski 1993). This model takes the following form:

$$g_i = \beta_0 + \mathbf{x}_i' \boldsymbol{\beta} + \rho\sum_{j=1}^{N} w_{ij}\, g_j + \sum_{j=1}^{N} w_{ij}\, \mathbf{x}_j' \boldsymbol{\gamma} + u_i; \qquad u_i = \lambda\sum_{j=1}^{N} w_{ij}\, u_j + \varepsilon_i \qquad (12.2)$$

where, for each $i = 1, ..., N$, $\sum_{j=1}^{N} w_{ij}\, g_j$ is the spatially lagged dependent variable, $\sum_{j=1}^{N} w_{ij}\, \mathbf{x}_j$ denotes the spatially lagged explanatory variables, $\sum_{j=1}^{N} w_{ij}\, u_j$ is the spatial lag of the disturbance terms. Here, w_{ij}s are the entries of the $N \times N$ spatial connectivity matrix.

Spatially lagged terms express the interactions among the observations in the dependent variable, in the disturbance term, and in the explanatory variables. The strength of these interactions depends on the value of the estimated parameters, represented by ρ, λ, and γ, respectively. While ρ is the spatial autoregressive coefficient, λ is the spatial autocorrelation coefficient associated to the error term. The vector γ expresses the effects of the spatial lag of covariates. The Manski model consists of a very general and complex specification. Hence, it represents a general form for subsequent linear spatial econometric models that can be derived starting from it (Elhorst 2014).

By assuming $\gamma = \mathbf{0}$, we obtain the SAC, also named SARAR or Cliff-Ord type model (Kelejian and Prucha 1998):

$$g_i = \beta_0 + \mathbf{x}_i^t \boldsymbol{\beta} + \rho\sum_{j=1}^{N} w_{ij} g_j + u_i; \qquad u_i = \lambda\sum_{j=1}^{N} w_{ij} u_j + \varepsilon_i \qquad (12.3)$$

This model contains both spatially lagged dependent variable and a spatially autocorrelated error term.

Equation (12.2) can be also simplified by assuming that no autocorrelation is present in the error term, thus $\lambda = 0$:

$$g_i = \beta_0 + \mathbf{x}_i^t \boldsymbol{\beta} + \rho\sum_{j=1}^{N} w_{ij} g_j + \sum_{j=1}^{N} w_{ij} \mathbf{x}_j^t \boldsymbol{\gamma} + \varepsilon_i; \qquad (12.4)$$

The spatial Durbin model (SDM; Anselin 1988), defined in equation (12.4), includes among the independent variables the spatially lagged dependent variable as well as the spatial lag of the covariates.

Further, assuming $\rho = 0$ in (12.2), we can define the Spatial Durbin Error Model (SDEM; LeSage 2014), which is specified as follows:

$$g_i = \beta_0 + \mathbf{x}_i^t \boldsymbol{\beta} + \sum_{j=1}^{N} w_{ij} \mathbf{x}_j^t \boldsymbol{\gamma} + u_i; \qquad u_i = \lambda\sum_{j=1}^{N} w_{ij} u_j + \varepsilon_i \qquad (12.5)$$

Two very well-known specifications can be obtained by assuming that interactions occur in the dependent variable or the disturbance term, respectively. In the first case, the spatial lag model (SLM) takes the following form:

$$g_i = \beta_0 + \mathbf{x}_i^t \boldsymbol{\beta} + \rho\sum_{j=1}^{N} w_{ij} g_j + \varepsilon_i \qquad (12.6)$$

In the second case, the spatial error model (SEM) is defined as:

$$g_i = \beta_0 + \mathbf{x}_i^t \boldsymbol{\beta} + u_i; \qquad u_i = \lambda\sum_{j=1}^{N} w_{ij} u_j + \varepsilon_i \qquad (12.7)$$

Besides model specification, the estimation has also extreme relevance. Generally, practitioners may select among maximum likelihood (ML), the instrumental variables/generalized method of moments (IV/GMM), and the Bayesian Markov Chain Monte Carlo (MCMC) method. ML estimation has been among the most often adopted techniques (Anselin 1988), and it relies on the assumption of normality. This assumption is not required by IV/GMM (Elhorst 2014). Moreover, these classes of estimators are useful in the case of one or more endogenous explanatory variables, other than the spatially lagged dependent variables (Fingleton and Le Gallo 2008). The Bayesian estimation method has been attracting more researchers over the last years as this approach may tackle for the presence of heteroscedastic disturbances (LeSage and Pace 2009). However, as ML is still one of the most exploited options in the empirical setting, it will be applied in this chapter.

Furthermore, an often-underrated issue is the interpretation of the estimated parameters. As stressed by LeSage and Pace (2009), the analysis of the spatial models that include dependence in the response variable should be based on the estimated impacts (LeSage and Fischer 2008; LeSage and Pace 2014).

In the standard regression model, a β coefficient parameter can be directly interpreted as the change produced on the dependent variable for one-unit change in the correspondent explanatory variable, while holding other variables in the model constant. In this standard case, the model parameters are estimated under the explicit assumption that the observations are independent; changes in values for one observation (in this case unit i) do not "spill-over" to affect values of other observations (for every $j \neq i$). Note that the previous simple interpretation is still appropriate for some spatial models with only spatially lagged errors (i.e., SEM, see LeSage and Pace 2009).

Unfortunately, for the other spatial model specifications (for example, SLM or SDM) this interpretation no longer holds, and the interpretation of the coefficients should be based on the use of impacts measure.

The direct impact is the effect that a change in an exploratory variable in a single unit causes on the dependent variable for the same unit. The indirect impact denotes the effect that a change of an explanatory variable in a given unit produces on the dependent variable in the neighbouring units. Finally, the total impact is defined as the sum of the direct and indirect effects. LeSage and Pace (2009) developed summary measures expressing the average direct, indirect, and total impact. A significative average indirect impact gives evidence of the presence of spatial spillovers.

12.3 Models estimation and results

As aforementioned, in the current chapter we estimate a model of agricultural conditional convergence (Deller et al. 2003). Additionally, we expand the model accounting for spatial effects as in many convergence studies (LeSage and Fischer 2008).

Data for the application includes 208 NUTS 2 European regions from 12 different countries (Austria, Belgium, Finland, France, Germany, Italy, Portugal, Spain, Sweden, the Netherlands, Poland, and the United Kingdom), which allow for a broad consideration of agricultural β-convergence in Europe. The period under analysis is 1992–2017. NUTS 2 units represent a relevant level for consideration of policies for subsidies in the field of structural funds of the European Commission (Le Gallo and Dall'erba 2006).

Data are obtained from the European Regional Database by Cambridge Econometrics and are now available free of charge from the new Annual Regional Database of the European Commission's Directorate General for Regional and Urban Policy (ARDECO, https://ec.europa.eu/knowledge4policy/territorial/ardeco-database_en) maintained by the European Commission's Joint Research Centre.

The variables considered in the model are: the average growth rate of GVA per-worker of the agricultural sector (g), the natural logarithm of the initial level of GVA per-worker of the agricultural sector (q), the saving rate (s) measured as an average, across the years, of the share of gross fixed capital formation on the GVA for the agricultural sector. Besides, we consider as explanatory variable $v_i = n_i + l + d$, where

n_i is the growth of the working population in agriculture, l is the rate of technological progress, d indicates the depreciation rate of capital, with $l + d = 0.05$ according to Mankiw et al. (1992) and also in line with others empirical evidences for OECD Countries (Vander Donckt and Chan 2019).

As a first step, we download data into the R environment. Particularly, our data file "data_agr.csv " is uploaded using the common base function read.csv:

```
>data <- read.csv("~/SP_AGR/data_agr.csv")
>attach(data)
```

This function uploads the data from a specified directory. We take a look at a preview of the data on our console:

```
>data
  CNTR_CODE FID        g        q        s        v
1     BE BE31 1.26221084 9.642145 -1.06685274 -2.398142
2     BE BE32 0.79582535 9.790938 -1.10254638 -2.473539
3     BE BE33 0.58986699 9.977081 -1.07771466 -2.497582
4     BE BE34 0.76883422 9.569205 -0.95068105 -2.548142
5     AT AT11 0.64210185 9.543297 -0.88999793 -2.532814
6     AT AT12 0.65556778 9.635459 -0.64657724 -2.629505
[...]
```

In the dataset, only six variables are reported, namely the country code (CNTR_CODE), the regional code (FID), the average growth rate of GVA (g), the natural logarithm of the starting point of GVA (q), the natural logarithm of the saving rate (s), and the variable $v = \ln(n_i + l + d)$ (v). Using R, the conditional β-convergence parameters are estimated through OLS as:

```
>ols<-lm(g~q+s+v)
# This function is used to fit linear model
>summary(ols)  # This function returns information about
the estimated   models, including estimated coefficients,
parameters inference, and fit.
Call:
lm(formula = g ~ q + s + v)
Residuals:
      Min       1Q   Median       3Q      Max
  -1.65473  -0.28949  0.04955  0.33390  2.84361
Coefficients:
              Estimate  Std. Error  t value  Pr(>|t|)
(Intercept)   4.88636     0.46236    10.568  < 2e-16  ***
q            -0.38850     0.04686    -8.290  1.5e-14  ***
s             0.04311     0.08263     0.522   0.6024
v             0.21765     0.12112     1.797   0.0738   .
---
Signif. codes: 0 '***' 0.001 '**' 0.01 '*' 0.05 '.' 0.1 ' ' 1
Residual standard error: 0.5385 on 204 degrees of freedom
Multiple R-squared: 0.3213,    Adjusted R-squared: 0.3113
F-statistic: 32.19 on 3 and 204 DF, p-value: < 2.2e-16
```

OLS results offer significant estimates of the intercept and the starting point of GVA. This coefficient is negative, showing the presence of conditional β-convergence. The F-statistic suggests that the complete model can be accepted, but representativeness is quite low (see multiple R-squared in the output). Furthermore, standard linear model does not take into account potential dependence occurring between the regional units under analysis. Therefore, we move a step ahead applying spatial regression models.

Before proceeding to a spatial estimation, we have to define a contiguity matrix that introduces spatial structure into our model. We upload the official shapefile of EU NUTS 2[1] regions new_agr_conv (".shp" extension can be omitted inside the layer of the function) modified to only contain map of the 208 NUTS 2 regions under analysis. We use readOGR at this aim, whose functions are made available by recalling the package rgdal.

```
>library (rgdal)
>map_1<-readOGR(dsn="~/...new_agr_conv.shp",layer = "new_
agr_conv")
#upload the shapefile from the desired directory.
```

After we uploaded the shapefile, we build a contiguity matrix that sets linkages between neighbour regions. To this end, we adopt the spdep library (Pebesma et al. 2015).

```
>library(spdep)
```

This library contains a wide range of functions that enable us to develop different alternatives of the contiguity matrix. In this application, we will consider a k-nearest neighbour matrix setting $k = 8$. This matrix ensures a fixed of neighbours to each of the considered regions so that it is especially useful while dealing with irregular patchworks. To build our contiguity matrix we follow three steps:

```
>coords<-coordinates (map_1)
#extract centroids coordinates from the shapefile
>kn_n<-knearneigh(coords,8)
#find k=8 nearest neigh for each unit in the map
>nb_n<-knn2nb(kn_n)
#develop a nb object based on the k=8 rule
```

Once we have a "nb" (nb_n), we need to obtain a "listw" object, which contains the spatial weights necessary to derive spatial lag terms. Therefore, we also have to set a parameter defining which kind of standardization we prefer. Row-standardised weights are computed by setting style="W" as an argument in the nb2listw function. Accordingly, the "listw" object can be further transformed to a weighting matrix using listw2mat.

```
>neigh<-nb2listw(nb_n,style="W")
#derive a listw object including row-standardized weights
>W<-listw2mat (neigh)
# spatial W corresponding to the listw "neigh"
```

---

[1] Shapefiles for EU regions are available at the following Eurostat link: https://ec.europa.eu/eurostat/web/gisco/geodata/reference-data/administrative-units-statistical-units/nuts.

Once **W** is obtained, we proceed to the estimation of different spatial models. Note that some of the spatial models adopted in the current chapter would allow different spatial structures to be used, for example using different spatial matrices for the lag of the dependent variable and another one for the lag of the errors (Parent and LeSage 2008). However, for the sake of simplicity, we use only one weight matrix for all estimated spatial specifications.

For the estimation process, we adopt the library `spatialreg`, which has been recently developed to enhance functions that were previously included in the `spdep` package. We follow a top-down strategy (i.e., we start from the most general spatial specification), so that we estimate the GNS model first. This model is estimated using the `sacsarlm` function and setting the argument `Durbin=TRUE`. This allows us to account for the lag of covariates, the spatial lag of the dependent variable, and the spatial term in the error. The `gns` object, including estimated parameters from the GNS model, is obtained as the following:

```
>library(spatialreg)
>form<-g~q+s+v #we set the formula for the model to be estimated
>gns<-sacsarlm(form, data = data, listw = neigh, Durbin = TRUE)
#estimation of the GNS - Manski's model using the "neigh" listw
specification of the weight matrix for calculating all spatial terms
> summary(gns) # This function returns information about the
estimated model, including estimated coefficients, parameters
inference, and fit.
Call: sacsarlm(formula = form, data = data, listw = neigh,
Durbin = TRUE)
Residuals:
      Min          1Q     Median          3Q          Max
-1.377500   -0.209724   0.024275   0.223416   1.999161
Type: sacmixed
Coefficients: (asymptotic standard errors)
             Estimate    Std. Error    z value   Pr(>|z|)
(Intercept)   0.164270    0.471802     0.3482    0.72771
q            -0.697936    0.049909   -13.9840    < 2e-16
s            -0.096162    0.082593    -1.1643    0.24431
v             0.217679    0.100029     2.1762    0.02954
lag.q         0.668844    0.066839    10.0068    < 2e-16
lag.s         0.120682    0.109568     1.1014    0.27071
lag.v        -0.309722    0.157768    -1.9631    0.04963
Rho: 0.77808
Asymptotic standard error: 0.086403
    z-value: 9.0052, p-value: < 2.22e-16
Lambda: -0.44279
Asymptotic standard error: 0.28376
    z-value: -1.5605, p-value: 0.11865
LR test value: 92.326, p-value: < 2.22e-16
Log likelihood: -118.2199 for sacmixed model
```

```
ML residual variance (sigma squared): 0.16282, (sigma: 0.40351)
Number of observations: 208
Number of parameters estimated: 10
AIC: 256.44, (AIC for lm: 338.77)
```

The output summary shows estimates for the covariate coefficients. Moreover, we have information about the estimated parameters for the lag of covariates ("lag.q","lag.s","lag.v"), the spatial autocorrelation parameter for the dependent variable (rho) and the spatial parameter in the error (lambda). While the first is positive and significant, the autocorrelation parameter for the residuals is negative and not significant. Another important piece of information contained into the summary is the AIC for the model, which is compared to the one of the standard linear regression also including the lag of covariates (known as SLX model, Gibbson and Overman 2012). The evidence suggests that the use of a spatial model including spatial autocorrelation terms represents a valid alternative, as it presents lower a level of the AIC and it implies better representativeness.

As aforementioned, to estimate the impact of covariates on the dependent variable in a spatial model, we cannot consider the coefficients by themselves. In fact, we have to estimate average direct, indirect, and total impacts. The average impacts can be easily estimated in spatialreg package, but in order to speed up calculation of the spatial derivatives, we use a vector of traces of powers for the spatial weights matrix, as suggested by (LeSage and Pace 2009). The code is as follows:

```
>W <- as(neigh,"CsparseMatrix")
>trn <- trW(W, type="mult")
#calculate traces of powers of the W matrix, type="mult" is
used for sparse matrices
```

The object trn includes stored traces of the row-standardised contiguity matrix.

Some arguments have to be introduced for the calculation and testing of impacts. In fact, to obtain inference for impacts, a simulation is carried out for deriving their empirical distributions. In the summary, we do not report MC sequence selecting short=TRUE, but we include results for test statistics and p-values from the MC drawn (zstats=TRUE). Testing is based on the specification of the variance-covariance matrix of parameters (Elhorst 2014). Conversely, setting useHESS=TRUE would lead to Hessian approximation even if asymptotic coefficient covariance matrix is available. In the application, we set the number of replications for the simulation to 1000 (R=1000).

```
> set.seed(1)
> gns_imp<-impacts(gns, tr=trn, R=1000)
> summary(gns_imp,zstats=TRUE, short=TRUE)
Impact measures (sacmixed, trace):
          Direct     Indirect        Total
q -0.67460675  0.5435833 -0.1310234
s -0.08765811  0.1980900  0.1104319
v  0.19165463 -0.6061878 -0.4145332
```

```
================================================================
Simulation results (asymptotic variance matrix):
================================================================
Simulated standard errors
        Direct      Indirect         Total
q   0.04726199   0.2695768     0.2774557
s   0.08356289   0.4498788     0.4595181
v   0.10264820   0.8599459     0.8967116
Simulated z-values:
        Direct      Indirect         Total
q   -14.274235    2.1429881   -0.3493487
s    -1.027303    0.4343635    0.2384378
v     1.879192   -0.6963506   -0.4526854
Simulated p-values:
        Direct      Indirect         Total
q      < 2e-16    0.032114      0.72683
s     0.304278    0.664025      0.81154
v     0.060218    0.486209      0.65078
```

Average direct and indirect impacts show opposite signs for the case of the starting point of GVA (q). Both average direct and indirect impacts for this variable are significant. Moreover, the impacts for the other variables show no significance.

The GNS can be simplified in a model that includes spatial terms both in the dependent variable and in the error term (i.e., SAC). Estimation of the SAC model using spatialreg requires the same function as GNS to be adopted, but specifying Durbin=FALSE. This argument excludes spatial lag of covariates in the function sacsarlm.

```
> sac<-sacsarlm(form, data=data, listw=neigh, Durbin=FALSE)
> summary(sac)
Call:sacsarlm(formula = form, data = data, listw = neigh,
Durbin=FALSE)
Residuals:
        Min           1Q         Median           3Q          Max
-1.3437e+00 -2.1875e-01 -3.8618e-06 2.1735e-01 1.9520e+00
Type: sac
Coefficients: (asymptotic standard errors)
               Estimate    Std. Error    z value     Pr(>|z|)
(Intercept)    7.626323      0.485092     15.7214      < 2e-16
q             -0.656028      0.048276    -13.5890      < 2e-16
s             -0.078850      0.078334     -1.0066      0.31413
v              0.207837      0.095869      2.1679      0.03017
Rho: -0.49306
Asymptotic standard error: 0.17568
    z-value: -2.8067, p-value: 0.0050059
Lambda: 0.83249
Asymptotic standard error: 0.055521
```

```
   z-value: 14.994, p-value: < 2.22e-16
LR test value: 84.703, p-value: < 2.22e-16
Log likelihood: -122.0317 for sac model
ML residual variance (sigma squared): 0.16449, (sigma: 0.40558)
Number of observations: 208
Number of parameters estimated: 7
AIC: 258.06, (AIC for lm: 338.77)
```

SAC estimates highlight differences when compared to the GNS estimation. Agricultural convergence is confirmed (parameter linked to q negative and significant), but the coefficient ρ is negative and significant. Conversely, the estimated Lambda is positive and extremely significant. The level of AIC is in line with the GNS. This circumstance may happen in spatial applications of the SAC and suggests how the practitioners have to be particularly aware of the different modelling options. The implied impacts for the SAC are calculated as:

```
> set.seed(1)
> sac_imp<-impacts(sac, tr=trn, R=1000)
> summary(sac_imp, zstats=TRUE, short=TRUE)
Impact measures (sac, trace):
           Direct           Indirect              Total
q  -0.6686748        0.22929148       -0.43938329
s  -0.0803704        0.02755936       -0.05281104
v   0.2118433       -0.07264199        0.13920132

=============================================================
Simulation results (asymptotic variance matrix):
=============================================================
Simulated standard errors
           Direct           Indirect              Total
q  0.04977331        0.06888887         0.07011345
s  0.08268387        0.03000625         0.05575383
v  0.10262259        0.03882023         0.07456165
Simulated z-values:
           Direct           Indirect              Total
q -13.4762228         3.2358284        -6.3874134
s  -0.9864038         0.9148781        -0.9704737
v   2.0549027        -1.7721611         1.9055873
Simulated p-values:
           Direct           Indirect              Total
q    < 2e-16          0.0012129        1.6872e-10
s   0.323935          0.3602557         0.331810
v   0.039888          0.0763678         0.056704
```

Average direct impact of the starting point of GVA is negative, while the spillover effects (i.e., average indirect impact) corresponding to that variable are positive. The average total impact is negative and significant.

Both GNS and SAC have received major attention in theoretical studies. In fact, one problem in the estimation of those two models is the simultaneous treatment of two different forms of autocorrelation in the dependent variable and in the error term. This feature emerges as a relevant issue in our application as this leads to puzzling results in the signs, magnitude, and significance of spatial autocorrelations coefficients. Those difficulties in the interpretations lead many researchers involved into applied research to the use of SDM as a flexible specification (LeSage 2014). Using the following code, the SDM model is estimated.

```
> sdm<-lagsarlm(form, data=data, listw = neigh, Durbin = TRUE)
> summary(sdm)
Call:lagsarlm(formula = form, data = data, listw = neigh,
Durbin = TRUE)
Residuals:
        Min          1Q       Median          3Q         Max
  -1.425927   -0.224802    0.022803    0.245974    2.111680
Type: mixed
Coefficients: (asymptotic standard errors)
              Estimate  Std. Error   z value    Pr(>|z|)
(Intercept)   0.492469    0.652192    0.7551     0.45019
q            -0.688417    0.050493  -13.6339    < 2.2e-16
s            -0.096976    0.080988   -1.1974     0.23114
v             0.223423    0.100016    2.2339     0.02549
lag.q         0.627029    0.076086    8.2410    2.22e-16
lag.s         0.117763    0.126706    0.9294     0.35267
lag.v        -0.337534    0.198413   -1.7012     0.08891
Rho: 0.61585, LR test value: 46.808, p-value: 7.8304e-12
Asymptotic standard error: 0.078282
    z-value: 7.8671, p-value: 3.5527e-15
Wald statistic: 61.891, p-value: 3.6637e-15
Log likelihood: -119.0999 for mixed model
ML residual variance (sigma squared): 0.17475, (sigma: 0.41803)
Number of observations: 208
Number of parameters estimated: 9
AIC: 256.2, (AIC for lm: 301.01)
LM test for residual autocorrelation
test value: 1.5131, p-value: 0.21866
```

In the SDM the convergence parameter is still negative and significant together with the positive coefficient corresponding to the spatial lag of the starting point (lag.q). The parameter ρ is very significant, indicating a strong level of spatial autocorrelation in the dependent variable. Another relevant detail can be obtained by looking at the level of the AIC, which is lower compared to both GNS and SAC. We then calculate impacts for the SDM.

```
> set.seed(1)
> sdm_imp<-impacts(sdm, tr=trn, R=1000)
> summary(sdm_imp,zstats=TRUE, short=TRUE)
Impact measures (mixed, trace):
```

```
        Direct     Indirect          Total
q -0.6669842   0.5071856 -0.15979865
s -0.0908507   0.1449613  0.05411057
v  0.2023210 -0.4993649 -0.29704397
================================================
Simulation results (asymptotic variance matrix):
================================================
Simulated standard errors
          Direct       Indirect         Total
q  0.04804364     0.1638502     0.1637816
s  0.07976174     0.3049376     0.3120589
v  0.10639250     0.5367829     0.5712575
Simulated z-values:
          Direct       Indirect         Total
q -13.883269      3.1224503    -0.9487560
s  -1.138115      0.4960668     0.1938461
v   1.890894     -0.9557415    -0.5458988
Simulated p-values:
          Direct       Indirect         Total
q     < 2e-16     0.0017935      0.34274
s    0.255072     0.6198473      0.84630
v    0.058638     0.3392028      0.58514
```

Average impacts for the SDM suggest the presence of direct and indirect effects in the starting point (q).

As the SDM, the SDEM includes lag of the covariates on the right-hand side of the equation. However, spatial autocorrelation parameter is included in this model in the error term. The SDEM model is estimated in spatialreg through the function errorsarlm as follows:

```
>sdem<-errorsarlm(form, data=data, listw=neigh, Durbin=TRUE)
>summary(sdem)
Call:errorsarlm(formula = form, data = data, listw = neigh,
Durbin = TRUE)

Residuals:
      Min           1Q     Median          3Q         Max
-1.417401  -0.224243   0.018388   0.248036   2.127209
Type: error
Coefficients: (asymptotic standard errors)
                Estimate    Std. Error    z value     Pr(>|z|)
(Intercept)     2.898325      1.272756     2.2772     0.022774
q              -0.666000      0.049261   -13.5198    < 2.2e-16
s              -0.076894      0.079690    -0.9649     0.334590
v               0.176850      0.103679     1.7058     0.088055
lag.q.          0.322527      0.118905     2.7125     0.006678
lag.s           0.220435      0.216651     1.0175     0.308933
lag.v          -0.568088      0.367614    -1.5453     0.122265
```

```
Lambda: 0.64371, LR test value: 44.522, p-value: 2.515e-11
Asymptotic standard error: 0.076699
    z-value: 8.3927, p-value: < 2.22e-16
Wald statistic: 70.437, p-value: < 2.22e-16
Log likelihood: -120.2427 for error model
ML residual variance (sigma squared): 0.1756, (sigma: 0.41904)
Number of observations: 208
Number of parameters estimated: 9
AIC: 258.49, (AIC for lm: 301.01)
```

In the summary, we observe that convergence parameter estimate (q) is negative and significant. Estimates corresponding to other variables show little significance, as for the case of investment. Equivalently to the SDM, the spatial lag of the starting point is positive and significant (lag.q). Representativeness for the SDEM and SDM are highly comparable, however the AIC criterion is slightly higher for the SDEM.

In the case of SDEM, average direct and indirect impact can be also estimated. However, it has to be noted that those impacts do not generate from the presence of global dependence, as it happens for other GNS, SAC, SDM, and SLM. Conversely, the average impacts based on the Durbin error model depend on the presence of local spillovers and are only linked to the spatial interactions embedded in the lag of covariates (LeSage 2014). Hence, since the main diagonal elements of **W** are zero and the row-sums are unity, average indirect impacts equal the coefficients obtained for the lag of covariates. In fact, as can be observed from the following, print impacts do not differ from the model parameters.

```
> set.seed(1)
> sdem_imp<-impacts(sdem, tr=trn, R=1000)
> summary(sdem_imp,zstats=TRUE, short=TRUE)
Impact measures (SDEM, estimable, n):
          Direct      Indirect        Total
q   -0.6660000    0.3225272   -0.3434728
s   -0.0768942    0.2204347    0.1435405
v    0.1768504   -0.5680876   -0.3912372
==========================================================
Standard errors:
          Direct      Indirect        Total
q    0.0492612    0.1189048    0.1209169
s    0.0796905    0.2166512    0.2340682
v    0.1036789    0.3676145    0.4087755
==========================================================
Z-values:
          Direct      Indirect        Total
q -13.5197678    2.712483   -2.8405692
s  -0.9649106    1.017464    0.6132421
v   1.7057505   -1.545335   -0.9570955
```

```
p-values:
           Direct        Indirect            Total
q         < 2e-16       0.0066781         0.0045033
s         0.334590      0.3089330         0.5397163
v         0.088055      0.1222652         0.3385190
```

Starting from the SDM specification, another model can be introduced. In fact, by setting the parameters for the lag of covariates to zero, we can further simplify to only include the lag of the dependent variable and estimate the SLM. This model can be as estimated in the R environment through the function `lagsarlm` but setting `Durbin=FALSE`.

```
> slm<-lagsarlm(form, data=data, listw=neigh, Durbin=FALSE)
> summary(slm)
Call: lagsarlm(formula = form, data=data, listw = neigh,
Durbin = FALSE)
Residuals:
       Min           1Q       Median            3Q          Max
-1.568155    -0.262469    0.027234     0.318184     2.693916
Type: lag
Coefficients: (asymptotic standard errors)
                Estimate     Std. Error      z value       Pr(>|z|)
(Intercept)     4.514144       0.517961       8.7152       < 2.2e-16
q              -0.365783       0.050946      -7.1798       6.979e-13
s               0.089555       0.078662       1.1385        0.25492
v               0.211082       0.114607       1.8418        0.06551
Rho: 0.38753, LR test value: 15.321, p-value: 9.0726e-05
Asymptotic standard error: 0.086376
    z-value: 4.4866, p-value: 7.2381e-06
Wald statistic: 20.129, p-value: 7.2381e-06
Log likelihood: -156.7228 for lag model
ML residual variance (sigma squared): 0.25965, (sigma: 0.50956)
Number of observations: 208
Number of parameters estimated: 6
AIC: 325.45, (AIC for lm: 338.77)
LM test for residual autocorrelation
test value: 64.604, p-value: 8.8818e-16
```

The output shows how estimated coefficient signs and significance are in line with the other spatial estimations. The estimated spatial autocorrelation parameter of the dependent variable is also positive and significant. The AIC indicates that, for the case of SLM, representativeness improves (i.e., 325.45) with respect to the standard linear model (i.e., 338.77).

Nevertheless, there are some features that have to be taken into account when comparing results of this model with others that include spatial lag of covariates. On the one hand, SLM is a much simpler model when compared to others, such as SAC and SDM. On the other hand, the magnitude of the convergence parameter, which is lower than other spatial models estimated so far, suggests how this model can potentially fall in bias linked to an omitted variable problem (Pace and LeSage 2010). This feature can be extremely relevant in the case of β-convergence. Empirical literature has stressed the fact that SLM model can suffer from omitted variables compared to SDM (LeSage and Pace 2009).

In the special case of agricultural convergence, the problem of omitted variables may occur as our model is far from considering all the determining factors in the determination of GVA growth. Features in soil productivity are among those not explicitly accounted for in the analysis (Chatterjee 2017). Moreover, variables for the share of agricultural helps and subsidies might also be included, as in Esposti (2007). Impacts calculated for the SLM are reported below.

```
> set.seed(1)
> slm_imp<-impacts(slm, tr=trn, R=1000)
> summary(slm_imp,zstats=TRUE, short=TRUE)
Impact measures (lag, trace):
             Direct         Indirect            Total
q   -0.37289861     -0.22433050       -0.5972291
s    0.09129661      0.05492274        0.1462194
v    0.21518776      0.12945389        0.3446417
========================================================
Simulation results (asymptotic variance matrix):
========================================================
Simulated standard errors
             Direct         Indirect            Total
q    0.05249288      0.07801573       0.09138657
s    0.08304004      0.06462254       0.14322379
v    0.12151185      0.09922825       0.20915452
Simulated z-values:
             Direct         Indirect            Total
q     -7.097981      -2.9654000        -6.608644
s      1.081894       0.9288991         1.046393
v      1.786209       1.3967017         1.700359
Simulated p-values:
             Direct         Indirect            Total
q    1.2659e-12      0.0030229       3.8785e-11
s     0.279300       0.3529414         0.295380
v     0.074065       0.1625033         0.089063
```

The average impacts suggest how direct, indirect, and total effects for the starting point of GVA are significant. In this direction, one difference emerges from a comparison with the SDM. In fact, average indirect impact for the starting point has a negative sign, a circumstance that likely occurs from not considering the lag of covariates.

Another modelling strategy that may be considered is the SEM. This model considers the presence of spatial autocorrelation in the error terms. In this case, direct, indirect, and total impacts do not need to be calculated, and the parameters may be interpreted as first order derivatives as in the simple linear model specification.

```
> sem<-errorsarlm(form, data=data, listw = neigh)
> summary(sem)
Call:errorsarlm(formula = form, data=data, listw = neigh)
Residuals:
      Min          1Q       Median          3Q         Max
-1.351419   -0.220657    0.010182    0.243193    2.070409
Type: error
Coefficients: (asymptotic standard errors)
             Estimate    Std. Error     z value    Pr(>|z|)
(Intercept)  7.474362      0.488810     15.2909      < 2e-16
q           -0.653169      0.049798    -13.1164      < 2e-16
s           -0.062397      0.081648     -0.7642      0.44474
v            0.242093      0.100219      2.4157      0.01571
Lambda: 0.76615, LR test value: 74.478, p-value: < 2.22e-16
Asymptotic standard error: 0.056621
    z-value: 13.531, p-value: < 2.22e-16
Wald statistic: 183.09, p-value: < 2.22e-16
Log likelihood: -127.144 for error model
ML residual variance (sigma squared): 0.18118, (sigma: 0.42565)
Number of observations: 208
Number of parameters estimated: 6
AIC: 266.29, (AIC for lm: 338.77)
```

When considering spatial models, SEM is often compared to SLM. This leads to a common dilemma between modelling spatial autocorrelation in the error term or in the dependent variable. In the case under investigation, the AICs for the two models show that the SEM outperforms the SLM. In fact, AIC for SEM is 266.29 compared to 325.45 of the SLM. This feature would lead practitioners to take into account SEM for the modelling of agricultural convergence as a simpler specification that obtains a sufficient fit. However, it has to be noted that both SEM and SLM do not perform as well as their Spatial Durbin counterparts (i.e., SDM and SDEM) in terms of representativeness. Hence, looking at the results suggests the use of spatial specifications other than SLM and SEM, a circumstance that has led many researchers to prefer the SDM (LeSage and Pace 2009).

Elhorst (2014) also strongly suggests testing whether the SDM specification degenerates into an SEM. In fact, the SDM brings a specification that may degenerate to a SEM when $\lambda\beta = -\gamma$. Hence, practitioners have to verify what is known as the common factor hypothesis, for which the SEM should be preferred to the SDM if the null hypothesis $\lambda\beta + \gamma = 0$ holds. Conversely, in the alternative hypothesis the best modelling options will still be SDM. The common factor test takes the form of a Likelihood Ratio between the unconstrained SDM and the SEM (Mur and Angulo 2009). In spatialreg, this test can be performed as:

```
> common_factor<-LR.sarlm(sdm, sem)
> print(common_factor)
    Likelihood ratio for spatial linear models
```

```
data:
Likelihood ratio = 16.088, df = 3, p-value = 0.001088
sample estimates:
Log likelihood of sdm Log likelihood of sem
            -119.0999                -127.1440
```

The difference in the log-likelihoods and the p-value suggest how the null hypothesis cannot be accepted, so that we can refer to the unconstrained SDM for modelling agricultural β-convergence.

Lastly, another relevant decision has to be taken between two non-nested options, SDM and SDEM. In our application, it can be seen that both specifications result in lower levels of the AIC. The intuitive difference between those two is that, in one case, spatial autocorrelation is modelled from a global perspective, including spatial lag of dependent variables, while in the other case, spatial dependence is in the error term. From a theoretical perspective, this gives rise to problems in terms of identification of spatial parameters (LeSage 2014). From an applied point of view, the SDM allows global spatial spillovers to be explicitly modelled where a change in the controls of each units affect agricultural growth of neighbour units. By this mechanism, the presence of spatial linkages and a spatial lag of the dependent variable determines a *cascade* of adjustments that leads to variations to all regions. For this reason, spillovers are called global (Le Gallo 2014). Conversely, by local spillovers a change in one region covariates solely affects the agricultural output of neighbours. This issue has a certain relevance in terms of policy implications, so that this is often considered in favour of SDM as the most appealing spatial specification.

12.4 Concluding remarks

In this chapter, we provide some examples in the modelling of spatial regression with R. Particularly, we estimate different spatial models specifications to test agricultural conditional β-convergence. The problem of modelling spatial interconnections in agricultural convergence is faced with reference to different forms of spatial dependence. In the estimation process, we start from the more general form, i.e., the GNS, to the simpler specifications, such as the SLM and the SEM. In this sense, some suggestions on how to select the better spatial model are offered with some justifications on the empirical importance of SDM. This circumstance has been studied with reference to European NUTS2 regions.

References

Akram, V., P.K. Sahoo and B.N. Rath. 2020. A sector-level analysis of output club convergence in case of a global economy. *Journal of Economic Studies* 47: 747–767.

Anselin, L. 1988. *Spatial Econometrics: Methods and Models*. Dordrecht: Kluwer Academic Publishers.

Anselin, L., R. Bongiovanni and J. Lowenberg-DeBoer. 2004. A spatial econometric approach to the economics of site-specific nitrogen management in corn production. *American Journal of Agricultural Economics* 86: 675–687.

Anselin, L. 2010. Thirty years of spatial econometrics. *Papers in Regional Science* 89: 3–25.

Barro, R.J. and X. Sala-i-Martin. 1995. *Economic Growth*. Boston: McGraw Hill.

Bivand, R.S. and R.J. Brunstad. 2003. Regional growth in Western Europe: An empirical exploration of interactions with agriculture and agricultural policy. pp. 351–373. *In*: Fingleton, B. (ed.). *European Regional Growth*. Berlin, Heidelberg: Springer.

Bivand, R. and G. Piras. 2015. Comparing implementations of estimation methods for spatial econometrics. *Journal of Statistical Software* 63: 1–36.

Cartone, A., P. Postiglione and G.J.D. Hewings. 2021. Does economic convergence hold? A spatial quantile analysis on European regions. *Economic Modelling* 95: 408–417.

Chatterjee, T. 2017. Spatial convergence and growth in Indian agriculture: 1967–2010. *Journal of Quantitative Economics* 15: 121–149.

Crescenzi, R., D. Luca and S. Milio. 2016. The geography of the economic crisis in Europe: National macroeconomic conditions, regional structural factors and short-term economic performance. *Cambridge Journal of Regions, Economy and Society* 9: 13–32.

Deller, S.C., B.W. Gould and B.L. Jones. 2003. Agriculture and rural economic growth. *Journal of Agricultural and Applied Economics* 35: 517–527.

Elhorst, J.P. 2014. *Spatial Econometrics: From Cross-Sectional Data to Spatial Panels*. Berlin, Heidelberg: Springer.

Ertur, C. and W. Koch. 2007. Growth, technological interdependence and spatial externalities: theory and evidence. *Journal of Applied Econometrics* 22: 1033–1062.

Esposti, R. 2007. Regional growth and policies in the European Union: Does the Common Agricultural Policy have a counter-treatment effect? *American Journal of Agricultural Economics* 89: 116–134.

Fingleton, B. and J. Le Gallo. 2008. Estimating spatial models with endogenous variables, a spatial lag and spatially dependent disturbances: Finite sample properties. *Papers in Regional Science* 87: 319–339.

Getis, A. 2009. Spatial weights matrices. *Geographical Analysis* 41: 404–410.

Gibbons, S. and H.G. Overman. 2012. Mostly pointless spatial econometrics? *Journal of Regional Science* 52: 172–191.

Kelejian, H.H. and I.R. Prucha. 1998. A generalized spatial two-stage least squares procedure for estimating a spatial autoregressive model with autoregressive disturbances. *The Journal of Real Estate Finance and Economics* 17: 99–121.

Le Gallo, J. and S. Dall'Erba. 2006. Evaluating the temporal and spatial heterogeneity of the European convergence process, 1980–1999. *Journal of Regional Science* 46: 269–288.

Le Gallo, J. 2014. Cross-section spatial regression models. pp. 1511–1533. *In*: Fischer, M.M. and P. Nijkamp (eds.). *Handbook of Regional Science*. Berlin, Heidelberg: Springer.

LeSage, J.P. 2014. What regional scientists need to know about spatial econometrics. *The Review of Regional Studies* 44: 13–32.

LeSage, J.P. and M.M. Fischer. 2008. Spatial growth regressions: Model specification, estimation and interpretation. *Spatial Economic Analysis* 3: 275–304.

LeSage, J.P. and R.K. Pace. 2009. *Introduction to Spatial Econometrics*. Boca Raton: Taylor & Francis.

LeSage, J.P. and R.K. Pace. 2014. The biggest myth in spatial econometrics. *Econometrics* 2: 217–249.

Lusigi, A., J. Piesse and C. Thirtle. 1998. Convergence of per capita incomes and agricultural productivity in Africa. *Journal of International Development: The Journal of the Development Studies Association* 10: 105–115.

Mankiw, N.G., D. Romer and D.N. Weil. 1992. A contribution to the empirics of economic growth. *The Quarterly Journal of Economics* 407–437.

Manski, C.F. 1993. Identification of endogenous social effects: The reflection problem. *Review of Economic Studies* 60: 531–542.

Mur, J. and A. Angulo. 2009. Model selection strategies in a spatial setting: Some additional results. *Regional Science and Urban Economics* 39: 200–213.

Pace, R.K. and J.P. LeSage. 2010. Omitted variable biases of OLS and spatial lag models. *pp.* 17–28. *In*: *Progress in Spatial Analysis*. Berlin, Heidelberg: Springer.

Panzera, D. and P. Postiglione. 2014. Economic growth in Italian NUTS 3 provinces. *The Annals of Regional Science* 53: 273–293.

Parent, O. and J.P. LeSage. 2008. Using the variance structure of the conditional autoregressive spatial specification to model knowledge spillovers. *Journal of Applied Econometrics* 23: 235–256.

Pebesma, E., R. Bivand and P.J. Ribeiro. 2015. Software for spatial statistics. *Journal of Statistical Software* 63: 1–8.

Rezitis, A.N. 2010. Agricultural productivity and convergence: Europe and the United States. *Applied Economics* 42: 1029–1044.

Sassi, M. 2010. OLS and GWR approaches to agricultural convergence in the EU-15. *International Advances in Economic Research* 16: 96–108.

Solow, R.M. 1956. A contribution to the theory of economic growth. *The Quarterly Journal of Economics* 70: 65–94.

Suhariyanto, K. and C. Thirtle. 2001. Asian agricultural productivity and convergence. *Journal of Agricultural Economics* 52: 96–110.

Vander Donckt, M. and P. Chan. 2019. The New FAO global database on agricultural investment and stock. *FAO Statistics Working Paper Series* 19–16.

CHAPTER 13
Spatial Panel Regression Models in Agriculture

Paolo Postiglione

13.1 Introduction

Panel models typically refer to data containing time series observations on a number of individuals. Therefore, observations in panel data include two different dimensions; a cross-sectional dimension, indicated by i, for $i = 1,..., N$, and a time series one, denoted by t, for $t = 1,..., T$.

The increasing use of panel data is presumably connected with the growing data set availability (Hsiao 2007). There are many advantages using panel data with respect to cross-sectional or time series data (Baltagi 2013; Hsiao 2014):

1. Panel data models ensure the possibility that individuals, firms, states, countries, regions or spatial units are heterogeneous: time-series and cross-sectional studies do not control for this typology of heterogeneity and, therefore, lead to biased estimates.

2. Panel data offer more informative data and results. From a statistical point of view, they provide more variability, less multicollinearity between variables, and more efficiency in the estimates. Conversely, time series analysis suffers from multicollinearity problems.

3. Panel data are more efficient when describing the dynamics of processes. Cross-sectional analyses, that are static by themselves, do not explain any temporal change of the phenomena under investigation.

4. Panel data allow us to test more complicated hypotheses on the real world than simple cross-sectional or time series data, because they model changes across time and/or individuals.

5. Panel data collected on micro units (i.e., individuals, firms, households) are more accurate and possible biases resulting from aggregations can be avoided.

"G. d'Annunzio" University of Chieti-Pescara, Department of Economic Studies.
Email: postigli@unich.it

However, panel models suffer some shortcomings. Baltagi (2013) and Hsiao (2014) evidence the possible existence of design and data collections problems related, for example, to coverage of the population and non-response bias. Furthermore, other possible limitations, such as distortions of measurement errors and selectivity problems, may affect the reliability of a panel model.

Panel models are largely used to analyse different and important topics in agricultural economics. A stochastic frontier production function model for panel data is described by Battese and Coelli (1992), where the firm effects are an exponential function of time. The model is verified using agricultural data for paddy farmers in an Indian village. Dawson (2005) defines a panel model for 62 less developed countries (LDC) for the years from 1974 to 1995 for studying the contribution of agricultural exports to economic growth. The results highlight significant structural differences in economic growth between low, lower-middle, and upper-income LDCs. Lio and Liu (2006) evidence that ICT can play a significant role in improving agricultural productivity, using panel data from 81 countries over the period 1995–2000. Hackl et al. (2007) analyse local compensation payments made to farmers by municipalities in Austrian tourist communities for preserving a typical agricultural landscape using unbalanced panel data. The results evidence that the probability of introducing compensation payments depends positively on the benefits of landscape amenities. Kim et al. (2012) analyse a panel of 3,140 Korean farm households between 2003 and 2007 and evaluate the presence and sources of productivity gains associated with specialization and diversification in rice production. The main findings are that benefits from diversification are positive, but non-statistically significant for Korean farms. Khandker and Koolwal (2016) use household panel data for Bangladesh over 20 years to investigate the impact of rural credit (i.e., microcredit and formal bank channels) on economic results for agricultural households. The authors do not find any influence of microcredit on crop income.

However, as evidenced by Baylis et al. (2011), these kinds of agricultural economic analysis have often an important spatial dimension. During the last decades, spatial econometric models for panel data have helped applied analysts in controlling for the presence of cross-sectional dependence, as well as heterogeneity across units in terms of spatial specific effects and time specific effects. In this respect, spatial panel models may provide an additional choice for the estimation of many agricultural economic models. Spatial panel models consider individuals that are associated with a particular position in space (see Section 13.2 for further details).

Recently, many contributions have used spatial panel methodologies in agriculture. Druska and Horrace (2004) define a spatial panel estimator that is the extension to the panel case of the cross-sectional one introduced by Kelejian and Prucha (1999). The model is applied in the stochastic frontier production framework to a panel of Indonesian rice farms. This data set is very popular in agricultural economic literature and forms the basis for our illustration, using a different spatial model, in Section 13.3. The paper by Baylis et al. (2011) contains an interesting discussion on possible spatial panel data applications in agriculture. In particular, they estimate a spatial panel model to evaluate the effect of temperature and precipitation on the log of farmland values using county-level data limited to the counties east of the 100th meridian to control for non-irrigated agriculture. Chakir and Le Gallo (2013) predict land use area in France, considering four land use classes: agriculture, forest, urban, and other use. The authors use a spatial panel model specification with a data set covering cross-sectional observations of NUTS 3 regions from 1992 to 2003. Tong et al. (2013) estimate a spatial Durbin panel data model for measuring the influence of transportation infrastructure on US agricultural output. In particular, the analysis employs a panel data set for 44 states in the United States over the period from 1981 to 2004. The empirical evidence suggests that road disbursement in a given state has positive direct effects on its own agricultural output. Zouabi et al. (2015) use several spatial panel specifications to evaluate the impact of climate on Tunisian agriculture. They define a Cobb-Douglas production function for agricultural products based on regional data. The analysis shows that Tunisian agriculture is strongly influenced by temperature and precipitation at the regional level. Wu et al. (2020) aim at studying the impact of industrial agglomeration and some policies introduced by the Chinese government on agricultural energy efficiency. To this end, the authors estimate different spatial panel models on 30 Chinese provinces from 2000 to 2016.

This chapter wants to contribute to the literature in the field of spatial panel, highlighting the huge potential of those modelling strategies for agricultural data. In particular, the main aim is to show how spatial panel data models can be estimated in R through the library `spml` (Millo and Piras 2012).

The chapter is structured as follows. In Section 13.2, the main specifications for spatial panel data models are illustrated and discussed. Section 13.3 contains the description of the panel approach to stochastic frontier analysis with a particular focus on agriculture. Additionally, the R codes for the estimation of a spatial stochastic frontier panel model are presented and commented. Finally, Section 13.4 concludes.

13.2 Models

Spatial panel data consider cross sectional of geo-referenced spatial observations (i.e., countries, regions, and points) repeated across several time periods. If the number of spatial observations is constant over time, the spatial panel is defined as balanced. In this chapter, we only focus on spatial balanced panels.

Panel data can be specified and analysed using static or dynamic models (Hsiao 2014; Croissant and Millo 2019). The static models include contemporaneous values of the dependent and independent variables. Dynamic panel models are defined as having a lagged (in time) dependent variable and/or lagged (in both time and space) dependent variable. Dynamic panels consider time dimension autocorrelation (allowing in the spatial case calculating short and long run impacts; Debarsy et al. 2012). However, dynamic panel sometimes require some conditions in terms of the number of time observations to have a proper estimation. Furthermore, spatial dynamic panel model has been not currently implemented in `splm`. Therefore, in this chapter, the attention will focus only on static panel data models that can be widely applied in many agricultural studies (Chakir and Le Gallo 2013).

Consider the General Nesting spatial model (GNS) for cross-sectional data defined in equation (12.2).

The extension of this model, in matrix notation, to a space-time model can be simply obtained pooling the data as:

$$\mathbf{y}_t = \beta_0 \mathbf{i} + \mathbf{X}_t \boldsymbol{\beta} + \lambda \mathbf{W} \mathbf{y}_t + \mathbf{W} \mathbf{X}_t \boldsymbol{\gamma} + \mathbf{u}_t; \qquad \mathbf{u}_t = \rho \mathbf{W} \mathbf{u}_t + \boldsymbol{\varepsilon}_t \qquad (13.1)$$

where, for each time period $t = 1,..., T$, β_0 is the intercept, $\mathbf{X}_t$ is the $N \times p$ matrix of explanatory variables, $\mathbf{y}_t$ is the $N \times 1$ vector of the dependent variable, $\mathbf{W}$ is the $N \times N$ matrix of spatial weights that measures the connectivity between spatial units, $\boldsymbol{\beta}$ and $\boldsymbol{\gamma}$ are the unknown vectors of parameters, λ is the spatial autoregressive coefficient, ρ is the spatial autocorrelation coefficient, $\boldsymbol{\varepsilon}_t$ is the vector of the independent and identically distributed (*iid*) errors with zero mean and variance σ_ε^2. The entry $w_{ij} \in \mathbf{W}$ is greater than 0 only if i and j are neighbours and 0 otherwise. The elements w_{ii} are conventionally set equal to 0. The spatial weight matrix $\mathbf{W}$ is often row standardised. Under row-standardisation, the elements of each row sum to unity. This procedure is used to create proportional weights in cases where features have an unequal number of neighbours. Model (13.1) can be estimated in a similar way as the cross-sectional model (12.2).

The main shortcoming for this approach is that the model (13.1) does not consider any spatial and temporal heterogeneity. To this end, it is possible to define an augmented model as:

$$\mathbf{y}_t = \beta_0 \mathbf{i} + \mathbf{X}_t \boldsymbol{\beta} + \lambda \mathbf{W} \mathbf{y}_t + \mathbf{W} \mathbf{X}_t \boldsymbol{\gamma} + \boldsymbol{\alpha} + \xi_t \mathbf{i} + \mathbf{u}_t; \qquad \mathbf{u}_t = \rho \mathbf{W} \mathbf{u}_t + \boldsymbol{\varepsilon}_t; \qquad (13.2)$$

where $\boldsymbol{\alpha} = (\alpha_1,..., \alpha_N)^T$ is the vector of the spatial specific effects and ξ_t is the time-period effect. The term α_i summarizes the effect of the omitted variables that are characteristic of each spatial unit and that are very difficult to measure (Elhorst 2014). Similarly, the omission of the time-period effect ξ_t, that controls for all spatial invariant variables, could affect the estimates in time series analysis (Arellano 2003; Elhorst 2014).

It is possible to specify model equation (13.2) by introducing some restrictions. In particular, by assuming $\boldsymbol{\gamma} = \mathbf{0}$, we obtain the SARAR model for panel data as:

$$\mathbf{y}_t = \beta_0 \mathbf{i} + \mathbf{X}_t \boldsymbol{\beta} + \lambda \mathbf{W} \mathbf{y}_t + \boldsymbol{\alpha} + \xi_t \mathbf{i} + \mathbf{u}_t; \qquad \mathbf{u}_t = \rho \mathbf{W} \mathbf{u}_t + \boldsymbol{\varepsilon}_t \qquad (13.3)$$

This model contains both the spatially lagged dependent variable and the spatially autocorrelated error term.

If, as in Elhorst (2003), we consider only one form of spatial dependence, we can define the spatial lag model (SLM) for panel data if only the spatial lag of the dependent variable is included, or the spatial error model (SEM) if only a spatially autocorrelated error term is considered. The SLM for panel data assumes the following form (Elhorst 2003):

$$\mathbf{y}_t = \beta_0 \mathbf{i} + \mathbf{X}_t \boldsymbol{\beta} + \lambda \mathbf{W} \mathbf{y}_t + \boldsymbol{\alpha} + \xi_t \mathbf{i} + \boldsymbol{\varepsilon}_t \qquad (13.4)$$

Conversely, the SEM for spatial data is defined as (Elhorst 2003):

$$\mathbf{y}_t = \beta_0 \mathbf{i} + \mathbf{X}_t \boldsymbol{\beta} + \boldsymbol{\alpha} + \xi_t \mathbf{i} + \mathbf{u}_t; \qquad \mathbf{u}_t = \rho \mathbf{W} \mathbf{u}_t + \boldsymbol{\varepsilon}_t \qquad (13.5)$$

The spatial Durbin model for panel data (SDM; Elhorst 2014) can be defined assuming $\rho = 0$ in equation (13.2):

$$\mathbf{y}_t = \beta_0 \mathbf{i} + \mathbf{X}_t \boldsymbol{\beta} + \lambda \mathbf{W} \mathbf{y}_t + \mathbf{W} \mathbf{X}_t \boldsymbol{\gamma} + \boldsymbol{\alpha} + \xi_t \mathbf{i} + \boldsymbol{\varepsilon}_t; \qquad (13.6)$$

Model (13.6) extends the SLM by allowing for a spatial association not only in the dependent variable, but also in the explanatory variables, obtaining more flexible spatial effects (LeSage and Pace 2009).

The specific effects α_i and ξ_t can be treated as fixed or random. The fixed effect model (FEM) introduces a dummy variable for each spatial unit and time period to model unobserved heterogeneity. Conversely, in the random effect model (REM), α_i and ξ_t are treated as random variable independently and identically distributed with zero mean and variance σ_α^2 and σ_ξ^2, respectively. In the REM, the unobserved spatial and time effects are implicitly assumed as uncorrelated with the other explanatory variables and, therefore, can be treated as components of the error term.

Two different specifications can be defined for the error term. In the first case, the spatial diffusion effect is considered only for the idiosyncratic error term and not for the random individual effect (Baltagi et al. 2003). Conversely, in the second case, the spatial correlation structure is applied both to the individual effects and the remaining component of the error term (Kapoor et al. 2007). These two different specifications for the error term imply different spatial spillover process managed by the two different variance-covariance matrices. For more details about these different model specifications of the spatial panel data models, see Baltagi (2013, Chapter 13) and Kelejian and Piras (2017, Chapter 15).

The models described can be essentially estimated following two different approaches. The first is based on the principle of maximum likelihood (ML), while the second considers the instrumental variable or generalised method of moments procedure (IV/GMM).

ML estimators need stronger conditions on the distribution of the error term (i.e., normality). The IV/GMM procedure is often preferred for its less computational effort and the possibility to estimate spatial models with an endogenous explanatory variable.

Elhorst (2003) describes the maximum likelihood (ML) estimator of the SLM and the SEM under the fixed and the random effects assumptions. Note that the SDM can be estimated as a spatial lag model with covariates $[\mathbf{x}_{it} \, w_{ij} \, \mathbf{x}_{jt}]$ instead of $\mathbf{x}_{it}$. Lee and Yu (2010) demonstrate that the ML estimator of the SLM and the SEM with spatial fixed effects, as defined in Elhorst (2003), provides an inconsistent parameter estimate of the variance if N is large and T is small, as well as inconsistent estimates of all parameters of the SLM and the SEM with spatial and time-period fixed effects, if both N and T are large. To correct for this problem, they propose a simple bias correction procedure (Lee and Yu 2010).

The IV/GMM estimation was introduced in the spatial case by Kelejian and Prucha (1999) for cross-sectional data, firstly extended to the panel case by Druska and Horrace (2004) and then by Kapoor et al. (2007) and Mutl and Pfaffermayr (2011).

In order to choose between fixed and random specific effects, a Hausman test can be used (Hausman 1978). Mutl and Pfaffermayr (2011) extend this procedure to the framework of spatial panel models. The Hausman statistic H is asymptotically distributed as a χ^2 with k degrees of freedom, where k is the number of covariates in the model. Under the null hypothesis, the fixed effects estimator is consistent but inefficient, while the random effects estimator is both consistent and efficient. Rejection of the null hypothesis supports the use of a fixed effects specification.

Lagrange multiplier (LM) tests have been introduced in literature to verify some spatial autocorrelation specifications (Anselin et al. 2008; Baltagi et al. 2003; Baltagi et al. 2007).

Anselin et al. (2008) defined LM tests for a spatially lagged dependent variable and spatial error correlation. The robust version of these tests was studied by Elhorst (2010). These represent the extension of the standard tests for cross-sectional model (Anselin 1988; Anselin et al. 1996) and are based on the residual vector of a pooled regression model without any spatial or time-specific effects (Elhorst 2014). The test LM_{lag} (LM_{err}) aims at verifying the null hypothesis $\lambda = 0$ ($\rho = 0$). These two tests are accompanied by their robust version. The RLM_{lag} tests the null hypothesis $\lambda = 0$ and it is robust to the potential presence of a spatial autoregressive term in the errors (i.e., $\rho \neq 0$); conversely, the RLM_{err} verifies the null hypothesis $\rho = 0$ and it is robust to the potential presence of a spatial autoregressive term in the model of the dependent variable (i.e., $\lambda \neq 0$). These LM tests have been used to discriminate between SLM and SEM models, as described in Florax et al. (2003).

Baltagi et al. (2003) derived marginal and conditional LM tests for all combinations of random effects and spatial error autocorrelation.

In particular, the following hypotheses are considered:

1) H_0^a: $\rho = 0$ and no random effects, under the alternative that at least one component is not zero (LMH).

2) H_0^b: no random effects, assuming $\rho = 0$, under the one-sided alternative that the random effects assumption is reasonable (LM1).

3) H_0^c: $\rho = 0$, assuming no random effects, under the two-sided alternative that the spatial autocorrelation coefficient $\rho \neq 0$ (LM2).

4) H_0^d: $\rho = 0$, assuming the possible existence of random effects, under the two-sided alternative that the spatial autocorrelation coefficient is $\rho \neq 0$ (CLMlambda).

5) H_0^e: no random effects, assuming $\rho \geq 0$ and the one-sided alternative that the random effects assumption is reasonable (CLMmu).

The conditional tests (4) and (5) are the most useful tests, since they test for one effect, and they are robust against the other alternative. Finally, Baltagi et al. (2007) introduced LM tests for spatial error autocorrelation, serial correlation, and random effects in spatial panel models.

As emphasized by LeSage and Pace (2009), for cross-sectional data, and by Debarsy et al. (2012) and Piras (2014), for panel models, the coefficients of spatial models are often wrongly interpreted in terms of the partial derivatives of the dependent variable with respect to the explanatory variables. This interpretation is still correct for the SEM model, but it does not hold for models with spatial dependence in the dependent variable or the covariates (i.e., GNS, SLM, SARAR, SDM). For these last models, the computation of the average direct, indirect, and total impacts for each variable in the model is requested.

In the aforementioned spatial models, changes in an explanatory variable in a particular unit i influence not only the dependent variable in i, but also the dependent variables in other units. The first is called direct impact, while the second is denoted as indirect impact. Since these impacts are different across units, average statistics have been introduced for each impact. Piras (2014) extends these measures, introduced by LeSage and Pace (2009) for cross-sectional data, defining average direct and indirect impacts for spatial panel models, taking into account the specificity of the dimension of the spatial weighting matrix in the panel case.

13.3 An empirical application in agriculture with R

13.3.1 The stochastic frontier production model

A production function defines the technological relationship between the level of inputs and the resulting level of outputs. The general implicit hypothesis is that all firms produce efficiently, and deviations from this situation are assumed to be random. Conversely, if we consider an approach based on the estimation of the production frontier, we assume that the limit of the production function is only reached by *the best performance* firms.

The stochastic frontier analysis (SFA) is widely used to estimate firm individual efficiency and has been introduced by Aigner et al. (1977) and Meeusen and van de Broeck (1977). The SFA is motivated by the conjectural idea that no economic firm can go beyond the frontier, and difference from this limit represents individual inefficiencies. This frontier describes the maximum output attainable by a unit (i.e., the farm) with given inputs (Greene 2008).

These models were first based on cross-sectional data, but similar models can be developed for panel data (Pitt and Lee 1981; Schmidt and Sickles 1984; Horrace and Schmidt 1996).

In this section, our aim is to describe a stochastic spatial panel frontier model to evaluate efficiency in agricultural farms.

The basic standard stochastic frontier model for panel data can be defined as (Greene 2005):

$$y_{it} = \beta_0 + \boldsymbol{\beta}'\mathbf{X}_{it} + \varepsilon_{it} - v_i \qquad (13.7)$$

where $i = 1,\dots, N$ are the farms, $t = 1,\dots, T$ is the time index, ε_{it} is the random component representing stochastic elements and any specific heterogeneity, and $v_i \geq 0$ is a measure of technical inefficiency, assumed to be time invariant. $\mathbf{X}_{it}$ and v_js are assumed as independent for $t,s = 1,\dots, T$ and $i, j = 1,\dots, N$. The variable y_{it} is the natural logarithm of the output and $\mathbf{X}_{it}$ is the vector of the natural logarithm of the inputs, for each i and t.

The general basic assumption is that farms lying on this frontier apply the most efficient production processes. In this case, this efficient production level can be estimated as:

$$y_{it}^* = \widehat{\beta_0} + \widehat{\boldsymbol{\beta}}'\mathbf{X}_{it} \qquad (13.8)$$

where y_{it}^* is the maximum output for each unit i (no inefficiency in this case).

The production of each firm i can be obtained as:

$$\hat{y}_{it} = \widehat{\beta_0} + \widehat{\boldsymbol{\beta}}'\mathbf{X}_{it} - v_i \qquad (13.9)$$

Assuming the natural logarithm specification for output and inputs, as in our case, the technical efficiency can be defined as:

$$\ln T E_i = \hat{y}_{it} - y_{it}^* = -v_i \tag{13.10}$$

and, hence, $T E_i = \exp(-v_i)$, with $0 \leq T E_i \leq 1$. If $v_i = 0$, then $T E_i = 1$, and production is said to be technically efficient.

It is worth noting that the estimation of model (13.7) differs from OLS estimation of a production frontier. In fact, the model (13.7) assumes the presence of two random elements: one is the usual random disturbance term (i.e., ε_{it} as in OLS), while the other (v_i) is an efficiency scaling term, see Greene (2008) for further details.

Now define $\alpha_i = \beta_0 - v_i$, so that the panel model (13.7) can be rewritten in the standard formulation as:

$$y_{it} = \alpha_i + \boldsymbol{\beta}' \mathbf{X}_{it} + \varepsilon_{it} \tag{13.11}$$

Any latent heterogeneity is either absent or included in α_i (Greene 2005).

Note that model (13.7) is formally identical to the panel data model without inefficiency terms v_i, but with farm specific effects, α_i, as (Behr 2015):

$$y_{it} = \beta_0 + \boldsymbol{\beta}' \mathbf{X}_{it} + \alpha_i + \varepsilon_{it} \tag{13.12}$$

Without distributional assumptions on ε_{it} and v_i, but allowing for correlation between α_i and $\mathbf{X}_{it}$, the model can be analysed through a fixed effects specification, as suggested by Schmidt and Sickles (1984).

Conversely, the random effects approach considers the model (13.7) with the additional distributional assumption that $\varepsilon_{it} \sim N(0, \sigma_\varepsilon^2)$ and $v_i \sim N^+(0, \sigma_v^2)$.

The panel data model (13.7) assumes a time invariant inefficiency effect. This hypothesis is considered unrealistic, since it theoretically excludes the possibility for farms to respond to inefficiencies. For further details about the panel approach to SFA, see Greene (2005) and Behr (2015).

The SF panel equations, spatially augmented according to the models outlined in Section 13.2, are estimated with R in the following paragraph.

13.3.2 Spatial panel estimations of stochastic frontier model

For the estimation and testing of spatial panel models, we use the R library `splm`. In this package, maximum likelihood and generalized moments estimators using fixed as well as random effects are considered (Millo and Piras 2012).

The empirical application concerns the analysis of production efficiency in the Indonesian rice farming through the `RiceFarms` dataset, available in `splm`. This topic has received a lot of attention in the literature (Lee and Schmidt 1993; Trewin et al. 1995; Horrace and Schmidt 1996; Druska and Horrace 2004; Feng and Horrace 2012; Millo 2014; Croissant and Millo 2019).

The `RiceFarms` dataset includes 171 observations of rice farms in Indonesia ($N = 171$), observed during six growing seasons ($T = 6$), between 1975 and 1983. In this case, we have large N and small T. The farms are located in six different villages of the Cimanuk River basin in West Java (Horrace and Schmidt 1996).

In our study, to estimate the spatial panel production frontier equation, we use output measured in kilograms of rice (`goutput`) and some inputs to the production of rice, as seed in kg (`seed`), urea in kg (`urea`), total labour in hours (`totlabor`), and land in hectares (`size`). All the inputs are considered in natural logarithm.

The preliminary steps in R for the spatial panel estimation are the following.

```
> library (spdep) #load library sdpep for spatial analysis
> library(splm) #load library splm for spatial panel data analysis
> data (RiceFarms) #load RiceFarms dataset
```

The `RiceFarms` dataset is a `data.frame` where the first two variables are the individual and time indexes. The index argument should be left to the default value (i.e., `NULL`). For further details see Figure 13.1 where the first rows of the `RiceFarms` dataset are reported.

id	time	size	seed	urea	totlabor	goutput
101001	1	3	90	900	2915	7980
101001	2	2	40	600	2155	4083
101001	3	1	100	700	1075	2650
101001	4	2	60	600	2091	4500
101001	5	3.572	105	400	3889	16300
101001	6	3.572	105	400	3519	17424
101017	1	1.42	50	120	810	3840
101017	2	1.42	20	100	855	2800
101017	3	0.428	15	150	460	950
101017	4	0.214	7	50	109	240
101017	5	0.428	15	100	230	1500
101017	6	0.428	15	100	180	2280

Figure 13.1. `RiceFarms` dataset included in `splm` library.

Note that panel data are often stacked first by cross-section and then by time period (as the dataset in Figure 13.1). Conversely, spatial panel data are usually ordered first by time period and then by cross-section (in this case successions of cross-sections are stacked). However, `splm` can effectively estimate also according to the standard conventions of panel data model (Millo and Piras 2012).

Data can be provided in two other forms to estimate a spatial panel model with `splm`: a `data.frame` and a character vector indicating the indexes variables or an object of the class `pdata.frame` of `plm`.

To estimate a spatial panel model, it is necessary to define a spatial weight matrix that measures connectivity between units. In the case of Indonesian farms, a contiguity matrix (`riceww`) is provided in `splm`, where, for each farm, all other farms from the same village are considered as neighbours.

The next steps of the analysis in R consist in loading the weight matrix and transforming it into a `listw` object using the `spdep` package as:

```
> data (riceww)
> rice_listw<-mat2listw(riceww)
```

The general model estimated in the application is defined in R as:

```
> mod<- log(goutput) ~ log(seed) + log(totlabor) + log(size)
+ log(urea)
```

Fixed and random effect models are estimated through ML using the function spml. The main arguments that need to be specified are:

- model that can assume the option within for fixed effects, random for random effects, and pooling for no-effects estimation.
- effect that can assume the option individual (i.e., α_i), time (i.e., ξ_t) and twoways (both) according to the effects introduced in the model.
- lag that can assume the attribute TRUE or FALSE. If TRUE, a spatial autoregressive term in the dependent variable is included in the model.
- spatial.error that can assume b (Baltagi et al. 2003), kkp (Kapoor et al. 2007), and none for no spatial error correlation.

First, we estimate a fixed effect spatial model using the option within in the argument model. In particular, for the sake of simplicity, we present the code for the estimation of the SEM and SARAR model, with the presence of only individual time invariant effects in the model (i.e., α_i).

```
> #FE-SEM Maximum Likelihood Estimation
> FE_SEM_ML<-spml(mod, data=RiceFarms, listw=rice_listw,
model=  "within",   effect=  "individual",   lag=FALSE,
spatial.error= "b")
> summary(FE_SEM_ML)
Spatial panel fixed effects error model

Call:
spml(formula = mod, data = RiceFarms, listw = rice_listw,
model = "within", effect = "individual", lag = FALSE,
spatial.error = "b")

Residuals:
      Min    1st Qu.     Median    3rd Qu.        Max
-1.053064  -0.185242   0.018638   0.201089   1.342150

Spatial error parameter:
     Estimate   Std. Error   t-value   Pr(>|t|)
rho  0.781880     0.025967     30.11   < 2.2e-16   ***

Coefficients:
               Estimate  Std. Error  t-value   Pr(>|t|)
log(seed)      0.102454    0.022491   4.5553  5.231e-06  ***
log(totlabor)  0.243276    0.026106   9.3188  < 2.2e-16  ***
log(size)      0.487660    0.027684  17.6150  < 2.2e-16  ***
log(urea)      0.107562    0.014722   7.3062  2.747e-13  ***
---
Signif. codes: 0 '***' 0.001 '**' 0.01 '*' 0.05 '.' 0.1 ' ' 1
```

Fixed effects can be extracted using the function `effects` as:

```
> effects(FE_SEM_ML)
Intercept:
               Estimate   Std. Error   t-value   Pr(>|t|)
(Intercept)   5.34524      0.16126     33.147    < 2.2e-16   ***
Spatial fixed effects:
        Estimate       Std. Error    t-value    Pr(>|t|)
1      0.06403816     0.20227826     0.3166     0.751559
2     -0.00634851     0.18403795    -0.0345     0.972482
3      0.08084476     0.18812360     0.4297     0.667383
4      0.10168740     0.18343106     0.5544     0.579330
5      0.45809336     0.18348351     2.4966     0.012537     *
6      0.07036177     0.19265234     0.3652     0.714942
7      0.19607748     0.19803795     0.9901     0.322125
8      0.12715208     0.18558482     0.6851     0.493254
9      0.20661835     0.18315346     1.1281     0.259271
10    -0.05006643     0.19268688    -0.2598     0.794992
11    -0.10554135     0.18735679    -0.5633     0.573219
12    -0.00604802     0.18912501    -0.0320     0.974489
13    -0.02735803     0.18884916    -0.1449     0.884816
14     0.13670470     0.19120320     0.7150     0.474627
15     0.09864297     0.19703814     0.5006     0.616632
...
```

The print method displays the type of effects (with significance levels) and the intercept.

The RE-SEM is estimated in R as:

```
> #RE-SEM Maximum Likelihood Estimation
>RE_SEM_ML<-spml(mod, data=RiceFarms, listw=rice_listw,
model= "random", effect= "individual", lag=FALSE,
spatial.error= "b")
> summary(RE_SEM_ML)
ML panel with , random effects, spatial error correlation
Call:
spreml(formula = formula, data = data, index = index, w =
listw2mat(listw),
w2 = listw2mat(listw2), lag = lag, errors = errors, cl = cl)
Residuals:
         Min        1st Qu.      Median      3rd Qu.         Max
-1.1283831   -0.2426953   0.0089203   0.2388153   1.3860407
Error variance parameters:
        Estimate    Std. Error    t-value    Pr(>|t|)
phi    0.201844     0.045220     4.4636    8.058e-06    ***
rho    0.762172     0.028678    26.5769    < 2.2e-16    ***
```

```
Coefficients:
               Estimate  Std. Error  t-value    Pr(>|t|)
(Intercept)    5.245044    0.178444  29.3932   < 2.2e-16  ***
log(seed)      0.114381    0.023073   4.9574  7.145e-07   ***
log(totlabor)  0.239830    0.026185   9.1590   < 2.2e-16  ***
log(size)      0.506221    0.027433  18.4531   < 2.2e-16  ***
log(urea)      0.136519    0.014017   9.7393   < 2.2e-16  ***
---
Signif. codes:
0 `***' 0.001 `**' 0.01 `*' 0.05 `.' 0.1 ` ' 1
```

As usual, the `summary` prints a short description of the model, a summary of the residuals, and the table of estimated coefficients. In particular, the estimate phi is $\varphi = \sigma_\alpha^2/\sigma_\varepsilon^2$ where σ_α^2 is the variance of the cross-sectional specific effect and σ_ε^2 is the variance of the error term. The `rho` coefficient represents the estimate of the parameter of spatial autocorrelation of the error term. The estimates of the two models are essentially similar in magnitude and all strongly significant.

A spatial Hausman test can be used to investigate whether the individual effects can be treated as fixed or random. In `splm` this test is implemented as:

```
> sphtest(mod, data=RiceFarms, listw=rice_listw,spatial.
model = "error", method = "ML", errors = "BSK")
    Hausman test for spatial models
data: x
chisq = 8.818, df = 4, p-value = 0.06581
alternative hypothesis: one model is inconsistent
```

The option BSK concerns the specification for the error term introduced by Baltagi et al. (2003). Conversely, one may choose the option KKP, that is the formula of the error term presented in Kapoor et al. (2007). The spatial Hausman test supports the null (acceptance of H_0), at a level of 5%, that the REM is the proper specification for this data set.

The SARAR panel model can be estimated adding a spatial lag of the dependent variable to the SEM specifications. The FE-SARAR model is estimated as:

```
> #FE-SARAR Maximum Likelihood Estimation
>FE_SARAR_ML<-spml(mod, data=RiceFarms, listw=rice_listw,
model=    "within",    effect=    "individual",    lag=TRUE,
spatial.error= "b")
> summary(FE_SARAR_ML)
Spatial panel fixed effects sarar model
Call:
spml(formula = mod, data = RiceFarms, listw = rice_listw,
model  =  "within",  effect  =  "individual",  lag  =  TRUE,
spatial.error = "b")
Residuals:
        Min     1st Qu.     Median    3rd Qu.         Max
-0.9790407  -0.1731642  0.0091114  0.1849727  1.4462310
```

```
Spatial error parameter:
        Estimate    Std. Error    t-value    Pr(>|t|)
rho    0.699546     0.049541       14.12    < 2.2e-16    ***
Spatial autoregressive coefficient:
        Estimate    Std. Error    t-value    Pr(>|t|)
lambda  0.22558      0.08711      2.5896    0.009608     **
Coefficients:
              Estimate    Std. Error    t-value    Pr(>|t|)
log(seed)     0.103043    0.022652      4.5489    5.393e-06    ***
log(totlabor) 0.243345    0.026212      9.2838    < 2.2e-16    ***
log(size)     0.490360    0.027882     17.5870    < 2.2e-16    ***
log(urea)     0.108451    0.014805      7.3254    2.383e-13    ***
---
Signif. codes: 0 '***' 0.001 '**' 0.01 '*' 0.05 '.' 0.1 ' ' 1
```

Conversely, the RE specification of the SARAR model is estimated as:

```
> #RE-SARAR Maximum Likelihood Estimation
> RE_SARAR_ML<-spml(mod, data=RiceFarms, listw=rice_listw,
model= "random", effect= "individual", lag=TRUE,
spatial.error="b")
> summary(RE_SARAR_ML)
ML panel with spatial lag, random effects, spatial error
correlation
Call:
spreml(formula = formula, data = data, index = index, w =
listw2mat(listw),
    w2 = listw2mat(listw2), lag = lag, errors = errors, cl = cl)
Residuals:
   Min   1st Qu.   Median    Mean   3rd Qu.    Max
 -0.340    0.548    0.796    0.788    1.027    2.179
Error variance parameters:
        Estimate    Std. Error    t-value    Pr(>|t|)
phi    0.203897     0.045620      4.4694    7.842e-06    ***
rho    0.722249     0.041366     17.4601    < 2.2e-16    ***
Spatial autoregressive coefficient:
          Estimate    Std. Error    t-value    Pr(>|t|)
lambda    0.117070    0.074214      1.5775    0.1147
Coefficients:
              Estimate    Std. Error    t-value    Pr(>|t|)
(Intercept)   4.443784    0.177746     25.0007    < 2.2e-16    ***
log(seed)     0.113262    0.023123      4.8982    9.672e-07    ***
log(totlabor) 0.242227    0.026222      9.2374    < 2.2e-16    ***
log(size)     0.506250    0.027491     18.4149    < 2.2e-16    ***
log(urea)     0.137081    0.014055      9.7528    < 2.2e-16    ***
---
Signif. codes: 0 '***' 0.001 '**' 0.01 '*' 0.05 '.' 0.1 ' ' 1
```

In addition to the SEM, for the SARAR case, the `lambda` coefficient is also presented in the summary. This value represents the estimate of the spatial autoregressive coefficient of the lag of the dependent variable.

The spatial Hausman test for checking the correct specification of the individual effect can be also implemented in this alternative way:

```
> sphtest(RE_SARAR_ML,FE_SARAR_ML)
    Hausman test for spatial models
data: formula
chisq = 1.7537, df = 4, p-value = 0.7809
alternative hypothesis: one model is inconsistent
```

Also, in this case, there is evidence that RE model is the appropriate specification for the model.

In the package `splm`, the routine for the GMM estimation of spatial panel data model is also available.

Under the normality of the errors, the GMM is asymptotically equivalent to the ML estimation (Kelejian and Prucha 1999). However, the computational effort of the GMM estimators is much lower.

In `splm`, the GMM procedure is implemented through the function `spgm`. In analogy to the ML procedure, the argument model can be specified as `within` (estimating a FEM) or `random` (estimating a REM). Furthermore, a model with a spatial lag of the dependent variable is estimated setting `lag` to TRUE. Finally, the SEM model is estimated adding the argument `spatial.error` that can assume only TRUE or FALSE, since only the KKP (Kapoor et al. 2007) option is available. In this chapter, we only describe the code for the FE-SARAR model. For further details, see Millo and Piras (2012), Croissant and Millo (2019).

The FE-SARAR panel model is estimated in R through the GMM method as:

```
> #FE-SARAR General Methods of Moments Estimation
> FE_SARAR_GM<-spgm(mod, data=RiceFarms, listw=rice_listw,
  model= "within", lag=TRUE, spatial.error=TRUE)
> summary(FE_SARAR_GM)
Spatial panel fixed effects GM model

Call:
spgm(formula = mod, data = RiceFarms, listw = rice_listw,
model = "within", lag = TRUE, spatial.error = TRUE)

Residuals:
 Min  1st Qu.  Median  Mean  3rd Qu.   Max
2.17    3.06    3.30   3.29    3.50   4.82
Estimated spatial coefficient, variance components and theta:
            Estimate
rho         0.654665
sigma^2_v   0.076146
Spatial autoregressive coefficient:
          Estimate   Std. Error   t-value    Pr(>|t|)
lambda    0.306275    0.078993    3.8772    0.0001057    ***
```

```
Coefficients:
               Estimate  Std. Error  t-value  Pr(>|t|)
log(seed)      0.102875   0.024999    4.1151  3.87e-05  ***
log(totlabor)  0.242545   0.028862    8.4036  <2.2e-16  ***
log(size)      0.490886   0.030758   15.9596  <2.2e-16  ***
log(urea)      0.108328   0.016368    6.6185  3.63e-11  ***
---
Signif. codes:
0 '***' 0.001 '**' 0.01 '*' 0.05 '.' 0.1 ' ' 1
```

The print of the summary for the GMM estimation is very similar to that obtained with the ML technique. Note that `sigma^2_v` is the variance components of the disturbance process. Finally, note that the standard errors of `rho` are not available and henceforth no inference can be performed on it.

If both the arguments `lag` and `spatial.error` are set to FALSE (i.e., `lag=FALSE`, `spatial.error=FALSE`), an endogenous variable (`endog`) should be specified together with a set of instruments (`instruments`).

Comparing the results of FE_SARAR_GM with those of FE_SARAR_ML, it is possible to note that the magnitude of the estimated coefficients is very similar, only some differences in the spatial coefficients can be appreciated.

In `splm`, some LM tests have been implemented. To discriminate between spatial lag and spatial error specification (Anselin et al. 2008; Elhorst 2010), we may use the function `slmtest` that is based on a pooling assumption, not allowing for any kind of individual effect. There are four available tests: LM_{lag}, LM_{err}, RLM_{lag}, RLM_{err}. Using our dataset, we obtain the following results:

```
> #SLM/SEM Tests
> slmtest(mod,data=RiceFarms, listw=rice_listw,test="lme")
     LM test for spatial error dependence
data: formula
LM = 1350, df = 1, p-value < 2.2e-16
alternative hypothesis: spatial error dependence
> slmtest(mod,data=RiceFarms, listw=rice_listw,test="lml")
     LM test for spatial lag dependence
data: formula
LM = 163.37, df = 1, p-value < 2.2e-16
alternative hypothesis: spatial lag dependence
> slmtest(mod,data=RiceFarms, listw=rice_listw,test="rlme")
    Locally robust LM test for spatial error dependence sub
spatial lag
data: formula
LM = 1198.5, df = 1, p-value < 2.2e-16
alternative hypothesis: spatial error dependence
> slmtest(mod,data=RiceFarms, listw=rice_listw,test="rlml")
    Locally robust LM test for spatial lag dependence sub
spatial error
data: formula
```

```
LM = 11.886, df = 1, p-value = 0.0005655
alternative hypothesis: spatial lag dependence
```

The interpretation of these four tests is very similar to that in the cross-sectional case. The first step is to run LM_{lag} and LM_{err} tests. If only LM_{lag} (LM_{err}) is significant, one should estimate the SLM (SEM). If both are significant, we need to check the robust LM tests. If only RLM_{lag} (RLM_{err}) is significant, the SLM (SEM) should be estimated. If both are significant, the suitable specification is the spatial regression model matching the most significant statistic. In our case, the RLM_{err} test gives slight evidence that the most appropriate model is the SEM.

As evidenced by Millo (2014) and Croissant and Millo (2019) who analysed the same dataset with a different spatial regression model, for Indonesian farms, spatial dependence can be more supported in the error terms rather than in the spatial lag of the dependent variable. Our results are in line with this narrative.

The LM tests for testing the presence of random effects and spatial error autocorrelation (Baltagi et al. 2003) are implemented in `splm` trough the function `bsktest`.

There are currently five options, corresponding to the tests described in Section 13.2: `LM1`, `LM2`, `LMJOINT`, `CLMlambda`, and `CLMmu`. The results for our dataset are:

```
> #Test Baltagi et al. (2003) on random effect and spatial error
autocorrelation
>bsktest(mod, data = RiceFarms, listw = rice_listw, test ="LMH"
    Baltagi, Song and Koh LM-H one-sided joint test
data: log(goutput) ~ log(seed) + log(totlabor) + log(size) +
log(urea)
LM-H = 1359.9, p-value < 2.2e-16
alternative hypothesis: Random Regional Effects and Spatial
autocorrelation
>bsktest(mod, data = RiceFarms, listw = rice_listw, test ="LM1"
    Baltagi, Song and Koh SLM1 marginal test
data: log(goutput) ~ log(seed) + log(totlabor) + log(size) +
log(urea)
LM1 = 3.1461, p-value = 0.001655
alternative hypothesis: Random effects
>bsktest(mod, data = RiceFarms, listw = rice_listw, test ="LM2"
    Baltagi, Song and Koh LM2 marginal test
data: log(goutput) ~ log(seed) + log(totlabor) + log(size) +
log(urea)
LM2 = 36.743, p-value < 2.2e-16
alternative hypothesis: Spatial autocorrelation
> bsktest(mod, data = RiceFarms, listw = rice_listw,
test ="CLMlambda"
    Baltagi, Song and Koh LM*-lambda conditional LM test
(assuming sigma^2_mu >= 0)
```

```
data: log(goutput) ~ log(seed) + log(totlabor) + log(size) +
log(urea)
LM*-lambda = 39.773, p-value < 2.2e-16
alternative hypothesis: Spatial autocorrelation
> bsktest(mod, data = RiceFarms, listw = rice_listw,
test ="CLMmu"
    Baltagi, Song and Koh LM*- mu conditional LM test (assuming
lambda may or may not be = 0)
data: log(goutput) ~ log(seed) + log(totlabor) + log(size) +
log(urea)
LM*-mu = 7.9576, p-value = 1.754e-15
alternative hypothesis: Random regional effects
```

The conditional tests CLMlambda and CLMmu are probably the most interesting ones in this approach, since they analyse one effect (i.e., spatial error autocorrelation or random effect) assuming the possible existence of the other (i.e., random effect or spatial error autocorrelation). These tests evidence if spatial error autocorrelation and random effects are appropriate for modelling RiceFarms.

As we noted in Section 13.2 for the spatial models including spatial lag of the dependent variable, it is suitable to calculate the impact measures. Since splm only copes with static panel model, the impacts for cross-sectional data developed in spatialreg can be easily extended to spatial panel modelling (Piras 2014). The major change concerns the spatial weight matrix. splm assumes for panel data analysis that the $N \times N$ weight matrix **W** is constant across time and, given the dimension of the panel (*NT*), we only need to transform this matrix in a block diagonal matrix whose blocks are the spatial weighting matrix itself. For further details about the estimation of impacts in a spatial panel model see Piras (2014).

The function impacts.splm estimates the impacts for static spatial model. We present the impacts for FE-SARAR model estimated via ML.

```
> set.seed(19)
> imp <- impacts(FE_SARAR_ML, listw = mat2listw(riceww,
style = "W"), time = 6, R=1000)
> summary(imp, zstats=TRUE, short=TRUE)
Impact measures (lag, trace):
                     Direct        Indirect          Total
log(seed)         0.1032875     0.02977090      0.1330584
log(totlabor)     0.2439224     0.07030656      0.3142290
log(size)         0.4915240     0.14167357      0.6331976
log(urea)         0.1087090     0.03133354      0.1400425
=================================================================
Simulation results ( variance matrix):
=================================================================
Simulated standard errors
                     Direct        Indirect          Total
log(seed)         0.02173241    0.01644761      0.03170894
log(totlabor)     0.02575848    0.03664277      0.04987035
```

```
log(size)        0.02809732   0.07338208   0.08351069
log(urea)        0.01504992   0.01634792   0.02471603
Simulated z-values:
                     Direct     Indirect       Total
log(seed)          4.731456     1.884578    4.220348
log(totlabor)      9.463674     2.017284    6.370294
log(size)         17.519905     2.033847    7.681775
log(urea)          7.218765     2.004282    5.721295
Simulated p-values:
                     Direct     Indirect       Total
log(seed)        2.2292e-06     0.059487  2.4393e-05
log(totlabor)    < 2.22e-16     0.043666  1.8867e-10
log(size)        < 2.22e-16     0.041967  1.5765e-14
log(urea)        5.2469e-13     0.045040  1.0572e-08
```

Note that only row-standardised weights are supported in this function, therefore the transformation `listw = mat2listw(riceww, style = "W")` is needed. The argument `time` is requested. As an additional argument, it is possible to include the option `R` that rules the number of simulations that are used to compute the distributions for the impact measures. The impact measures are all extremely significant and the interpretation of these summary statistics is the same of those of cross-sectional regression analysis.

13.4 Conclusions

This chapter discusses the great advantages of spatial panel modelling for the statistical analysis of agricultural variables. In particular, the use of the library `splm`, available in R for spatial panel data analysis, is presented and illustrated on a data set containing information related to Indonesian rice farms. The results of the different spatial panel specifications, estimated in this chapter, confirm previous analyses on the same data set (Druska and Horrace 2004; Millo 2014; Croissant and Millo 2019).

The library `splm` is a fairly complete package for this spatial analysis. Future improvements should include routines for the estimation of spatial dynamic panel models that, at this moment, are not available.

References

Aigner, D., C.A. Knox Lovell and P. Schmidt. 1977. Formulation and estimation of stochastic frontier production function models. *Journal of Econometrics* 6: 21–37.

Anselin, L. 1988. *Spatial Econometrics: Methods and Models*. Dordrecht: Kluwer Academic Publishers.

Anselin, L., A.K. Bera, R. Florax and M.J. Yoon. 1996. Simple diagnostic tests for spatial dependence. *Regional Science and Urban Economics* 26: 77–104.

Anselin, L., J. Le Gallo and H. Jayet. 2008. Spatial panel econometrics. pp. 624–660. *In*: Matyas, L. and P. Sevestre (eds.). *The Econometrics of Panel Data, Fundamentals and Recent Developments in Theory and Practice*. Berlin: Springer.

Arellano, M. 2003. *Panel Data Econometrics*. Oxford, UK: Oxford University Press.

Baltagi, B.H., S.H. Song and W. Koh. 2003. Testing panel data regression models with spatial error correlation. *Journal of Econometrics* 117: 123–150.

Baltagi, B.H., S.H. Song, B.C. Jung and W. Koh. 2007. Testing for serial correlation, spatial autocorrelation and random effects using panel data. *Journal of Econometrics* 140: 5–51.

Baltagi, B.H. 2013. *Econometric Analysis of Panel Data. 5th Edition.* New York, United States: John Wiley & Sons.

Battese, G.E. and T.J. Coelli. 1992. Frontier production functions, technical efficiency and panel data: With application to paddy farmers in India. *Journal of Productivity Analysis* 3: 153–169.

Baylis, K., N. Paulson and G. Piras. 2011. Spatial approaches to panel data in agricultural economics: A climate change application. *Journal of Agricultural and Applied Economics* 43: 325–338.

Behr, A. 2015. *Production and Efficiency Analysis with R.* Berlin: Springer.

Chakir, R. and J. Le Gallo. 2013. Predicting land use allocation in France: A spatial panel data analysis. *Ecological Economics* 92: 114–125.

Croissant, Y. and G. Millo. 2019. *Panel Data Econometrics with R.* Hoboken, NJ: John Wiley & Sons Ltd.

Dawson, P.J. 2005. Agricultural exports and economic growth in less developed countries. *Agricultural Economics* 33: 145–152.

Debarsy, N., C. Ertur and J.P. LeSage. 2012. Interpreting dynamic space-time panel data models. *Statistical Methodology* 9: 158–71.

Druska, V. and W.C. Horrace. 2004. Generalized moments estimation for spatial panel data: Indonesian rice farming. *American Journal of Agricultural Economics* 86: 185–98.

Elhorst, J.P. 2003. Specification and estimation of spatial panel data models. *International Regional Science Review* 26: 244–268.

Elhorst, J.P. 2010. Spatial panel data models. pp. 377–407. *In*: Fischer, M.M. and A. Getis (eds.). *Handbook of Applied Spatial Analysis.* Berlin: Springer.

Elhorst, J.P. 2014. *Spatial Econometrics: From Cross-Sectional Data to Spatial Panels.* Berlin: Springer.

Feng, Q. and W.C. Horrace. 2012. Alternative technical efficiency measures: Skew, bias and scale. *Journal of Applied Econometrics* 27: 253–268.

Florax, R.J.G.M., H. Folmer and S.J. Rey. 2003. Specification searches in spatial econometrics: The relevance of Hendry's methodology. *Regional Science and Urban Economics* 33: 557–579.

Greene, W. 2005. Fixed and random effects in stochastic frontier models. *Journal of Productivity Analysis* 23: 7–32.

Greene, W.H. 2008. The econometric approach to efficiency analysis. pp. 92–250. *In*: Fried, H.O., C.A.K. Lovell and S.S. Schmidt (eds.). *The Measurement of Productive Efficiency and Productivity Growth.* Chap. 2, New York: Oxford University Press.

Hackl, F., M. Halla and G.J. Pruckner. 2007. Local compensation payments for agri-environmental externalities: A panel data analysis of bargaining outcomes. *European Review of Agricultural Economics* 34: 295–320.

Hausman, J. 1978. Specification tests in econometrics. *Econometrica* 46: 1251–1271.

Horrace, W.C. and P. Schmidt. 1996. Confidence statements for efficiency estimates from stochastic frontier models. *Journal of Productivity Analysis* 7: 257–282.

Hsiao, C. 2007. Panel data analysis—advantages and challenges. *TEST* 16: 1–22.

Hsiao, C. 2014. *Analysis of Panel Data* (3rd ed). Cambridge: Cambridge University Press.

Kapoor, M., H.H. Kelejian and I.R. Prucha. 2007. Panel data model with spatially correlated error components. *Journal of Econometrics* 140: 97–130.

Kelejian, H.H. and I.R. Prucha. 1999. A generalized moments estimator for the autoregressive parameter in a spatial model. *International Economic Review* 40: 509–533.

Kelejian, H.H. and G. Piras. 2017. *Spatial Econometrics.* Academic Press.

Khandker S.R. and G.B. Koolwal. 2016. How has microcredit supported agriculture? Evidence using panel data from Bangladesh. *Agricultural Economics* 47: 157–16.

Kim, K., J.-P. Chavas, B. Barham and J. Foltz. 2012. Specialization, diversification, and productivity: A panel data analysis of rice farms in Korea. *Agricultural Economics* 43: 687–700.

Lee, Y.H. and P. Schmidt. 1993. A production frontier model with flexible temporal variation in technical efficiency. pp. 237–255. *In*: Fried, H.O. and S.S. Schmidt (eds.). *The Measurement of Productive Efficiency: Techniques and Applications.* Oxford, UK: Oxford University Press.

Lee, L.F. and J. Yu. 2010. Estimation of spatial autoregressive panel data models with fixed effects. *Journal of Econometrics* 154: 165–185.

LeSage, J.P. and K.R. Pace. 2009. *Introduction to Spatial Econometrics*. Boca Raton: Chapman & Hall/ CRC.

Lio, M. and M.C. Liu. 2006. ICT and agricultural productivity: evidence from cross-country data. *Agricultural Economics* 34: 221–228.

Meeusen, W. and J. van de Broeck. 1977. Efficiency estimation from Cobb–Douglas production functions with composed errors. *International Economic Review* 18: 435–444.

Millo, G. and G. Piras. 2012. splm: spatial panel data models in R. *Journal of Statistical Software* 47: 1–43. http://www.jstatsoft.org/v47/i01/.

Millo, G. 2014. Maximum likelihood estimation of spatially and serially correlated panels with random effects. *Computational Statistics & Data Analysis* 71: 914–933.

Mutl, J. and M. Pfaffermayr. 2011. The Hausman test in a Cliff and Ord panel model. *Econometrics Journal* 14: 48–76.

Piras, G. 2014. Impact estimates for static spatial panel data models in R. *Letters in Spatial and Resource Sciences* 7: 213–23.

Pitt, M.M. and L.F. Lee. 1981. The measurement and sources of technical inefficiency in the Indonesian weaving industry. *Journal of Development Economics* 9: 43–64.

Schmidt, P. and R.C. Sickles. 1984. Production frontiers and panel data. *Journal of Business & Economic Statistics* 2: 367–374.

Tong, T., T.H.E. Yu, S.H. Cho, K. Jensen and D. De La Torre Ugarte. 2013. Evaluating the spatial spillover effects of transportation infrastructure on agricultural output across the United States. *Journal of Transport Geography* 30: 47–55.

Trewin, R., L. Weiguo, S. Erwidodo and S. Bahri. 1995. Analysis of the technical efficiency over time of west javanese rice farms. *Australian Journal of Agricultural and Resource Economics* 39: 143–163.

Wu, J., Z. Ge, S. Han, L. Xing, M. Zhu, J. Zhang and J. Liu. 2020. Impacts of agricultural industrial agglomeration on China's agricultural energy efficiency: A spatial econometrics analysis. *Journal of Cleaner Production* 260: 121011. https://doi.org/10.1016/j.jclepro.2020.121011.

Zouabi, O. and N. Peridy. 2015. Direct and indirect effects of climate on agriculture: an application of a spatial panel data analysis to Tunisia. *Climatic Change* 133: 301–320.

Index

About the Editors

Paolo Postiglione

Paolo Postiglione is a Professor of Economic Statistics at University of Chieti-Pescara (Italy). He has been a visiting researcher at Regional Economics Applications Laboratory of University of Illinois at Urbana-Champaign, at Regional Research Institute of West Virginia University, and received a Ph.D. in Statistics from the University of Chieti-Pescara in 1998.

Currently, he is the Principal Investigator for University "G. d'Annunzio" of Chieti-Pescara for the Horizon 2020 Project "Integrative Mechanisms for Addressing Spatial Justice and Territorial Inequalities in Europe" (IMAJINE), H2020-SC6-REV-INEQUAL-2016. His research interests mainly concern regional quantitative analysis, spatial statistics and econometrics, regional economic convergence, models for spatial non-stationary data, agricultural statistics, and spatial sampling. He is author of a book edited by Springer, several articles on peer review journals and other publications on these topics.

Roberto Benedetti

Roberto Benedetti is a Professor of Economic Statistics at University of Chieti-Pescara (Italy). He obtained his PhD in Methodological Statistics in 1994 from "La Sapienza" University of Rome (Italy). From 1994 to 2001, he was employed at Italian National Statistical Institute as Research Director as the head of the Agricultural Statistical Service.

He was visiting researcher at the National Centre for Geographic Information Analysis of the University of California at Santa Barbara, at Regional Economics Applications Laboratory of University of Illinois at Urbana-Champaign, at Centre for Statistical and Survey Methodology of University of Wollongong.

His current research interests focus on agricultural statistics, sample design, small area estimation, and spatial data analysis. On these topics, he published a book edited by Springer and many articles on referred journals.

Federica Piersimoni

Federica Piersimoni is Senior Researcher at Processes Design and Frames Service within the Methodological Department of the Italian National Statistical Institute, joined in 1996. In the same institution, she spent more than ten years at the Agricultural Statistical Service within the Economic Department. She was a visiting researcher at Regional Economics Applications Laboratory of University of Illinois at Urbana-Champaign, and at Centre for Statistical and Survey Methodology of University of

Wollongong. In 1999 she received a Specialization Degree in Operational Research and Decision Theory from the University of Rome "La Sapienza" and received a Ph.D. in Statistics from the University of Chieti-Pescara in 2014. Her main research interests concern disclosure control, and sample design, on which topics she published a book edited by Springer and journal papers.